LONDON MATHEMATICAL SOCIETY LECTURE NOTE SERIES

Managing Editor: Professor J.W.S. Cassels, Department of Pure Mathematics and Mathematical Statistics, University of Cambridge, 16 Mill Lane, Cambridge CB2 1SB, England

The titles below are available from booksellers, or, in case of difficulty, from Cambridge University Press.

London Mathematical Society Lecture Note Series. 216

Stochastic Partial Differential Equations

Edited by

Alison Etheridge
University of Edinburgh

Published by the Press Syndicate of the University of Cambridge
The Pitt Building, Trumpington Street, Cambridge CB2 1RP
40 West 20th Street, New York, NY 10011-4211, USA
10 Stamford Road, Oakleigh, Melbourne 3166, Australia

© Cambridge University Press 1995

First published 1995

Printed in Great Britain at the University Press, Cambridge

Library of Congress cataloging in publication data available

British Library cataloguing in publication data available

ISBN   0 521 48319 0 paperback

Contents

Foreword

In March 1994 a two week workshop on Stochastic Partial Differential Equations was held at the University of Edinburgh under the auspices of the International Centre for Mathematical Sciences. The meeting attracted an international audience including researchers from all aspects of the field aswell as participants with backgrounds in biology, physics, engineering and finance. This volume is a collection of articles contributed by participants at that meeting. Some report work presented at the meeting and some are surveys, but all are accessible to a wide audience. Although the subject matter reflects the diversity of the field, all the contributions are closely related to the central theme of stochastic partial differential equations.

There are many people who contributed to the meeting. Only a few of them are mentioned here, but I am deeply indebted to them all. The original proposal for the workshop would never have been completed without the invaluable input of Terry Lyons. I should like to thank him and the other members of the Scientific Committee, Peter Donnelly and Steve Evans. A great deal of the administration was handled by the International Centre for Mathematical Sciences and I should especially like to thank Lucy Young, not only for her hard work but also for her patience with my inefficiency. Thanks also to Sandra Bashford for her sustained cheerfulness and limitless supplies of coffee. The Science and Engineering Research Council provided funding under grant number GR/H94092.

The production of the present volume could not have taken place without the help of Roger Astley at CUP and of course the contributors themselves. The referees must remain anonymous, but they have my heartfelt thanks.

AME
March 1995

Participants

A Alabert	Universitat Autònoma de Barcelona
O Brockhaus	Universität Bonn
Z Brzezniak	Ruhr-Universität Bochum
A Cairns	Heriot-Watt University
T Chan	Heriot-Watt University
N Cutland	University of Hull
I Davies	University College Swansea
D Dawson	Carleton University
B Djehiche	Royal Institute of Technology, Stockholm
R Doney	University of Manchester
P Donnelly	Queen Mary and Westfield College, London
D Elworthy	University of Warwick
T Erhardsson	Royal Institute of Technology, Stockholm
A Etheridge	University of Edinburgh
S Evans	University of California, Berkeley
B Fisher	MIT
K Fleischmann	Institute of Applied Analysis and Stochastics, Berlin
J Gaines	University of Edinburgh
J Geiger	Universität Frankfurt
N Goncharuk	University of Warwick
L Gorostiza	Cinvestav, Mexico
S Harris	University of Bath
R Hudson	University of Nottingham
I Kaj	Uppsala Universitet
J Kennedy	University of Oxford
T Knudsen	Technical University of Denmark
M Kreer	NatWest Markets
J-A Lopes-Mimbela	Universität Frankfurt
J Lunt	NatWest Markets
O Lyne	University of Bath
T Lyons	Imperial College, London
G Lythe	DAMTP, University of Cambridge
X Mao	University of Strathclyde
S Maybank	University of Oxford

D Mollison	Heriot-Watt University
J Norris	DPMMS, University of Cambridge
N O'Connell	Dublin Institute of Advanced Study
B Øksendal	University of Oslo
L Overbeck	Univerisité Paris VI
E Perkins	University of British Columbia
F Rezakhanlou	University of California, Berkeley
M Röckner	Universität Bonn
E Rodrigues	Harvard University
F Russo	Université de Provence
A Rybko	Institute for Information and Transmission Sciences, Moscow
M Sanz-Sole	Universitat de Barcelona
A Schied	Universität Bonn
B Schmuland	University of Alberta
R Streater	King's College, London
A Truman	University College Swansea
A Veretennikov	Institute for Information and Transmission Sciences, Moscow
J Verzani	University of Washington, Seattle
A Wakolbinger	Universität Frankfurt
J Warren	University of Bath
A Wulfsohn	University of Warwick
H-T Yau	Courant Institute
B Zegarlinski	Imperial College, London
H Zhao	University College Swansea

STOCHASTIC DIFFERENTIAL EQUATIONS WITH BOUNDARY CONDITIONS AND THE CHANGE OF MEASURE METHOD

Aureli Alabert

Universitat Autònoma de Barcelona, Departament de Matemàtiques.
08193-Bellaterra, Catalonia.

1. Introduction

The definition of several types of stochastic integrals for anticipating integrands put the basis for the development, in recent years, of an anticipating stochastic calculus. It is natural to consider, as an application, some problems that can be stated formally as stochastic differential equations, but that cannot have a sense within the theory of non-anticipating stochastic integrals. For example, this is the case if we impose to an s.d.e. an initial condition which is not independent of the driving process, or if we prescribe boundary conditions for the solution.

In this paper, we will try to survey the work already done concerning s.d.e. with boundary conditions, and to explain in some detail a method based in transformations and change of measure in Wiener space. An alternative approach is sketched in the last Section.

Transformations on Wiener space provide a natural method, among others, to prove existence and uniqueness results for nonlinear equations. At the same time, a Girsanov type theorem for not necessarily adapted transformations allows to study properties of the laws of the solutions from properties of the solution to an associated linear equation. A natural first question about these laws is to decide if they satisfy some kind of Markov (or conditional independence) property.

In Section 2, we introduce stochastic differential equations with boundary conditions, the particular instances that have been studied, and the kind of results obtained concerning conditional independence properties of the solutions. The short Section 3 outlines the idea of the method of transformations and change of measure. In Section 4, we recall the necessary elements of Wiener space analysis in order to enounce the Girsanov type theorem we want to apply. In Section 5, we describe the use of suitable transformations to obtain existence and uniqueness results for nonlinear equations. Section 6 is devoted to explain how the change of measure induced by the same transformations allows to derive characterizations of conditional independence properties. Finally, in Section 7 we describe briefly other alternatives to this method and the situations to which they have been applied.

1

Consider the following general problem:

$$
\left.
\begin{aligned}
(2.1) \qquad dX_t = f(t, X_t)\, dt + \sum_{i=1}^{k} \sigma_i(t, X_t) \circ dW_t^i \quad , \quad 0 \le t \le 1 \\
h(X_0, X_1) = 0
\end{aligned}
\right\}
$$

where $f, \sigma_i \colon [0,1] \times \mathbb{R}^d \longrightarrow \mathbb{R}^d$, $h \colon \mathbb{R}^{2d} \longrightarrow \mathbb{R}^d$, and $\{W_t,\ 0 \le t \le 1\}$ is a k-dimensional Wiener process ($k \le d$). The customary initial condition for X_0 is replaced by the boundary condition $h(X_0, X_1) = 0$.

By a *solution* to (2.1) we mean a d-dimensional continuous process X_t verifying the system

$$
\left.
\begin{aligned}
(2.2) \qquad X_t = \int_0^t f(s, X_s)\, ds + \sum_{i=1}^{k} \int_0^t \sigma_i(s, X_s) \circ dW_s^i \quad , \quad 0 \le t \le 1 \\
h(X_0, X_1) = 0
\end{aligned}
\right\}
$$

But here, unlike the initial-value problem, we cannot expect in general the existence of a solution adapted to the Wiener process, since the boundary condition makes X_0 depend on X_1, which in turn will depend on the whole Wiener process, through the integral equation. Therefore, the stochastic integral in (2.2) has to be understood as an anticipating stochastic integral. With the circle we denote, as usual, the Stratonovich anticipating integral.

As stated in the Introduction, two problems have been tackled concerning such equations: First, of course, the problem of existence and uniqueness of a solution; and secondly, to find sufficient and necessary conditions on the coefficients to have some conditional independence property for the solution.

Which kind of conditional independence property should be expected? The classical Markov Process property will not hold in general because a random variable X_t can hardly make independent the past and the future of the process, given that the first and last variables are linked by the boundary condition.

It turns out that the relevant property to study is the Markov Field property. This is the natural Markov property for random fields, and therefore it is more clearly formulated with a general parameter set:

A random field $\{X_t,\ t \in T\}$, with $T \subset \mathbb{R}^k$, is a *Markov Field* (M.F., for short) if and only if for every Borel and bounded set D (with $\overline{D} \subset T$), the families of random variables $\{X_t,\ t \in \overline{D}\}$ and $\{X_t,\ t \in \overline{D^c}\}$ are conditionally independent given $\{X_t,\ t \in \partial D\}$. Here ∂D denotes the boundary of D. It is enough to check this property for open sets D. Translated to the one parameter case, with $T = [0,1]$, the property can be stated as follows: $\{X_t,\ t \in [0,1]\}$ is a Markov Field if and only if the families $\{X_u,\ u \in [s,t]\}$ and $\{X_u,\ u \in]s, t[^c\}$ are independent given X_s and X_t. It is obvious that every Markov Process is a Markov Field, but the converse is not true.

Let us now take a look to the particular equations studied so far, and the kind of results obtained concerning the Markov Field property. Our aim here is only to sketch these results. We refer the reader to the original references for the precise statements. Particularly, we remark that some technical hypothesis on the coefficients of the equations are needed, first of all, to obtain existence and uniqueness theorems, and then, to characterize the Markov Field property.

A) **First order equations**: Equations like (2.1) are first order equations. Within this setting, the first work was done by Ocone and Pardoux [19] (1989), who considered *linear equations*:

$$(2.3) \qquad \left. \begin{array}{c} dX_t = (AX_t + a(t))\,dt + \sum_{i=1}^{k}(B_i X_t + b_i(t)) \circ dW_t^i \quad , \quad 0 \le t \le 1 \\ F_0 X_0 + F_1 X_1 = F \end{array} \right\}$$

(affine drift, diffusion coefficient and boundary condition), where A, $B_1, \ldots,$ B_k, F_0, F_1 are $d \times d$-matrices of constants, $F \in I\!\!R^d$, and $a(t), b_1(t), \ldots, b_k(t)$ are d-dimensional processes.

Their main result states that each of the following are sufficient conditions to have a M.F.

a) $B_1 = \cdots = B_k = 0$ (Gaussian case).

b) $a = b_1 = \cdots = b_k = 0$ and $\Phi_t \cdot \Phi_s^{-1}$ is a diagonal matrix, $\forall t, s$, where Φ_t is the matrix solution of

$$(2.4) \qquad \left. \begin{array}{c} d\Phi_t = A\Phi_t\,dt + \sum_{i=1}^{k} B_i \Phi_t \circ dW_t^i \quad , \quad 0 \le t \le 1 \\ \Phi_0 = I \end{array} \right\}$$

Notice that, in particular, in dimension one and with linear drift and diffusion, the solution is always a M.F. Further results can be given for special forms of the boundary condition (see [19]).

Nualart and Pardoux [16] considered the non-linear equation

$$(2.5) \qquad \left. \begin{array}{c} dX_t = f(X_t)\,dt + \sigma\,dW_t \quad , \quad 0 \le t \le 1 \\ h(X_0, X_1) = 0 \end{array} \right\}$$

and proved that, in dimension 1 ($d = 1$), X is a M.F. iff f is affine. For $d > 1$, they showed examples with nonlinear f for which X is a M.F. (or even a Markov Process) and others where X is not a M.F. With certain special structures of the function f one can recover dichotomy results (see Ferrante [8] and Ferrante–Nualart [9]). ing

Donati-Martin [5] took the next step considering a linear diffusion coefficient:

$$(2.6) \qquad \left. \begin{array}{c} dX_t = f(X_t)\,dt + \sigma X_t \circ dW_t \quad , \quad 0 \le t \le 1 \\ F_0 X_0 + F_1 X_1 = F \end{array} \right\}$$

In dimension 1, the solution is a M.F. iff f takes the form $f(x) = Ax + Bx \log x$, with $|B| < 1$.

B) **Second order equations**: In [17], Nualart and Pardoux studied the following second order stochastic differential equation in dimension one, with Dirichlet type boundary conditions:

$$(2.7) \qquad \left. \begin{array}{l} \ddot{X}_t = f(X_t, \dot{X}_t) + \dot{W}_t \quad , \quad 0 \le t \le 1 \\ X_0 = a, \ X_1 = b \end{array} \right\}$$

A solution to (2.7) will be a $\mathcal{C}^1$ process verifying

$$(2.8) \qquad \left. \begin{array}{c} \dot{X}_t = \dot{X}_0 + \displaystyle\int_0^t f(X_s, \dot{X}_s)\,ds + W_t \quad , \quad 0 \le t \le 1 \\ X_0 = a, \ X_1 = b \end{array} \right\}$$

In this problem, we cannot expect to have any type of conditional independence property for X, because in a $\mathcal{C}^1$ process the positions X_t do not keep enough information to make independent the past and the future or the interior and exterior of an interval.

However, the two-dimensional stochastic process $\{(X_t, \dot{X}_t),\ t \in [0,1]\}$ is a M.F. iff f is affine. The same result is true if we change Dirichlet to Neumann boundary conditions (see Nualart [14]).

C) **Partial differential equations**: The following parabolic stochastic partial differential equation with periodicity conditions has been considered by Nualart and Pardoux [18]:

$$(2.9) \qquad \left. \begin{array}{rl} \dfrac{\partial X_{t,y}}{\partial t} - \dfrac{\partial^2 X_{t,y}}{\partial y^2} = f(X_{t,y}) + \dfrac{\partial^2 W_{t,y}}{\partial t \partial y} & , \quad (t,y) \in [0,1]^2 \\ X(t,0) = X(t,1) = 0 & , \quad t \in [0,1] \\ X(0,y) = X(1,y) & , \quad y \in [0,1] \end{array} \right\}$$

where $\dfrac{\partial^2 W}{\partial t \partial y}$ is a space-time White Noise. The condition f affine is again necessary and sufficient for the Markov Field property of the $C_{0,0}([0,1])$-valued process X_t.

Donati-Martin [6], [7], considered the elliptic equation with Dirichlet boundary condition:

$$(2.10) \qquad \left. \begin{array}{l} \Delta X_t = f(X_t) + \dot{W}_t \quad , \quad t \in T \\ X_{|\delta T} = 0 \end{array} \right\}$$

where Δ is the Laplacian, T is a bounded domain of $I\!\!R^k$ ($k \le 3$), and $\dot{W}$ represents a White Noise in $I\!\!R^k$. Here, the condition f affine is equivalent to a slightly weaker conditional independence property: For any Borel and bounded set D (with $\overline{D} \subset U \subset T$, for some open set U), the σ-fields $\sigma\{X_t,\ t \in \overline{D}\}$ and $\sigma\{X_t,\ t \in \overline{D^c}\}$ are conditionally independent given the σ-field $\bigcap_{\varepsilon>0} \sigma\{X_t,\ t \in (\partial D)_\varepsilon\}$, where $(\partial D)_\varepsilon$ denotes an ε-neighbourhood of ∂D. X is said to be a *Germ Markov Field* (G.M.F.). □

All these results concerning nonlinear equations have been obtained using a common method, which is based in an argument of change of measure in Wiener space. Our aim is to explain this method, and to illustrate it with some examples. In a non-anticipating context, a change of measure for s.d.e. would rely in the celebrated (Cameron–Martin–Maruyama)–Girsanov Theorem. In our case, an extended version of this theorem, essentially due to Ramer [20] and Kusuoka [12], and allowing for anticipating transformations, is the basic tool to use.

3. Idea of the change of measure method

The idea of the change of measure to study nonlinear anticipating s.d.e. is analogous to that of the classical Girsanov theorem for non-anticipating ones. Starting with a nonlinear equation on a probability space $(\Omega, \mathcal{F}, P)$, with solution process X_t, one considers another measure Q on $(\Omega, \mathcal{F})$, and a linear equation conveniently related with the original one. These measure and linear equation should be chosen in such a way that the law of the solution Y_t under Q coincide with the law of X_t under P.

Then, anything we can prove concerning the law of Y_t under Q produces automatically the same result for the law of X_t under P, which is the process we are interested in. In other words, we switch to a simpler process (possibly in explicit form), at the price of dealing with a more complicated measure, given by its Radon–Nikodým derivative with respect to P.

4. Some elements of analysis in Wiener space

Let (B, H, P) be a Wiener space. That is, H is an infinite-dimensional real separable Hilbert space, equipped with the Gauss cylinder measure μ, B is the completion of H with respect to a measurable norm, and P is the extension of μ to a measure on the Borel σ-field of B. P is called the *Wiener measure* on B. On the other hand, given a Banach space B and a Gaussian centered measure P on B, with supp $P = B$, there exists a unique Hilbert space $H \subset B$ such that (B, H, P) is a Wiener space. We will call H the *Cameron–Martin space* relative to B and P. See for example Kuo [11] for details.

This is the definition of an *abstract Wiener space*. The *classical Wiener space* is the particular instance in which we take as B the space $C_0([0,1])$ of continuous functions vanishing at zero with the supremum norm, H is the subspace of functions with derivatives in $L^2([0, 1])$, and P is the measure induced by a one-dimensional standard Wiener process.

Many interesting mappings and functionals on Wiener space, such as solutions to s.d.e., are not Fréchet differentiable in general; therefore the classical infinite-dimensional calculus is of little help. For this reason, several infinite-dimensional calculi well adapted to these functionals have been introduced. We recall here the definition of derivation and other basic features of the Malliavin infinite-dimensional calculus.

Let E be a real separable Hilbert space, $F: B \to E$ a mapping and $\omega_0 \in B$. We say that F is H-differentiable at ω_0 iff there exists $\nabla F(\omega_0) \in \mathcal{L}(H; E)$ (a linear continuous mapping from H into E) such that

$$\lim_{\substack{h \in H \\ ||h||_H \to 0}} \frac{||F(\omega_0 + h) - F(\omega_0) - [\nabla F(\omega_0)](h)||_E}{||h||_H} = 0 \quad .$$

Clearly, if F is Fréchet differentiable at ω_0, then F is H-differentiable at ω_0 and $\nabla F(\omega_0)$ coincides with the Fréchet differential restricted to H.

A *smooth E-valued cylinder functional on B* is a mapping $F: B \to E$ of the form

$$F(\omega) = \sum_{j=1}^{m} f_j((\ell_1, \omega), \ldots, (\ell_n, \omega)) e_j \quad ,$$

where $\ell_1, \ldots, \ell_n \in B^*$ (the topological dual of B), $e_j \in E$, and f_j are $\mathcal{C}^\infty$ functions on $\mathbb{R}^n$ with polynomial growth, together with all their derivatives. Denote by $\mathcal{S}(E)$ the set of these functionals.

Since an element $F \in \mathcal{S}(E)$ is clearly Fréchet differentiable on B, its H-differential exists for every ω and

$$\nabla F(\omega) = \sum_{j=1}^{m} \sum_{i=1}^{n} \partial_i f_j((\ell_1, \omega), \ldots, (\ell_n, \omega)) \ell_i \otimes e_j \quad ,$$

considered as an element of $\mathcal{L}(H; E)$. Of course, $(\ell_i \otimes e_j, h)$ means $(\ell_i, h) \cdot e_j$, and is an element of the algebraic tensor product $H \otimes E$ (after identification of H and H^*). Its completion by the inner product $\langle \ell \otimes e, \ell' \otimes e' \rangle := \langle \ell, \ell' \rangle_H \cdot \langle e, e' \rangle_E$ is the space of Hilbert–Schmidt operators from H to E, which we will denote by the same symbol $H \otimes E$.

Up to this point, we have only taken into account the topological structure of B. Now, using the measure P, one can prove that $\forall F \in \mathcal{S}(E)$, $\forall p \geq 1$, $F \in L^p(B; E)$ and $\nabla F \in L^p(B; H \otimes E)$, and that $\mathcal{S}(E)$ is dense in $L^p(B; E)$ (see Ikeda and Watanabe [13], Remark 8.2). Moreover, the mapping

$$\nabla: L^p(B; E) \longrightarrow L^p(B; H \otimes E) \quad ,$$

with domain $\mathcal{S}(E)$, is closable. Denoting by $I\!\!D^{1,p}(E)$ the closure of $\mathcal{S}(E)$ under the graph norm

$$\|F\|_{I\!\!D^{1,p}(E)} := \|F\|_{L^p(B;E)} + \|\nabla F\|_{L^p(B;H\otimes E)} \quad,$$

we obtain a continuous mapping $\nabla: I\!\!D^{1,p}(E) \to L^p(B; H\otimes E)$, called the *gradient operator*. Recursively, one can define higher order gradient operators ∇^k ($\nabla^k F$ will be an element of $L^p(B; H^{\otimes k} \otimes E)$) and obtain the Sobolev spaces $I\!\!D^{k,p}(E)$.

The operator ∇ is local in the following sense: If, for some measurable set A, $F: B \to E$ verifies $F(\omega) = 0$, for a.a. $\omega \in A$, then $\nabla F: B \to H \otimes E$ verifies the same property. This fact justifies the following definition: The random variable $F: B \to E$ belongs to $I\!\!D_{\text{loc}}^{k,p}(E)$ iff there exist a sequence $\{B_n\}_{n\in I\!\!N}$ of measurable sets converging to B and a sequence $\{F_n\}_{n\in I\!\!N}$ of elements of $I\!\!D^{k,p}(E)$ such that $F_n = F$ on B_n. For $F \in I\!\!D_{\text{loc}}^{k,p}(E)$, the gradient ∇F is defined as $\nabla F(\omega) = \nabla F_n(\omega)$, if $\omega \in B_n$.

The following different concept of differentiability will be used in the Theorem below.

Definition. Let $F: B \to H$ be a random variable with values in the Cameron–Martin space H. The mapping F is H–$\mathcal{C}^1$ if for all $\omega \in B$, there exists a Hilbert–Schmidt operator $\mathcal{K}(\omega)$ such that
1) $\|F(\omega + h) - F(\omega) - [\mathcal{K}(\omega)](h)\|_H = o(\|h\|_H)$, as $\|h\|_H \to 0$, a.s.
2) The mapping $h \mapsto \mathcal{K}(\omega + h)$ from H to $H \otimes H$ is continuous, a.s. $\quad\square$

If F is H–$\mathcal{C}^1$, then $F \in I\!\!D_{\text{loc}}^{1,2}(H)$ (see Kusuoka [12] or Nualart [15]). On the other hand, $F \in I\!\!D_{\text{loc}}^{1,2}(H)$ implies that $\nabla F(\omega)$ is Hilbert–Schmidt, a.s. Therefore, $\nabla F(\omega)$ is the only candidate for the operator $\mathcal{K}(\omega)$ in the definition.

For any Hilbert-Schmidt operator $\mathcal{K}$ on a Hilbert space H, its Carleman–Fredholm determinant, denoted by $\det_2(I_H + \mathcal{K})$, is defined by

$$\det_2(I_H + \mathcal{K}) := \prod_{i=1}^{\infty}(1 + \lambda_i)e^{-\lambda_i} \quad,$$

where $\{\lambda_i\}_{i=1}^{\infty}$ is the family of (complex) eigenvalues of $\mathcal{K}$, counted with their multiplicity. For the properties of this quantity and its role in the theory of integral equations see, for instance, Cochran [4].

The Ramer–Kusuoka Theorem can be stated as follows:

Theorem (Ramer [20], Kusuoka [12]).
Let $F: B \to H$ be an H–$\mathcal{C}^1$ map. Assume:
  a) The transformation $T: B \to B$ given by $T(\omega) = \omega + F(\omega)$ is bijective.
  b) The operator $I_H + \nabla F(\omega): H \to H$ is invertible, a.s.
Then:

The measure $Q := P \circ T$ (the image probability of the Wiener measure by T^{-1}) is equivalent to P and

$$\frac{dQ}{dP}(\omega) = \big| \det\nolimits_2(I_H + \nabla F(\omega)) \big| \exp \big\{ -(\nabla^* F)(\omega) - \frac{1}{2} \|F(\omega)\|_H^2 \big\} \quad ,$$

where ∇^* is the adjoint of the gradient operator, considered here as an unbounded operator $L^2(B) \to L^2(B; H)$. □

Remark. ∇^* enjoys a local property which is similar to that of ∇, and ensures that, for $F \in \mathbb{D}^{1,2}_{\text{loc}}(H)$, $\nabla^* F$ is well-defined. □

Remark. There exist stronger versions of this theorem, but we will not make use of them. Particularly, Üstünel and Zakai ([23] [24]) have obtained representations for the density of Q without hypothesis *a)* and with less regularity on F. □

Sometimes it is useful, for the purpose of representation, to realize the Cameron–Martin space H as an L^2 space. Let $(T, \mathcal{B}, \mu)$ be a separable measure space. Denote $\tilde{H} = L^2(T, \mathcal{B}, \mu)$, and let $i: \tilde{H} \to H$ be an isomorphism. Given a random variable $F: B \to H^{\otimes n}$ (for some $n = 0, 1, 2, \dots$), define $G: B \to \tilde{H}^{\otimes n}$ by the equality $F = i^{\otimes n} \circ G$, where $i^{\otimes n}$ is the natural isomorphism between $\tilde{H}^{\otimes n}$ and $H^{\otimes n}$ induced by i. Define, also,

$$DG(\omega) := (i^{\otimes n})^{-1} \circ \nabla F(\omega) \circ i \quad \in \tilde{H}^{\otimes n+1} \quad .$$

If $F \in \mathbb{D}^{1,2}(H^{\otimes n})$, then clearly $G \in \mathbb{D}^{1,2}(\tilde{H}^{\otimes n})$ and we have $DG \in L^2(B; \tilde{H}^{\otimes n+1}) \simeq L^2(B \times T; \tilde{H}^{\otimes n})$.

The operator

$$D: \mathbb{D}^{1,2}(\tilde{H}^{\otimes n}) \longrightarrow L^2(B \times T; \tilde{H}^{\otimes n}) \quad ,$$

which transforms random variables into processes (indexed by the elements of T), is called the *Malliavin derivative operator*.

The statement of the Ramer–Kusuoka Theorem can be translated, using the Malliavin derivative, into a form which is usually more amenable to computations. Let $F: B \to H$, as in the Theorem. Then $DG(\omega) = i^{-1} \circ \nabla F(\omega) \circ i$ is a Hilbert–Schmidt operator on $\tilde{H}$, which has the same eigenvalues of $\nabla F(\omega)$, and the invertibility of $I_H + \nabla F(\omega)$ is equivalent to that of $I_{\tilde{H}} + DG(\omega)$.

Assume now $v: B \to \mathbb{R}$, $v \in L^2(B)$. In this case, $Dv(\omega) = \nabla v(\omega) \circ i$. If we denote by j the isomorphism between $L^2(B; \tilde{H})$ and $L^2(B; H)$ given by $j(F)(\omega) = i(F(\omega))$, we can write $D = j^{-1} \circ \nabla$ (as unbounded operators on $L^2(B)$) and the relation $D^* = \nabla^* \circ j$ holds for the adjoints. The adjoint of the Malliavin derivative operator D is usually denoted by δ and is called the *Skorohod integral*. We have then, for $F: B \to H$ and $F = i \circ G$,

$$\delta G = \nabla^* F \quad .$$

The formula for the density of Q with respect to P becomes

$$(4.1) \quad \frac{dQ}{dP}(\omega) = \left| \det{}_2(I_{\tilde{H}} + DG(\omega)) \right| \exp \left\{ - (\delta G)(\omega) - \frac{1}{2}\|G(\omega)\|_{\tilde{H}}^2 \right\} \quad .$$

It is convenient to keep in mind both interpretations of $DG(\omega)$. In fact, the process $\{D_t G(\omega), \ t \in T\}$ is the kernel of the integral operator $DG(\omega)$:

$$[DG(\omega)](h) = \int_T D_t G(\omega) h(t) \, dt, \quad h \in \tilde{H} \quad .$$

In the case of the classical Wiener space, taking $T = [0,1]$ with the Lebesgue measure, we have $i(h) = \int_0^{\cdot} h(t) \, dt$.

5. Transformations. Existence and Uniqueness of solutions

As stated before, transformations in Wiener space provide a natural method to achieve existence and uniqueness results for nonlinear s.d.e. We are going to describe here this procedure, and to apply it to some concrete examples. Another natural approach involves the use of stochastic flows (see, for instance, [5], Theorems 3.1 and 4.1).

Suppose we have a nonlinear equation of the following form:

$$\text{(N)} \qquad \left. \begin{array}{c} p(D)X_t = f(X_t) + \sigma(X_t) \circ \dot{W}_t \quad , \quad t \in T \subset \mathbb{R}^k \\ h(X_{|\delta T}) = 0 \end{array} \right\}$$

where $p(D)$ is a linear differential operator with constant coefficients, and we assume that σ is linear or constant. For simplicity of notation we make f, σ and h depend only on X and not on any of its derivatives (cf. equation (2.7)) or the time parameter t, but this is also possible. We assume also that the dimension is equal to one.

$\dot{W}_t$ represents a White Noise. That means: $\dot{W}$ is a centered Gaussian family $\{\dot{W}(A), \ A \in \mathcal{B}(\mathbb{R}^k)\}$ with covariance $E[\dot{W}(A)\dot{W}(B)]$ equal to the Lebesgue measure of $A \cap B$. Its (random) distribution function W_t (normalized with $W_0 = 0$) is a standard Wiener process with k-dimensional parameter. We can (and shall) assume that the probability space $(\Omega, \mathcal{F}, P)$ in which we are working is the canonical space of the Wiener process. Then $W_t(\omega) = \omega(t)$, $t \in T$.

Step 1: We can associate to (N) the linear equation

$$\text{(L)} \qquad \left. \begin{array}{c} p(D)Y_t = \alpha Y_t + \sigma(Y_t) \circ \dot{W}_t \quad , \quad t \in T \\ h(Y_{|\delta T}) = 0 \end{array} \right\}$$

for some convenient $\alpha \in \mathbb{R}$ (which very often can be taken to be zero). Usually, it is not difficult to find some explicit expression for the solution Y_t.

Step 2: Denote by Σ the set of trajectories of the process Y_t. We must identify this set and check that the mapping from Ω into Σ, defined by $\omega \mapsto Y(\omega)$, is bijective.

Step 3: Define a transformation $T: \Omega \to \Omega$ by $T(\omega) = \omega + F(\omega)$, with $F: \Omega \to H$. We must choose F in order to have the following:

   a) If $Z: \Omega \to \Omega$ satisfies $T(Z(\omega)) = \omega$, then $X(\omega) := Y(Z(\omega))$ solves (N).

   b) If X solves (N), then there exists such a Z.

F is usually found by inspection or by a formal manipulation of (N) and (L). The following fact is immediate:

Proposition. If T is exhaustive, there exists a solution to (N) whose paths belong to Σ. If, moreover, T is injective, the solution is unique, *within* the class of processes with paths in Σ. $\quad\square$

Example 1. Let us consider first an equation of the type (2.5):

(5.1)
$$\left.\begin{array}{c} \dot{X}_t = f(X_t) + \sigma \dot{W}_t \quad , \quad 0 \le t \le 1 \\ h(X_0, X_1) = 0 \end{array}\right\}$$

We assume here $d = 1$ and $\sigma > 0$. The linearized equation (L) is

(5.2)
$$\left.\begin{array}{c} \dot{Y}_t = \alpha Y_t + \sigma \dot{W}_t \quad , \quad 0 \le t \le 1 \\ h(Y_0, Y_1) = 0 \end{array}\right\}$$

The solution with initial condition Y_0 can be easily computed and (5.2) becomes equivalent to the system

(5.3)
$$\left.\begin{array}{c} Y_t = e^{\alpha t}\left(Y_0 + \displaystyle\int_0^t \sigma e^{-\alpha s}\, dW_s\right) \quad , \quad 0 \le t \le 1 \\ h\left(Y_0, e^{\alpha}\left(Y_0 + \displaystyle\int_0^1 \sigma e^{-\alpha s}\, dW_s\right)\right) = 0 \end{array}\right\}$$

Taking into account that the random variable $\int_0^1 e^{-\alpha s}\, dW_s$ is absolutely continuous and has the whole real line as support, (5.3) will have a solution iff $\forall z \in \mathbb{R}$ (a.e.), $h(y, e^{\alpha}y + z) = 0$ has a solution $y = g(z)$. And this happens when $h(X_0, X_1) = 0$ can be written as $X_0 = g(X_1 - e^{\alpha}X_0)$. Thus, this should be the case to take advantage of considering equation (L). For example, choosing α properly, the case of affine boundary conditions ($aX_0 + bX_1 + c = 0$) is fully covered. But observe also that with periodicity conditions ($X_0 = X_1$) we cannot take the simplest equation ($\alpha = 0$).

   The solution to (L) is then

$$Y_t = e^{\alpha t}\left[g\left(e^{\alpha}\int_0^1 \sigma e^{-\alpha s}\, dW_s\right) + \int_0^t \sigma e^{-\alpha s}\, dW_s\right] \quad .$$

It is immediate to prove that $\Sigma = \{\xi \in C([0,1]) : \xi_0 = g(\xi_1 - e^{\alpha}\xi_0)\}$ and that $Y: \Omega \to \Sigma$ is a bijection. This follows from the fact that any continuous function vanishing at zero is a trajectory of $\int_0^t e^{-\alpha s}\,dW_s$ (this integral has a sense pathwise).

To find the random variable F notice first that $T(\omega) = \omega + F(\omega)$ implies $T^{-1}(\omega) = \omega - F(T^{-1}(\omega))$. Then, using that $Y(T^{-1}(\omega))$ solves (N) and, at the same time, solves (L) when $W(\omega)$ is substituted by $W(T^{-1}(\omega))$, we find

$$F(\omega)_t = \frac{1}{\sigma}\int_0^t \left(\alpha Y_s(\omega) - f(Y_s(\omega))\right)ds \quad .$$

The bijectivity of T is true under some monotonicity or Lipschitz conditions on f (see Nualart and Pardoux [16], Propositions 2.2 and 2.5 and subsequent remarks).

Example 2. Consider the second order equation

$$(5.4) \qquad\qquad \ddot{X}_t = f(X_t) + \dot{W}_t \quad .$$

The solution of the corresponding equation (L)

$$(5.5) \qquad\qquad \ddot{Y}_t = \alpha Y_t + \dot{W}_t \quad ,$$

with initial conditions Y_0 and $\dot{Y}_0$, is, for $\alpha > 0$, and putting $\lambda = \sqrt{\alpha}$,

$$Y_t = Y_0 \cosh \lambda t + \dot{Y}_0 \frac{1}{\lambda}\sinh \lambda t + \int_0^t \cosh\left(\lambda(t-s)\right)W_s\,ds \quad .$$

For $\alpha < 0$, we get the same expression with the hyperbolic functions replaced by the corresponding trigonometric functions, and $\lambda = \sqrt{-\alpha}$. For $\alpha = 0$, we get simply

$$Y_t = Y_0 + \dot{Y}_0 t + \int_0^t W_s\,ds \quad .$$

A general boundary condition will have the form $h(X_0, \dot{X}_0, X_1, \dot{X}_1) = (0,0)$. As in Example 1, it is easy to find which form it must take (depending on α) to have a solution to (5.5). We would find that, for Dirichlet boundary conditions ($X_0 = 0$, $X_1 = 0$), one can take $\alpha = 0$, which is convenient to simplify later computations. But for Neumann ($\dot{X}_0 = 0$, $\dot{X}_1 = 0$) or periodicity ($X_0 = X_1$, $\dot{X}_0 = \dot{X}_1$) conditions, one must take $\alpha \neq 0$.

We must define

$$F(\omega)_t = \int_0^t \left(\alpha Y_s(\omega) - f(Y_s(\omega))\right)ds \quad ,$$

and it can be shown that if $y \mapsto \alpha y - f(y)$ is locally Lipschitz, nonincreasing, and with linear growth, then T is bijective.

Whatever boundary condition we prescribe, Y will be a bijection between Ω and the set Σ of $\mathcal{C}^1$ functions verifying the boundary conditions. This can be checked as in Example 1. The following example behaves differently concerning this point.

Example 3. Let us add a linear diffusion coefficient, depending only on $\dot{X}_t$, to the previous equation,

$$(5.6) \qquad\qquad \ddot{X}_t = f(X_t) + \sigma \dot{X}_t \circ \dot{W}_t \quad ,$$

and impose the boundary conditions $X_0 = 0$, $X_1 = 1$. Then, taking $\alpha = 0$ we can solve the associated equation (L) and get the solution

$$Y_t = \left(\int_0^1 e^{\sigma W_s}\, ds \right)^{-1} \cdot \int_0^t e^{\sigma W_s}\, ds \quad .$$

In this case, the set Σ consists of all $\mathcal{C}^1([0,1])$ functions ξ with $\xi' > 0$ in $[0,1]$ and $\xi_0 = 0$, $\xi_1 = 1$. Therefore, one is bound to prove existence and uniqueness of solutions to (5.6) within the class of processes with such trajectories. We must take

$$F(\omega)_t = \int_0^t -\frac{f(Y_s(\omega))}{\sigma \dot{Y}_s(\omega)}\, ds \quad .$$

This equation will be studied by Alabert and Nualart in a forthcoming paper. We remark that if we impose $X_0 = 0$, $X_1 = 0$, the solution to (L) is $Y_t \equiv 0$, and the bijectivity of $Y \colon \Omega \to \Sigma$ fails.

Example 4. Consider the partial differential equation (2.10), and assume ∂T is a smooth hypersurface. We must precise first what is meant by a solution to this equation. If the righthand side were a continuous function g, the solution would have the representation

$$X_t = \int_T -K(t,s)g(s)\, ds$$

for a symmetric kernel K. It is natural then to define a solution to (2.10) as a continuous process X_t verifying the integral equation

$$(5.7) \qquad X_t = \int_T -K(t,s)f(X_s)\, ds + \int_T -K(t,s)\, \dot{W}(ds) \quad .$$

This concept of solution coincides, on the other hand, with the one obtained thinking of (2.10) as an equation between distributions (see Buckdahn and Pardoux [3]).

Notice that $V_t = \int_T -K(t,s)\dot{W}(ds)$ defines a Gaussian measure $\dot{V}$ on the Banach space B of continuous functions on T vanishing at ∂T. We can substitute this term in (5.7). The auxiliary linear equation

(5.8)
$$\left.\begin{aligned} \Delta Y_t &= \dot{W}_t \quad, \quad t \in T \\ Y_{|\delta T} &= 0 \end{aligned}\right\}$$

has precisely the solution $Y_t = V_t$. Therefore, in this case, we can take $\Omega = \Sigma = B$, and Y is the identity mapping.

The restriction $k \leq 3$ is due to the fact that otherwise $K(t,s)$ is not square-integrable and the linear equation fails to have a solution (see Rozanov [21]).

F must be defined as

(5.9)
$$F(v)_t = \int_T K(t,s)f(v(s))\,ds \quad, \quad v \in B \quad,$$

and the bijectivity of T is obtained under the condition that f is continuous and non-decreasing (Donati-Martin [6], Lemma 3.1).

6. CHANGE OF MEASURE AND MARKOV PROPERTIES

In this Section we will describe the use of the transformations $T(\omega) = \omega + F(\omega)$, with $F: \Omega \to H$, to obtain conditions for a Markov type property to hold. Recall that we choose T in order to have $X_t = T^{-1}(Y)_t$, for X_t and Y_t solving equations (N) and (L), respectively. Therefore, defining $Q = P \circ T$, the law of Y under Q coincides with the law of X under P.

We decompose the procedure in several steps, as in the previous Section, to clarify the exposition. The objective is to obtain an equivalence between the Markov Field property and a measurability condition which depends only on the structure of the Carleman–Fredholm determinant $\det_2(I_{\tilde{H}} + DG(\omega))$. It is hoped that this measurability condition could be then translated into an analytical condition on the coefficients of the equation.

We assume we have performed steps 1 to 3 of Section 5, and that T is bijective.

Step 4: Check that, in the case of the linearized equation (L), the process we are interested in (the solution Y_t, the couple $(Y_t, \dot{Y}_t)$, or whatever) is a M.F. under the original probability P. Since possibly the process is in explicit form, there should be no difficulties in proving this fact by some direct method. Ocone–Pardoux [19], Rozanov [21], and Russek [22], for instance, give general results in this direction.

Step 5: Verify the hypothesis of the Ramer–Kusuoka Theorem and obtain $J(\omega) := \dfrac{dQ}{dP}(\omega)$. This involves, in principle, the computation of $\det_2(I_{\tilde{H}} +$

$DG(\omega)$), which is always rather cumbersome, but sometimes can be avoided, as in Example 4 below.

Step 6: Let us introduce the following notation: Given a subset D of the parameter space, such that $\overline{D} \subset T$, set

$$\mathcal{F}_D^i := \sigma\{Y_t,\ t \in \overline{D}\}, \quad \mathcal{F}_D^e := \sigma\{Y_t,\ t \in \overline{D^c}\}, \quad \mathcal{F}_D^b := \sigma\{Y_t,\ t \in \partial D\}.$$

We will write E_P and E_Q to denote the expectation taken with respect to probabilities P and Q, respectively. The following Lemma is an immediate consequence of the definition of conditional expectation and the Radon-Nikodým Theorem.

Lemma 1. For every random variable ξ, $\mathcal{F}_D^e$-measurable and Q-integrable,

$$(6.1) \qquad \mathrm{E}_Q\left[\xi/\mathcal{F}_D^i\right] = \frac{\mathrm{E}_P\left[\xi J/\mathcal{F}_D^i\right]}{\mathrm{E}_P\left[J/\mathcal{F}_D^i\right]} \qquad \square$$

The quotient is indeed well defined because the invertibility of $I_{\tilde{H}} + DG(\omega)$, which is assumed a.s., is equivalent to $\det_2(I_{\tilde{H}} + DG(\omega)) \neq 0$. Therefore, $J > 0$, a.s.

Now we have the equivalencies: X is a M.F. under P iff Y is a M.F. under Q iff $\forall D$ and $\forall \xi$, $\mathcal{F}_D^e$-measurable, $\mathrm{E}_Q\left[\xi/\mathcal{F}_D^i\right]$ is $\mathcal{F}_D^b$-measurable, iff the same property is true for the quotient in (6.1).

Thus, we get a measurability condition formulated again in terms of the original probability P, the σ-fields generated by the process Y and (at the cost of) the density J.

Step 7: Suppose at first that $J(\omega)$ can be written, for every Borel set D, as $J = L_D^e L_D^i$, with L_D^e and L_D^i random variables $\mathcal{F}_D^e$ and $\mathcal{F}_D^i$-measurable, respectively. Then, the Markov Field property holds: Indeed, using that Y is a M.F. under P (step 4), and this factorization for J, the quotient in (6.1) is clearly $\mathcal{F}_D^b$-measurable.

In all examples known, the exponential part of (4.1) can be factorized in this way, and it only remains to consider the Carleman–Fredholm determinant. In the adapted case, $\det_2(I_{\tilde{H}} + DG(\omega))$ is always equal to 1 and the factorization holds. Therefore, in the classical Wiener space, J reduces to Girsanov formula (recall that for adapted G, the Skorohod integral coincides with the Itô integral). This fact is literally "trivial": Zakai and Zeitouni show in [25] three different arguments.

Assume then that there is a partial factorization $J = ZL_D^e L_D^i$, for some Z. This simplifies the measurability condition:

$$\frac{\mathrm{E}_P\left[\xi J/\mathcal{F}_D^i\right]}{\mathrm{E}_P\left[J/\mathcal{F}_D^i\right]} = \frac{\mathrm{E}_P\left[\xi Z L_D^e/\mathcal{F}_D^i\right]}{\mathrm{E}_P\left[Z L_D^e/\mathcal{F}_D^i\right]},$$

and taking $\xi = \eta \cdot (L_D^e)^{-1}$, with η an $\mathcal{F}_D^e$-measurable random variable, we get that the Markov Field property is equivalent to the fact that

$$(6.2) \qquad \Lambda_\eta := \frac{\mathrm{E}_P\left[\eta Z / \mathcal{F}_D^i\right]}{\mathrm{E}_P\left[Z / \mathcal{F}_D^i\right]} \quad \text{is } \mathcal{F}_D^b\text{-measurable.} \quad \square$$

From this point onwards, the objective is to translate this characterization into an analytical one. The way to do this depends very much on the concrete problem under study, but in most cases the arguments employed make a fundamental use of the following Lemma:

Lemma 2. Let $F \in \mathbb{D}_{\mathrm{loc}}^{1,2}$, and M a closed subspace of $L^2(T)$. Denote by $\mathcal{F}_M$ the σ-field generated by the Gaussian random variables $\{\int_T h\,dW, \ h \in M\}$. Let $A \in \mathcal{F}_M$ and assume that $1_A F$ is $\mathcal{F}_M$-measurable. Then $DF(\omega) \in M$, for $\omega \in A$, a.s. $\quad \square$

For the proof, see for instance [15]. We illustrate steps 4 to 7 above with the examples that follow. In the second one, moreover, we sketch the derivation of the final analytical characterization, without going into details.

Finally, we notice that for the Germ Markov Field property, everything in this Section remains valid after changing the definition of $\mathcal{F}_D^b$.

Example 2 *(Continued)*. Consider again the second order equation (5.4), with Neumann boundary conditions ($\dot{X}_0 = 0, \dot{X}_1 = 0$). Take for instance $\alpha = 1$ in equation (L).

We have the solution

$$Y_t = -\frac{W_1 + \int_0^1 \sinh(1-s)W_s\,ds}{\sinh 1} \cdot \cosh t + \int_0^t \cosh(t-s)W_s\,ds \quad .$$

The process $(Y_t, \dot{Y}_t)$ is a M.F. (in fact, a Markov Process). This can be seen computing, for $t > s$, $\mathrm{E}\left[\psi(Y_t, \dot{Y}_t)/(Y_r, \dot{Y}_r), 0 \le r \le s\right]$ for a bounded measurable ψ by means of a regular version of the conditional probability. The result is a function of Y_s and $\dot{Y}_s$.

If $\bar{f}(y) := \alpha y - f(y)$, besides the conditions stated in Section 5, is of class $\mathcal{C}^2$, the hypothesis of the Ramer–Kusuoka Theorem are satisfied:

F is H–$\mathcal{C}^1$ because in fact it is a Fréchet continuously differentiable mapping $\Omega \to H$. We have $G(\omega)_s := (i^{-1} \circ F)(\omega)_s = Y_s(\omega) - f(Y_s(\omega))$ (cf. Section 4). The derivative operator satisfies a chain rule analogous to that of the ordinary derivative. This allows to compute the kernel of $DG(\omega)$:

$$D_t G(\omega)_s = \bar{f}'(Y_s) \cdot \left[-\frac{\cosh s}{\sinh 1} \cdot \cosh(1-t) - \sinh(1-t) \cdot 1_{\{t \ge s\}}\right] \quad .$$

From the Fredholm alternative, we have to check that the unique solution in $L^2([0,1])$ of the integral equation

$$(6.3) \qquad h(s) + \int_0^1 D_t G(\omega)_s h(t)\,dt = 0$$

is $h \equiv 0$.

Set $g(t) = \int_0^t \sinh(t - s)h(s)\,ds$. From (6.3), $g(t)$ must satisfy

$$g'(t) = g'(1) \cdot \int_0^t \Phi_{22}(t,s) \cdot \bar{f}'(Y_s) \cdot \frac{\cosh s}{\sinh 1}\,ds \quad ,$$

where $\Phi(t,s) = \Phi_t \cdot \Phi_s^{-1}$, and Φ_t is the solution to

$$\left. \begin{array}{cc} d\Phi_t = M_t \Phi_t\,dt & , \quad 0 \le t \le 1 \\ \Phi_0 = I & \end{array} \right\}$$

for $M_t = \begin{pmatrix} 0 & 1 \\ 1 - \bar{f}'(Y_t) & 0 \end{pmatrix}$.

The integrand is nonpositive, and we deduce $g'(1) = 0$, hence $h \equiv 0$. Therefore, we can apply the Theorem, and follow steps 6 and 7, to get the condition (6.2), with $Z = \det_2(I_{\tilde{H}} + DG(\omega))$. The same computations can be done taking any $\alpha > 0$, and the following analytical condition is derived (we refer to Nualart [14] for the final computations):

Theorem: If f is $\mathcal{C}^2$, with lineal growth, and $f' \ge \alpha$, for some $\alpha > 0$, then (5.4) has a unique solution X_t, and the process $\{(X_t, \dot{X}_t), 0 \le t \le 1\}$ is a M.F. iff $f'' = 0$.

Example 4 *(Continued).* In the previous Section we have set the Banach space $B = \{\omega \in C(\overline{T}) : \omega_{|\delta T} = 0\}$, with the supremum norm. Let μ be the law of $V: \Omega \to B$, which is a Gaussian measure on B. Instead of the Cameron–Martin space H relative to B and μ, we will choose an L^2 space to work in. Take $\tilde{H} = L^2(T)$. The isomorphism $i: \tilde{H} \to H$ is given by

$$h \mapsto \int_T K(t,\cdot)h(t)\,dt \quad .$$

Indeed, $i: \tilde{H} \to B$ is continuous with a dense image, and moreover every $\ell \in B^*$ is a random variable normally distributed, with mean zero and variance $\|\ell\|_H$. This characterizes (B, H, μ) as an abstract Wiener space and can be checked by computing the characteristic function of ℓ.

Recall the definition of the random variable F in (5.9). Since $F(\omega) = i(f \circ \omega)$, we have $G(\omega) = f \circ \omega$. Assuming $f \in \mathcal{C}^1$ and $f' > 0$, the hypothesis of the Ramer–Kusuoka Theorem are satisfied: It is immediate to compute $\nabla F(\omega)$ using Fréchet differentials:

$$[\nabla F(\omega)](h) = \int_T K(t,\cdot)f'(\omega_t)h(t)\,dt \quad .$$

Then, from the relation between $DG(\omega)$ and $\nabla F(\omega)$, we find the kernel

$$D_t G(\omega)_s = f'(\omega_s)K(t,s) \quad .$$

Moreover the mapping from $\tilde{H}$ into $L^2(T \times T)$ given by $h \mapsto f'(\omega_s + h_s)K(t,s)$ is obviously continuous.

We know that T is bijective. Finally, we need the invertibility of $I_{\tilde{H}} + DG(\omega)$. Equivalently, the integral equation

$$h_s + \int_T f'(\omega_s)K(t,s)h(t)\,dt = 0$$

must have $h \equiv 0$ as the unique solution in $L^2(T)$ (Fredholm alternative). Indeed, multiplying by $\frac{h(s)}{f'(\omega_s)}$ and integrating over T, we get

$$\int_T \frac{h^2(s)}{f'(\omega_s)}\,ds + \langle h, \mathcal{K}(h) \rangle_{\tilde{H}} = 0 \quad ,$$

where $\mathcal{K}(h) = \int_T K(t,\cdot)h(t)\,dt$. But it is known that $\langle h, \mathcal{K}(h) \rangle_{\tilde{H}} \geq 0$ (see, for instance, [3]). Both terms must be zero, from which $h \equiv 0$. Therefore, formula (4.1) holds true for the density $\frac{dQ}{dP}$.

The exponential part of (4.1) factorizes as explained in Step 7 (see [6]). Therefore, following steps 4 to 7, we arrive to the condition (6.2), with $\mathcal{F}_D^b = \bigcap_{\varepsilon > 0} \sigma\{Y_t,\ t \in (\partial D)_\varepsilon\}$ and $Z = \det_2(I_{\tilde{H}} + DG(\omega))$.

In the sequel, we are going to avoid the references to the set D and the space $\tilde{H}$, to simplify notation. Set also $\mathcal{F}^{b,\varepsilon} = \sigma\{Y_t,\ t \in (\partial D)_\varepsilon\}$. Let us assume that $f \in \mathcal{C}^2$ and $f' > 0$. We want to prove that X is a G.M.F. iff $f'' \equiv 0$.

One of the implications is obvious: If $f'' \equiv 0$, then Z is deterministic and the factorization holds. Conversely, suppose X is a G.M.F. We notice that $Z \geq 0$, because the eigenvalues of the operator with kernel $f'(\omega_s)K(t,s)$ are nonnegative. This is also a consequence of $\langle h, \mathcal{K}(h) \rangle \geq 0$.

For every $\varepsilon > 0$, Λ_η is $\mathcal{F}^{b,\varepsilon}$-measurable. We can assume $\eta \geq 0$ and smooth (in the sense of Section 4). Then $\Lambda_\eta \in I\!D_{\text{loc}}^{1,2}$, and by Lemma 2 above,

$$D.\Lambda_\eta \in \text{span}\{K(t,\cdot),\ t \in (\partial D)_\varepsilon\} \subset \tilde{H} \quad .$$

Let $\phi \in \mathcal{C}_{\text{comp}}^\infty(D - (\partial D)_\varepsilon)$. Then

(6.3) $\langle \phi., \Delta D.\Lambda_\eta \rangle = 0 \quad \text{and} \quad \langle \phi., \Delta D.\eta \rangle = 0 \quad .$

The operators Δ and D commute with the conditional expectations. Then, from (6.3), and using the chain rule for the operator D, we get

$$\mathrm{E}\left[Z/\mathcal{F}^i\right]\mathrm{E}\left[\eta\langle\phi., \Delta D.Z\rangle/\mathcal{F}^i\right] - \mathrm{E}\left[\eta Z/\mathcal{F}^i\right]\mathrm{E}\left[\langle\phi., \Delta D.Z\rangle/\mathcal{F}^i\right] = 0 \quad .$$

This implies that for every $\mathcal{F}^i$-measurable random variable ζ,

$$\mathrm{E}\left[\zeta\eta\,\mathrm{E}\left[Z/\mathcal{F}^i\right]\langle\phi., \Delta D.Z\rangle\right] = \mathrm{E}\left[\zeta\eta Z\,\mathrm{E}\left[\langle\phi., \Delta D.Z\rangle/\mathcal{F}^i\right]\right] \quad ,$$

which in turn gives

$$\mathrm{E}\left[Z/\mathcal{F}^i\right]\langle\phi.,\Delta D.Z\rangle = Z\,\mathrm{E}\left[\langle\phi.,\Delta D.Z\rangle/\mathcal{F}^i\right] \quad,$$

and we deduce that

(6.4) $\dfrac{1}{Z}\langle\phi.,\Delta D.Z\rangle$ is $\mathcal{F}^i$-measurable.

The Malliavin derivative of Z can be written

$$D_t Z = Z\cdot\mathrm{Tr}\left[(I + DG(\omega))^{-1}\circ(-DG(\omega))\circ D_t(DG(\omega))\right] \quad,$$

where Tr is the trace of the nuclear operator in brackets. Denoting by L the kernel of the Hilbert–Schmidt operator $(I + DG(\omega))^{-1}\circ(-DG(\omega)) = (I + DG(\omega))^{-1} - I$, from (6.4) we deduce that

$$f''(\omega_s)\int_T L(t,s)K(s,t)\,dt$$

is $\mathcal{F}^i$-measurable, for $s\in D-(\partial D)_\varepsilon$.

We can apply again Lemma 2, and repeat the arguments above, now with a function $\psi\in\mathcal{C}^\infty_{\mathrm{comp}}(D^c-(\partial D)_\varepsilon)$, to obtain

$$f''(\omega_s)f''(\omega_{\bar s})\left[K\cdot(I+DG(\omega))^{-1}\right](s,\bar s)\left[K\cdot(I+DG(\omega))^{-1}\right](\bar s,s) = 0 \quad,$$

for every $s\in D-(\partial D)_\epsilon$ and $\bar s\in D^c-(\partial D)_\epsilon$, from which it can be drawn that $f''(\omega_s) = 0$. This implies obviously $f'' = 0$. Thus we get:
Theorem: If $f\in\mathcal{C}^2$, with $f' > 0$, then (2.10) has a unique solution, which is a G.M.F. iff $f'' = 0$.   □

7. Other methods and equations

The most serious objection that can be raised against the change of measure method is to have to deal with the Carleman–Fredholm determinant of a certain integral operator. In example 4 of the last Section its explicit computation has been avoided, but this is not always possible.

With this in mind, some different procedures have been developed recently. In Alabert and Nualart [2], a new approach was proposed and applied to the following second order difference equation:

$$\text{(7.1)}\qquad \left.\begin{array}{c}X_{n+2} - 2X_{n+1} + X_n = f(X_{n+1}) + \xi_n \quad,\quad 0\le n\le N-2\\ X_0 = 0,\ X_N = 0\end{array}\right\}$$

where the variables $\{\xi_n\}_{n=0}^{N-2}$ are given and assumed to be independent. In case ξ_n are absolutely continuous with a strictly positive density on $I\!R$, then the pair (X_n, X_{n+1}) is a Markov Process iff f is affine. But if ξ_n have discrete laws, then (X_n, X_{n+1}) is *always* a Markov Process.

The method used in [2] is based in an application of the co-area formula from Geometric Measure Theory, and leads in a natural way to raise the following general question: Given two independent random variables Z_1 and Z_2, are they conditionally independent given some function $g(Z_1, Z_2)$?

Ferrante and Nualart [10] applied more direct arguments to

$$(7.2) \qquad \left. \begin{array}{c} X_{n+1} - X_n = f(X_n) + \sigma(X_n)\xi_n \quad , \quad 0 \le n \le N-1 \\ aX_0 + bX_N = c \end{array} \right\}$$

where f and σ are nonlinear and the noise is assumed positive and absolutely continuous. In this case, X_n is a M.F. iff

$$\begin{cases} x + f(x) = \beta x^\gamma \\ \sigma(x) = \alpha x^\gamma \end{cases}$$

for some $\alpha > 0$, $\beta > 0$ and $0 < \gamma \le 1$.

In Alabert, Ferrante and Nualart [1], the arguments used in these difference equations were refined and applied to

$$(7.3) \qquad \left. \begin{array}{c} dX_t = f(t, X_{t-})\,\mu(dt) + dW_t \quad , \quad 0 \le t \le 1 \\ X_0 = \psi(X_1) \end{array} \right\}$$

where μ is any finite positive measure. The change of measure scheme does not work in this case. X_t is a M.F. iff one of following conditions holds:

a) $\psi' \equiv 0$.

b) $\forall t \in \operatorname{supp}\mu$, $f(t, \cdot)$ is affine.

c) ψ' is constant, and *b)* holds except possibly in one only point t_0.

It turns out that the investigation of the Markov Field property can be reduced to the following particular form of the general question cited above: Given two independent σ-fields $\mathcal{F}_1$ and $\mathcal{F}_2$ of a probability space, and given two random variables determined as the solution of a system of the form

$$\left. \begin{array}{c} X = g_1(Y, \omega) \\ Y = g_2(X, \omega) \end{array} \right\}$$

where $g_i(y, \cdot)$ is $\mathcal{F}_i$-measurable $(i = 1, 2)$, under what conditions on g_1 and g_2 are $\mathcal{F}_1$ and $\mathcal{F}_2$ conditionally independent given X and Y? The answer is given in [1], Theorem 2.1.

Finally, we remark that sometimes it is possible to make a change of variables to get rid of a nonlinear diffusion coefficient, thus making easier the investigation of conditional independence properties. For instance, assume b, σ and ψ are $\mathcal{C}^1$ functions, with $\sigma > 0$. Then the problem

$$(7.4) \qquad \left. \begin{array}{c} dX_t = b(X_t)\,dt + \sigma(X_t) \circ dW_t \quad , \quad 0 \le t \le 1 \\ X_0 = \psi(X_1) \end{array} \right\}$$

can be transformed in

$$\left. \begin{array}{c} dX_t = f(X_t)\,dt + dW_t \quad , \quad 0 \le t \le 1 \\ X_0 = \varphi(X_1) \end{array} \right\}$$

and one finds that the solution to (7.4) is a M.F. iff either
  1) $\psi' \equiv 0$, or
  2) $b(x) = A\sigma(x) + B\sigma(x) \displaystyle\int_c^x \frac{1}{\sigma(t)}\,dt$, for some constants A, B, c.
(see also [1] for a more detailed discussion).

References

[1] A. Alabert, M. Ferrante, D. Nualart: *Markov field property of stochastic differential equations,* to appear in The Annals of Probability.

[2] A. Alabert, D. Nualart: *Some remarks on the conditional independence and the Markov property,* in *Stochastic Analysis and Related Topics,* Progress in Probability **31**, Birkhäuser (1992).

[3] R. Buckdahn, E. Pardoux: *Monotonicity methods for White Noise driven quasi-linear SPDEs,* in *Diffusion Processes and Related Problems in Analysis,* Progress in Probability **22**, Birkhäuser (1990).

[4] J. Cochran: *The Analysis of Linear Integral Equations,* McGraw-Hill (1972).

[5] C. Donati-Martin: *Equations différentielles stochastiques dans $I\!R$ avec conditions aux bords,* Stochastics and Stochastics Reports, **35** (1991).

[6] C. Donati-Martin: *Quasi-linear elliptic stochastic partial differential equations. Markov property,* Stochastics and Stochastics Reports, **41** (1992).

[7] C. Donati-Martin, D. Nualart: *Markov property for elliptic stochastic partial differential equations,* Preprint.

[8] M. Ferrante: *Triangular stochastic differential equations with boundary conditions,* Rendiconti Sem. Mat. Univ. Padova, **90** (1993).

[9] M. Ferrante, D. Nualart: *Markov Field property for stochastic differential equations with boundary conditions,* to appear in Stochastics and Stochastics Reports.

[10] M. Ferrante, D. Nualart: *On the Markov property of a stochastic difference equation*, Stochastic Processes and their Applications, **52** (1994).

[11] H. H. Kuo: *Gaussian Measures in Banach Spaces*, Lecture Notes in Mathematics, **463**, Springer-Verlag (1975).

[12] S. Kusuoka: *The nonlinear transformation of Gaussian measure on Banach space and its absolute continuity*, J. Fac. Sciences Univ. Tokio, Sec I.A., **29** (1982).

[13] T. Ikeda, S. Watanabe: *Stochastic differential equations and diffusion processes*, Second Edition, North Holland (1989).

[14] D. Nualart: *Nonlinear transformations of the Wiener measure and applications*, in *Stochastic Analysis, liber amicorum for Moshe Zakai*, Academic Press (1991).

[15] D. Nualart: *Markov fields and transformations of the Wiener measure*, in *Proceedings of the Oslo–Silivri Workshop on Stochastic Analysis*, Stochastics Monographs, **8**, Gordon and Breach (1993).

[16] D. Nualart, E. Pardoux: *Boundary value problems for stochastic differential equations*, The Annals of Probability, **19** (1991).

[17] D. Nualart, E. Pardoux: *Second order stochastic differential equations with Dirichlet boundary conditions*, Stochastic Processes and their Applications, **39** (1991).

[18] D. Nualart, E. Pardoux: *Markov field properties of solutions of White Noise driven quasi-linear parabolic PDE's*, Stochastics and Stochastics Reports, **48** (1994).

[19] D. Ocone, E. Pardoux: *Linear stochastic differential equations with boundary conditions*, Probability Theory and Related Fields, **82** (1989).

[20] R. Ramer: *On nonlinear transformations of Gaussian measures*, J. Functional Analysis, **15** (1974).

[21] Y. A. Rozanov: *Markov Random Fields*, Springer-Verlag (1982).

[22] A. Russek: *Gaussian n-Markovian processes and stochastic boundary value problems*, Z. Wahrscheinlichkeitstheorie verw. Gebiete, **57** (1980).

[23] A. S. Üstünel, M. Zakai: *Applications of the Degree Theorem to absolute continuity on Wiener space*, Probability Theory and Related Fields, **95** (1993).

[24] A. S. Üstünel, M. Zakai: *Transformations of the Wiener measure under non-invertible shifts*, Probability Theory and Related Fields, **99** (1994).

[25] M. Zakai, O. Zeitouni: *When does the Ramer formula look like the Girsanov formula?*, The Annals of Probability, **20** (1992).

The Martin Boundary of the Brownian Sheet

Oliver Brockhaus
Institut für Angewandte Mathematik
Universität Bonn

1 Introduction

Let P on $\Omega := C((0,\infty)^d, I\!R)$ denote the distribution of a d-parameter Brownian sheet and define $X_t(\omega) := \omega(t)$. The purpose of the paper is a description of the set $\mathcal{G}(\Pi)$ of all probability measures on Ω compatible with the family of conditional probabilities

$$\Pi = (P[\,.\mid X_t, t \notin R\,]),$$

with R ranging over all d-dimensional rectangles in $(0,\infty)^d$ with sides parallel to the axes. In particular, we show in Theorem 1 that the extremal elements of the convex set $\mathcal{G}(\Pi)$ are exactly the translations P^m of P by functions $m \in M \subset \Omega$,

$$M = \{t \mapsto \sum_{i=1}^d x_t^i + t_i y_t^i \mid x^i, y^i \in \Omega \text{ do not depend on the } i\text{-th coordinate}\}.$$

Thus M may be viewed as the Martin boundary of the Brownian sheet. This result extends Föllmer's result on the Martin boundary of infinite dimensional Brownian motion, cf. [F2], as well as an earlier result of the author, cf. [B]. Remark that a smooth function m is in M if and only if $Lm = 0$, with

$$L = \partial_{t_1}^2 \partial_{t_2}^2 \ldots \partial_{t_d}^2.$$

Alternatively, we give in Theorem 3 a description of $\mathcal{G}(\Pi)$ in terms of a stochastic differential equation in d parameters which generalizes results of T.Jeulin, M.Yor and the author, cf. [JY1] [JY2] [B]. Our approach can be applied to processes connected with the Brownian sheet by a rescaling of space and time. In particular, we obtain a characterization of the Martin boundary of the d-parameter Ornstein-Uhlenbeck process, cf. Röckner [Rö], Royer and Yor [RoY] for one-parameter versions of this result.

2 Definitions

Let $\Omega := C(T, I\!R)$, $T := (0, \infty)^d$ and define $X_t(\omega) := \omega(t)$. We consider the family of subfields of the Borel field $\mathcal{F} := \sigma(X_t, t \in T)$ defined by

$$\mathcal{F}_{s,t} := \sigma(X_t, t \in T - R_{s,t}) \qquad (s, t \in T),$$

with $R_{s,t}$ denoting the open rectangle $(s_1, t_1) \times \ldots \times (s_d, t_d) \subset T$ (or the empty set, if $s_i \geq t_i$ for some i). Let P on $(\Omega, \mathcal{F})$ denote the distribution of a d-parameter **Brownian sheet**, i.e., the distribution of a continuous Gaussian process $(B_t, t \in T)$ with mean 0 and covariance

$$E[\, B_s B_t \,] = (s_1 \wedge t_1) \cdot \ldots \cdot (s_d \wedge t_d) \qquad (s, t \in T).$$

On $(\Omega, \mathcal{F}, P)$ we define a d-parameter **Brownian bridge** $X^{s,t,\omega}$ with boundary values $(X_u(\omega), u \in T - R_{s,t})$ by

$$X^{s,t,\omega} := X^{s,t,\omega,1} \circ X^{s,t,\omega,2} \circ \ldots \circ X^{s,t,\omega,d},$$

with $X^{s,t,\omega,k} : \Omega \to \Omega$ given by

$$X_u^{s,t,\omega,k} := \left\{ \begin{array}{c} X_u + \frac{s_k - u_k}{t_k - s_k}(X_{u^k, t_k} - X_{u^k, t_k}(\omega)) \\ + \frac{u_k - t_k}{t_k - s_k}(X_{u^k, s_k} - X_{u^k, s_k}(\omega)) \quad : \quad u \in R_{s,t} \\ X_u(\omega) \quad : \quad else \end{array} \right.$$

and $u^k, v := (u_1, \ldots, u_{k-1}, v, u_{k+1}, \ldots, u_d)$.

The reason to call $X^{s,t,\omega}$ a Brownian bridge is the fact that $\pi_{s,t} f$, with

$$\pi_{s,t} f(\omega) := E^P[\, f(X^{s,t,\omega}) \,],$$

is a version of $E^P[\, f \mid \mathcal{F}_{s,t} \,]$ for all bounded and $\mathcal{F}$-measurable functions f. One can show that $\Pi := (\mathcal{F}_{s,t}, \pi_{s,t}, s, t \in T)$ is a specification in the sense of Föllmer [F1]. We denote the set of probability measures Q on $(\Omega, \mathcal{F})$ compatible with Π in the sense that

$$Q[\,. \mid \mathcal{F}_{s,t}\,] = \pi_{s,t} \qquad (s, t \in T)$$

by $\mathcal{G}(\Pi)$. Following [F1], we call the set $\mathcal{G}(\Pi)^e$ of extremal elements in the convex set $\mathcal{G}(\Pi)$ the **Martin boundary** of Π.

We conclude the section with two remarks on the Brownian bridge $X^{s,t,\omega}$.

Remark 1 Since $X_u^{s,t,\omega,k}(\omega) = X_u(\omega)$,

$$\begin{array}{rcl} X_u^{s,t,\omega,k} - X_u^{s,t,\omega,k}(\omega) & = & X_u - X_u(\omega) + \frac{s_k - u_k}{t_k - s_k}(X_{u^k, t_k} - X_{u^k, t_k}(\omega)) \\ & & + \frac{u_k - t_k}{t_k - s_k}(X_{u^k, s_k} - X_{u^k, s_k}(\omega)) \end{array}$$

for $u \in R_{s,t}$. This yields, by induction, the formula

$$X_u^{s,t,\omega} = X_u(\omega) + \sum_{\substack{I,J \subset \{1,\ldots,d\} \\ I \cap J = \emptyset}} (\prod_{i \in I} \frac{s_i - u_i}{t_i - s_i})(\prod_{j \in J} \frac{u_j - t_j}{t_j - s_j})(X_c - X_c(\omega)) \qquad (1)$$

for $u \in R_{s,t}$, with $c \in T$ defined by

$$c_i := \begin{cases} s_i : i \in J \\ t_i : i \in I \\ u_i : else \end{cases}.$$

Remark 2 If ω is smooth, then $X^{s,t,\omega} - X^{s,t,0}$ is the uniquely defined continuous solution of

$$\begin{cases} Lf = 0 & on\ R_{s,t} \\ f = \omega & on\ T - R_{s,t} \end{cases}, \qquad (2)$$

with $L := \partial_{t_1}^2 \partial_{t_2}^2 \ldots \partial_{t_d}^2$.

3   Description of the Martin boundary of Π

In order to determine $\mathcal{G}(\Pi)^e$ we define a set $M \subset \Omega$ by

$$M := \{\ t \mapsto \sum_{i=1}^d x^i(t) + t_i y^i(t) \mid$$
$$x^i, y^i \in \Omega \text{ do not depend on the } i\text{-th coordinate}, i = 1, \ldots, d\ \}(3)$$

Lemma 1 M is closed in Ω.

PROOF: Let us state the additional condition

$$x^i(t) = y^i(t) = 0 \qquad\qquad if\ t \in T\ and\ \exists j < i : t_j \in \{1,2\}. \qquad (4)$$

The idea of the proof is to show that the mapping

$$\Phi : ((x^i, y^i),\ i = 1, \ldots, d) \mapsto (\sum_{i=1}^d x^i(t) + t_i y^i(t),\ t \in T)$$

is an homeomorphism from the set of elements in Ω^{2d} satisfying (3) and (4) onto M. For brevity's sake, we write $((x^i, y^i))$ instead of $((x^i, y^i), i = 1, \ldots, d)$.

Φ is one to one: Let $m(t) = \sum_{i=1}^d x^i(t) + t_i y^i(t)$ $(t \in T)$, with $((x^i, y^i))$ satisfying (3) and (4). Now $\sum_{i=1}^j x^i(t) + y^i(t) = m(t)$ respectively $\sum_{i=1}^j x^i(t) + 2y^i(t) = m(t)$, if $t_j = 1$ respectively $t_j = 2$, allows to successively calculate x^j and y^j, $j = 1, \ldots, d$, from m. Remark that, in addition, this argument yields the continuity of the inverse function Φ^{-1}.

Φ is onto: Let $m \in M$. We construct inductively for $n = d, d-1, \ldots, 0$ pre-images $((x_n^i, y_n^i))$ of m satisfying (3) and

$$x_n^i(t) = y_n^i(t) = 0 \qquad if \ t \in T, \ n < i \ and \ \exists j < i : t_j \in \{1, 2\} :$$

Assume $((x_n^i, y_n^i))$ to be already constructed. Define $x_{n-1}^i := x_n^i$, $y_{n-1}^i := y_n^i$, $i > n$, as well as

$$x_{n-1}^n(t) := x_n^n(t) + \sum_{\substack{I, J \subset \{1, \ldots, n-1\} \\ I \cap J = \emptyset, \ I \cup J \neq \emptyset}} (\prod_{i \in I}(1 - t_i))(\prod_{j \in J}(t_j - 2))x_n^n(c), \qquad (5)$$

with $c \in T$ given by

$$c_i := \begin{cases} 1 : i \in J \\ 2 : i \in I \\ t_i : else \end{cases} .$$

In the same way, y_{n-1}^n is defined by (5) with x replaced by y. By (1) with $\omega = 0$ it is easy to see that

$$x_{n-1}^n(t) = y_{n-1}^n(t) = 0 \qquad if \ t \in T \ and \ \exists j \leq n-1 : t_j \in \{1, 2\}.$$

Each summand under the Σ-sign in (5) is affine in at least one of the co-ordinates $t_1, t_2, \ldots, t_{n-1}$. Therefore it is possible to define x_{n-1}^i, y_{n-1}^i, $i = 1, \ldots, n-1$, such that

$$\sum_{i=1}^n x_{n-1}^i(t) + t_i y_{n-1}^i(t) = \sum_{i=1}^n x_n^i(t) + t_i y_n^i(t) \qquad (t \in T).$$

This ends the proof. $\square$

Corollary 1 *M is the closure in $C(T, I\!R)$ of the set of all $f \in C^{2d}(T, I\!R)$ satisfying*

$$\partial_{t_1}^2 \ldots \partial_{t_d}^2 f = 0.$$

Now we state our main result.

Theorem 1 *Let Q denote a probability measure on $(\Omega, \mathcal{F})$. The following two statements are equivalent:*

1. $Q \in \mathcal{G}(\Pi)$

2. There exists a decomposition

$$X_t = B_t + \sum_{i=1}^d (X_t^i + t_i Y_t^i) \qquad (t \in T) \qquad (6)$$

of the coordinate mapping such that B is a Q-Brownian sheet Q-independent of the second summand. For any i, the pair (X^i, Y^i) does not depend on the i-th coordinate.

In particular, $\{P^m \mid m \in M\}$ is the Martin boundary of Π.

PROOF: We show $\mathcal{G}(\Pi)^e = \{P^m \mid m \in M\}$:

$\mathcal{G}(\Pi)^e \subset \{P^m \mid m \in M\}$: Let $Q \in \mathcal{G}(\Pi)^e$ and $R_{s_n,t_n} \uparrow T$. Due to Dynkin, cf. [D], there exists $\omega \in \Omega$ such that $\pi_{s_n,t_n}(\omega, .)$ converges weakly to Q. Writing $X_u^{s_n,t_n,\omega}$ as

$$X_u^{s_n,t_n,\omega} = (X_u^{s_n,t_n,\omega} - X_u^{s_n,t_n,0}) + (X_u^{s_n,t_n,0} - X_u) + X_u,$$

(1) shows that the second summand converges P-almost surely to 0. Therefore it remains to prove that the locally uniform limit of the sequence $(m_n := X^{s_n,t_n,\omega} - X^{s_n,t_n,0}, n \in I\!N)$ is an element of M. But this is true due to Lemma 1 together with the remark that there exists an element $\tilde{m}_n$ of M such that $m_n = \tilde{m}_n$ on R_{s_n,t_n}.

$\{P^m \mid m \in M\} \subset \mathcal{G}(\Pi)$: Let $m \in M$. Then one has $\tau_m(X^{s,t,\omega}) = X^{s,t,(\tau_m\omega)}$: Indeed, for $m_t = x_t^i + t_i y_t^i$, with $x^i, y^i \in \Omega$ not depending on the i-th coordinate,

$$(\tau_m X^{s,t,\omega})_u - X_u^{s,t,(\tau_m\omega)}$$

$$= \sum_{\substack{I,J\subset\{1,\ldots,d\} \\ I\cap J=\emptyset}} (\prod_{i\in I}\frac{s_i - u_i}{t_i - s_i})(\prod_{j\in J}\frac{u_j - t_j}{t_j - s_j})m_c$$

$$= X_u^{s,t,0}(m) = 0$$

due to (1) and since $X^{s,t,i,0}(m) = 0$. But this implies

$$E^{P^m}[\,(\pi_{s,t}f)\,g\,] = \int_\Omega P(d\omega)E^P[\,f(X^{s,t,\tau_m\omega})\,]\,(g \circ \tau_m)(\omega)$$

$$= \int_\Omega P(d\omega)E^P[\,(f \circ \tau_m)(X^{s,t,\omega})\,]\,(g \circ \tau_m)(\omega)$$

$$= E^P[\,(\pi_{s,t}(f \circ \tau_m))\,(g \circ \tau_m)\,]$$

$$= E^{P^m}[\,f\,g\,]$$

for bounded and $\mathcal{F}$- respectively $\mathcal{F}_{s,t}$-measurable functions f and g.   □

Remark 3 This theorem is related with Föllmer's result on the Martin boundary of infinite dimensional Brownian motion, cf. [F2]: His approach is based on a specification of the two-parameter Brownian sheet with respect to the family of rectangles

$$R_t^1 := (0,t) \times (0,\infty) \qquad (t > 0)$$

in $T = (0,\infty)^2$. Remark that there is no room for a convergence of the sets R_t^1 to T except in one direction. Therefore the Martin boundary is given by the closed set $M^1 \subset M$ consisting of functions m of type

$$(t_1, t_2) \mapsto t_2 y_{t_1} \qquad (y \in C((0,\infty), I\!R))$$

which is clearly isomorphic to $C((0,\infty), I\!R)$. In a first attempt to generalize this result we studied a specification based on rectangles

$$R_t^2 := (0, t_1) \times (0, t_2) \qquad (t_1, t_2 > 0),$$

cf. [B]. The Martin boundary $M^2 \subset M$ in this situation is given by functions

$$(t_1, t_2) \mapsto t_1 y_{t_2}^1 + t_2 y_{t_1}^2 \qquad (y^1, y^2 \in C((0,\infty), I\!R)),$$

while $x^1 = x^2 = 0$.

4 Transformations of the Brownian sheet

Our approach generalizes to processes which are related with the Brownian sheet by a rescaling of space and time. More precisely, let $\rho^k, \tau^k \in C((a_k, b_k),(0,\infty))$, with $-\infty \leq a_k < b_k \leq \infty$, and let, in addition, the τ^k be increasing bijections, $k = 1, \ldots, d$. We define the transformed process X^{tr} by

$$X_u^{tr} := \rho_u X_{\tau_u} \qquad (u \in R_{a,b}),$$

with $\rho_u := \rho_{u_1} \cdot \ldots \cdot \rho_{u_d}$ and $\tau_u := (\tau_{u_1}, \ldots, \tau_{u_d})$. Let $\Omega^{tr}, \mathcal{F}^{tr}$ and $\mathcal{F}_{s,t}^{tr}$ denote the images of $\Omega, \mathcal{F}$ and $\mathcal{F}_{s,t}$, respectively, under this mapping. Then, for f and g bounded and $\mathcal{F}_{s,t}^{tr}$- respectively $\mathcal{F}^{tr}$-measurable,

$$\begin{aligned} E^{P^{tr}}[\, f\, g\,] &= E^P[\, f(\rho_t X_{\tau_t}, t \in R_{a,b})\, g(\rho_t X_{\tau_t}, t \in R_{a,b} - R_{s,t})\,] \\ &= E^P[\,(\pi_{\tau_s, \tau_t}(f \circ X^{tr}))\, g \circ X^{tr}\,]. \end{aligned}$$

Therefore, $\pi_{s,t}^{tr} f$ defined by

$$(\pi_{s,t}^{tr} f)(X^{tr}(\omega)) := \pi_{\tau_s, \tau_t}(f \circ X^{tr})(\omega)$$

is a version of $E^{P^{tr}}[\, f \mid \mathcal{F}_{s,t}\,]$. In addition,

$$\Pi^{tr} := (\pi_{s,t}^{tr}, \mathcal{F}_{s,t}, s, t \in R_{a,b})$$

is a specification on $(\Omega^{tr}, \mathcal{F}^{tr})$. Obviously, the set of probability measures $\mathcal{G}_{\Pi^{tr}}$ is given by the images of probability measures in $\mathcal{G}_\Pi$ under the transformation.

The most famous example of a transformed process is the d-parameter **stationary Ornstein-Uhlenbeck process** with parameter m. This process can be obtained by a transformation defined by

$$\rho_t := e^{-m(t_1 + \ldots + t_d)} \qquad \tau_t := e^{2m(t_1 + \ldots + t_d)},$$

cf. [ReY]. Therefore, the Martin boundary of the d-parameter stationary Ornstein-Uhlenbeck process is given by

$$M^{tr} := \{\ t \mapsto \sum_{i=1}^d e^{-mt_i} x_t^i + e^{mt_i} y_t^i$$

$$\mid x^i, y^i \in \Omega \text{ not dependent on the } i\text{-th coordinate }\},$$

which is the closure in $C(I\!R^d, I\!R)$ of the set of all smooth solutions f of $L^{tr}f = 0$, with

$$L^{tr} := (\partial_{t_1}^2 - m^2) \ldots (\partial_{t_d}^2 - m^2).$$

In the one-parameter case $d = 1$ this result is due to Royer and Yor [RoY]. For $d = 2$ there are related results by Föllmer [F2] and Röckner [Rö], cp. Remark 3.

5 Characterization of $\mathcal{G}(\Pi)$ by a stochastic differential equation

The following theorem is due to T.Jeulin and M.Yor, cf. [JY2].

Theorem 2 *Let $d = 1$ and let Q denote a probability measure on $(\Omega, \mathcal{F})$. Then the following two statements are equivalent:*

1. $Q \in \mathcal{G}(\Pi)$

2. For all $0 < s < t$, the process

$$\tilde{X}^{s,t} := (X_u - X_s - \int_s^u dv \frac{X_v - X_s}{v - s}, \ s \le u \le t)$$

is a Q-Brownian motion Q-independent of $\mathcal{F}_{s,t}$. In addition, we assume the integral to be defined Q-almost surely for all $s > 0$.

PROOF: *1.$\Rightarrow$2.*: Due to Theorem 1, the coordinate process has a decomposition $X_t = B_t + \sum_{i=1}^d (X_t^i + t_i Y_t^i)$ w.r.t. Q, with (X, Y) a pair of random variables Q-independent of B and B $Q-$Brownian motion. Since the Wiener measure P on Ω satisfies *2.*, cf. [JY1], the assertion follows by $\tilde{X}^{s,t} = \tilde{B}^{s,t}$.

2.$\Rightarrow$1.: Since $\tilde{X}^{s,t}$ is a Brownian motion w.r.t. Q, one can apply Itô's formula or the identity of $L^2(T, dx)$-functions

$$\frac{1_{(u,t]}(r)}{r - s} = \frac{1_{(s,t]}(r)}{t - s} - \frac{1_{(s,u]}(r)}{u - s} + \int_u^t \frac{dv}{(v - s)^2} 1_{(s,v]}(r) \qquad (r \in T)$$

to obtain

$$\int_u^t \frac{d\tilde{X}_v^{s,t}}{v - s} = \frac{\tilde{X}_t^{s,t}}{t - s} - \frac{\tilde{X}_u^{s,t}}{u - s} + \int_u^t \frac{dv}{(v - s)^2} \tilde{X}_v^{s,t}$$

$$= \frac{X_t}{t - s} - \frac{X_u}{u - s} - \frac{1}{t - s} \int_s^t dv \frac{X_v - X_s}{v - s} + \frac{1}{u - s} \int_s^u dv \frac{X_v - X_s}{v - s}$$

$$+ \int_u^t \frac{dv}{(v - s)^2} X_v - \int_u^t \frac{dv}{(v - s)^2} \int_s^v dr \frac{X_r - X_s}{r - s}$$

$$= \frac{1}{s - u} X_u^{s,t,0} \qquad (s \le u \le t).$$

Hence

$$
\begin{aligned}
Q[\,f(X_u)\mid \mathcal{F}_{s,t}\,](\omega) &= Q[\,f(X_u^{s,t,\omega})\mid \mathcal{F}_{s,t}\,](\omega)\\
&= Q[\,f(\underbrace{X_u^{s,t,0}}_{(s-u)\int_u^t \frac{d\tilde{X}_v^{s,t}}{v-s}} + (X_u^{s,t,\omega} - X_u^{s,t,0}))\mid \mathcal{F}_{s,t}\,](\omega)\\
&= P[\,f(X_u^{s,t,0} + (X_u^{s,t,\omega} - X_u^{s,t,0}))\,]\\
&= \pi_{s,t}f(\omega),
\end{aligned}
$$

which ends the proof of the theorem. $\qquad\qquad\qquad\qquad\qquad\qquad$ □

Our purpose in this section is a generalization of Theorem 2 to $d > 1$. The definition of $\tilde{X}^{s,t} \in C(R_{s,t}, I\!R)$ is straightforward:

$$
\tilde{X}^{s,t} := \tilde{X}^{s,t,1} \circ \tilde{X}^{s,t,2} \circ \ldots \circ \tilde{X}^{s,t,d}, \tag{7}
$$

with

$$
\tilde{X}_u^{s,t,k} := X_u - X_{u^k,s_k} - \int_{s_k}^{u_k} dv\, \frac{X_{u^k,v} - X_{u^k,s_k}}{v - s_k}, \qquad (u \in R_{s,t})
$$

and $u^k, v := (u_1, \ldots, u_{k-1}, v, u_{k+1}, \ldots, u_d)$. Now we state the result:

Theorem 3 *Let $d \geq 1$ and let Q denote a probability measure on $(\Omega, \mathcal{F})$. Then the following two statements are equivalent:*

1. $Q \in \mathcal{G}(\Pi)$

2. *For all $0 < s_i < t_i$, $i = 1, \ldots, d$, the process $\tilde{X}^{s,t}$ is a Q-Brownian motion Q-independent of $\mathcal{F}_{s,t}$. In addition, the integrals in $\tilde{X}^{s,t}$ are assumed to be defined Q-almost surely.*

PROOF: The proof follows the lines of Theorem 2:

1.⇒2.: The application $\tilde{X}^{s,t,k}$ maps Brownian sheets to Brownian sheets (with the k-th coordinate restricted to $[s_k, t_k]$). Therefore $\tilde{X}^{s,t}$ is a Brownian sheet on $R_{s,t}$ w.r.t. P. Since the definition of $\tilde{X}^{s,t}$ in (7) does not depend on the order of the $\tilde{X}^{s,t,k}$, $k = 1, \ldots, d$, one obtains the independence of $\tilde{X}^{s,t}$ and $(X_u, u_i \notin [s_i, t_i]), i = 1, \ldots, d$, hence the independence of $\tilde{X}^{s,t}$ and $\mathcal{F}_{s,t}$, w.r.t. P by Theorem 2. In order to show *2.* for an arbitrary $Q \in \mathcal{G}(\Pi)$ remark that $\tilde{X}^{s,t} = \tilde{B}^{s,t}$, if X and B are related by formula (6).

2.⇒1.: Since $\tilde{X}^{s,t}$ is a Brownian sheet w.r.t. Q, one can apply an identity of $L^2(T, dx)$-functions as in Theorem 2 to obtain

$$
\prod_{i=1}^d (s_i - u_i) \int_{R_{u,t}} (\prod_{i=1}^d (v_i - s_i))^{-1} d\tilde{X}_v^{s,t} = X_u^{s,t,0} \qquad (u \in R_{s,t}).
$$

The same argument as in Theorem 2 ends the proof of the theorem. $\qquad\quad$ □

References

[B]   O.Brockhaus, *Sufficient statistics for the Brownian sheet*, Séminaire de Probabilités XXVII, Lecture Notes in Math. 1557, 44-52, Springer (1993).

[D]   E.B.Dynkin, *Sufficient Statistics and extreme points*, Annals of Prob. 6, 705-730 (1978).

[F1]   H.Föllmer, *Phase transition and Martin boundary*, Séminaire de Probabilités IX, Lecture Notes in Math. 465, 305-317, Springer (1975).

[F2]   H.Föllmer, *Martin Boundaries on Wiener Space*, Diffusion Processes and Related Problems in Analysis, volume I, Editor M.Pinsky, Progress in Probability 22, 3-16, Birkhäuser (1991).

[JY1] T.Jeulin, M.Yor, *Une décomposition non-canonique du drap brownien*, Séminaire de Probabilités XXVI, Lecture Notes in Math.1526, 322-347, Springer (1992).

[JY2] T.Jeulin, M.Yor, *Private communication* (1993).

[ReY] D.Revuz, M.Yor, *Continuous Martingales and Brownian Motion*, Springer (1991).

[Rö]   M.Röckner, *On the parabolic Martin boundary of the Ornstein-Uhlenbeck operator on Wiener space*, Annals of Prob. 20, 1063-1085 (1992).

[RoY] G.Royer, M.Yor, *Représentation intégrale de certaines mesures quasi-invariantes sur* $C(I\!R)$; *Mesures extrémales et propriété de Markov*, Ann.Inst.Fourier 26, 2, 7-24 (1976).

Neocompact sets and stochastic Navier-Stokes equations.

Nigel J. Cutland
School of Mathematics, University of Hull.
Hull HU6 7RX, UK.

H. Jerome Keisler
Department of Mathematics, University of Wisconsin.
Madison, WI 537506, USA.

Abstract

We give a detailed exposition of the use of neocompact sets in proving existence of solutions to stochastic Navier-Stokes equations. These methods yield new results concerning optimality of solutions.

1   Introduction

In this paper we give a detailed exposition of the way in which the recent work of S. Fajardo and H. J. Keisler [6] can be used to establish existence of solutions to stochastic Navier-Stokes equations. Fajardo & Keisler [6] develop general methods for proving existence theorems in analysis, with the aim of embracing the many particular existence theorems that can be proved rather easily using nonstandard analysis. The machinery developed centres round the notion of a *neocompact set* - which is a weakening of the notion of a compact set of random variables with values in a metric space M - and the notion of a *rich adapted probability space*, in which any countable chain of nonempty neocompact sets has a nonempty intersection.

In the papers [2, 3] Capiński & Cutland used nonstandard methods to greatly simplify some known existence proofs for the deterministic Navier-Stokes equations and (using similar methods) solved a longstanding problem concerning existence of solutions to general stochastic Navier-Stokes equations. The aim here is to show how the main results of these papers can be obtained using the neocompactness methods developed in [6]. In addition, these methods yield additional information concerning the nature of the set of solutions and existence of optimal solutions.

For the convenience of the reader we begin with a brief summary of the notions and results we shall need from [6].

2   Neocompact sets.

In this section we review the notion of a neocompact set from the paper [6]. Neocompact sets share many of the useful properties of compact sets. We shall introduce the notion in two contexts – probability spaces and adapted spaces.

If $\Omega = (\Omega, P, \mathcal{G})$ is a probability space and (M, ρ_M) is a complete metric space, then $L^0(\Omega, M)$ will denote the metric space of all measurable functions from Ω into M with the metric of convergence in probability,

$$\rho(x,y) = \inf\{\epsilon > 0 \,:\, P[\rho_M(x(\omega), y(\omega)) < \epsilon] \geq \epsilon\}.$$

For any set A we write $A^\epsilon = \{x \,:\, \rho(x, A) \leq \epsilon\}$.

We let Meas(M) be the space of all Borel probability measures on M with the Prohorov metric, and let law : $L^0(\Omega, M) \to$ Meas(M) be the function mapping each random variable x to the measure on M induced by x. This function is continuous.

Definition 2.1 *Let Ω be a probability space and let M, M' denote complete separable metric spaces. A set $B \subset L^0(\Omega, M)$ is called* **basic** *if either*

(1) *B is compact, or*

(2) *$B = \{x \,:\, law(x) \in C\}$ for some compact set $C \subset$ Meas(M).*

By the family of **neocompact sets** *over Ω we mean the collection of all subsets of $L^0(\Omega, M)$ obtained by repeated application of the following rules:*

(a) *Every basic set is neocompact.*

(b) *Finite unions of neocompact sets are neocompact.*

(c) *Finite and countable intersections of neocompact sets are neocompact.*

(d) *Finite cartesian products of neocompact sets are neocompact (where we identify $L^0(\Omega, M) \times L^0(\Omega, M')$ with $L^0(\Omega, M \times M')$ in the natural way).*

(e) *If $C \subset L^0(\Omega, M \times M')$ is neocompact, then the set*

$$\{x \,:\, (\exists y)(x, y) \in C\}$$

is neocompact (and the analogous property holds for each factor in a finite Cartesian product).

(f) *If $C \subset L^0(\Omega, M \times M')$ is neocompact and $D \subset L^0(\Omega, M')$ is basic neocompact and nonempty, then the set*

$$\{x \,:\, (\forall y \in D)(x, y) \in C\}$$

is neocompact (and the analogous property holds for each factor in a finite Cartesian product).

It is not hard to see that the family of compact sets is closed under all of the rules (a)–(f). In fact, the family of compact sets is closed under arbitrary intersections, and condition (f) holds for arbitrary nonempty sets D. One of the reasons that compact sets are useful in proving existence theorems is that they have the following property:

If C is a set of compact sets such that any finite subset of C has a nonempty intersection, then C has a nonempty intersection.

We define a *rich probability space* as one in which the neocompact sets have a weaker form of this property, called the *countable compactness property*.

Definition 2.2 *A collection C of sets has the* **countable compactness property** *if the intersection of any countable decreasing chain $C_1 \supset C_2 \supset \cdots$ of nonempty sets in C is nonempty.*

Definition 2.3 *A probability space Ω is said to be* **rich** *if it is atomless and for any complete separable metric space M, the collection of neocompact sets in $L^0(\Omega, M)$ has the countable compactness property.*

We now turn to neocompact sets in adapted spaces. By an *adapted space* we shall mean a structure $\Omega = (\Omega, P, \mathcal{G}, \mathcal{G}_t)$ where t runs over the dyadic rationals and the $\mathcal{G}_t$ are σ-subalgebras of $\mathcal{G}$ which increase in t. For each real s, we let $\mathcal{F}_s$ be the P-completion of $\bigcap_{t>s} \mathcal{G}_t$.

Definition 2.4 *Let Ω be an adapted space. The families of* **basic** *sets and* **neocompact** *sets for the adapted space are defined exactly as for the case of a probability space except that we add to the family of basic sets all sets B of the form*

(3) $B = \{x \in L^0(\Omega, M) \,:\, \mathrm{law}(x) \in C$ *and x is $\mathcal{G}_t$-measurable*$\}$,

where C is compact in $\mathrm{Meas}(M)$ and t is a dyadic rational.

Definition 2.5 *An adapted space $\Omega = (\Omega, P, \mathcal{G}, \mathcal{G}_t)$ is said to be* **rich** *if the probability space $(\Omega, P, \mathcal{G}_0)$ is atomless, Ω admits a Brownian motion, and for any complete separable metric space M, the collection of neocompact sets in $L^0(\Omega, M)$ has the countable compactness property.*

The following fact is implicit in the paper [12] and will be proved explicitly in [7].

Theorem 2.6 *Rich probability spaces and rich adapted spaces exist.*

This is proved by showing that the adapted Loeb spaces, which were the underlying spaces used in [2] and [3], are rich. Every rich adapted space is also rich as a probability space.

The analogues of closed sets and continuous functions are defined in terms of neocompact sets in the following way.

Definition 2.7 *A set $C \subset L^0(\Omega, M)$ is* **neoclosed** *if $C \cap D$ is neocompact for every neocompact set D.*

A function f mapping a neoclosed set $C \subset L^0(\Omega, M)$ into $L^0(\Omega, N)$ is **neocontinuous** *if for every neocompact set $D \subset C$, the graph of f restricted to D is neocompact.*

It is shown in the paper [6] that for a rich adapted space, every neoclosed set is closed, and every neocontinuous function is continuous. Parallel to the classical case, images of neocompact sets under neocontinuous functions are neocompact, and preimages of neoclosed sets under neocontinuous functions are neoclosed. We shall identify N with the set of constant functions from Ω into N. With this identification, each closed subset of N becomes a neoclosed subset of $L^0(\Omega, N)$, and we obtain the notions of a neocontinuous function from $L^0(\Omega, M)$ into N, and a neocontinuous function from a closed subset of M into $L^0(\Omega, N)$.

In the paper [6], several important sets and functions are shown to be neocompact, neoclosed, or neocontinuous for a rich adapted space.

For example, it is shown that the set of Brownian motions on Ω is neocompact, the set of stopping times between 0 and 1 for Ω is neocompact, and the set of all $\mathcal{F}_t$-adapted stochastic processes with values in M is neoclosed.

Some examples of neocontinuous functions which will be used in this paper are the distance function ρ for $L^0(\Omega, M)$, the function $x(\cdot) \mapsto f(x(\cdot))$ where $f : M \to N$ is continuous, the expected value function $x(\cdot) \mapsto E(x(\cdot))$ restricted to a uniformly integrable subset of $L^0(\Omega, \mathbb{R})$, and the stochastic integral function $x \mapsto \int x \, db$ where b is a Brownian motion and x is in a bounded set of adapted processes.

Moreover, compositions of neocontinuous functions are neocontinuous.

In the paper [6] the notion of *neo-lower semicontinuity*, abbreviated *neo-lsc*, is defined as a useful generalisation of the classical notion of lowersemicontinuity. Recall that a function $f : M \to \overline{\mathbb{R}}$ is lower semicontinuous (lsc) if whenever $x_n \to x$ in M then $\underline{\lim}_{n \to \infty} f(x_n) \geq f(x)$; equivalently, f is lsc if for every compact $C \subseteq M$, the upper graph

$$\{(x, r) \in C \times \overline{\mathbb{R}} : f(x) \leq r\}$$

is compact. This is the definition that is generalised in [6]. Regard the set $\overline{\mathbb{R}} = [-\infty, \infty]$ as a compact metric space with the metric given by $\rho(r, s) = |\arctan(r) - \arctan(s)|$ and we have:

Definition 2.8 *Let* $D \subseteq L^0(\Omega, M)$; *a function* $f : D \to L^0(\Omega, \overline{\mathbb{R}})$ *is* **neo-lsc** *if for every neocompact set* $C \subseteq D$ *the upper graph*

$$\{(x, y) \in C \times L^0(\Omega, \overline{\mathbb{R}}) : f(x) \leq y \text{ a.s.}\}$$

is neocompact.

If $f : D \to \overline{\mathbb{R}}$ it is easy to check that f is neo-lsc if and only if the upper graph $\{(x, r) \in C \times \overline{\mathbb{R}} : f(x) \leq r\}$ is neocompact for each neocompact C.

It is shown in [6] that the expectation operator restricted to positive random variables is neo-lsc, and that a composition $g \circ f$ of two neo-lsc functions is neo-lsc provided that g is monotone. Also, if $f : M \to \overline{\mathbb{R}}$ is lsc then the function $g : L^0(\Omega, M) \to L^0(\Omega, \overline{\mathbb{R}})$ defined by $g(x)(\omega) = f(x(\omega))$ is neo-lsc.

We shall need the following important lemma from [6].

Lemma 2.9 *(Closure Under Diagonal Intersections.) Let* Ω *be either a rich probability space or a rich adapted space. Let* A_n *be neocompact in* $L^0(\Omega, M)$ *for each* $n \in \mathbb{N}$, *and let* $\epsilon_n \searrow 0$. *Then the set* $A = \bigcap_n ((A_n)^{\epsilon_n})$ *is neocompact in* $L^0(\Omega, M)$.

The key result from [6] which we shall apply in this paper is the following theorem.

Theorem 2.10 *(Approximation Theorem.) Let* Ω *be either a rich probability space or a rich adapted space. Let* A *be neoclosed in* $L^0(\Omega, M)$ *and let* $f : L^0(\Omega, M) \to L^0(\Omega, N)$ *be neocontinuous. Let* $B \subset L^0(\Omega, M)$ *and* $D \subset L^0(\Omega, N)$ *be neocompact. Suppose that for each* $\epsilon > 0$

$$(\exists x \in A \cap B^\epsilon) f(x) \in D^\epsilon.$$

Then

$$(\exists x \in A \cap B) f(x) \in D.$$

It is easy to check that the analogous result holds with compact, closed, and continuous in place of neocompact, neoclosed, and neocontinuous. We shall use this classical analogue as a warmup in Section 4, and then apply the Approximation Theorem for rich probability spaces and rich adapted spaces later in this paper.

3   The Navier-Stokes equations.

The classical Navier-Stokes equations describe the evolution in time of the velocity field $u : D \to \mathbb{R}^n$ of an incompressible fluid in a domain $D \subseteq \mathbb{R}^n$, so we are thinking of a function $u(x, t) : D \times [0, T] \to \mathbb{R}^n$ given by:

$$\begin{cases} \dfrac{\partial u}{\partial t} - \nu \Delta u + \langle u, \nabla \rangle u + \nabla p = f \\ \operatorname{div} u = 0 \end{cases} \tag{1}$$

(where $\langle , \rangle$ denotes the inner product in $\mathbb{R}^n$). For convenience we will assume a fixed finite time horizon T but there is no difficulty in extending to the time set $[0, \infty)$. The domain D is bounded with boundary of class C^2 and we will work with the homogeneous Dirichlet boundary condition $u|_{\partial D} = 0$. Of course in important applications, $n = 3$ but from the mathematical point of view we can allow $n \leq 4$. In this equation, p denotes the pressure, and f denotes the external forces.

The usual setting for these equations involves the function spaces $\mathbf{H}$, $\mathbf{V}$ which are obtained by closing the set $\{u \in C_0^\infty(D, \mathbb{R}^n) : \operatorname{div} u = 0\}$ in the norms $|\cdot|$ and $|\cdot| + \|\cdot\|$ respectively, where

$$|u| = (u, u)^{\frac{1}{2}}; \quad (u, v) = \sum_{j=1}^n \int_D u^j(\xi) v^j(\xi) \mathrm{d}\xi,$$

$$\|u\| = ((u, u))^{\frac{1}{2}}; \quad ((u, v)) = \sum_{j=1}^n \left(\frac{\partial u}{\partial \xi_j}, \frac{\partial v}{\partial \xi_j} \right).$$

$\mathbf{H}$, $\mathbf{V}$ are Hilbert spaces. We fix an orthonormal basis $(e_n)_{n \in \mathbb{N}}$ for $\mathbf{H}$ consisting of eigenvectors of the operator $-\Pi \circ \Delta$ (where Π is the orthogonal projection from $L^2(D, \mathbb{R}^n)$ onto $\mathbf{H}$) with eigenvalues $(\lambda_n)_{n \in \mathbb{N}}$ (see sec. 2.5 of [15]). For $u \in \mathbf{H}$ we write $u_k = (u, e_k)$. We let $\mathbf{V}'$ be the dual space of $\mathbf{V}$ with respect to the $|\cdot|$ norm, and let $(\cdot, \cdot)$ denote the duality between $\mathbf{V}'$ and $\mathbf{V}$ extending the scalar product in $\mathbf{H}$. In the equation (1) it is usual to take the force $f \in L^2(0, T; \mathbf{V}')$, and then the equation is understood as a Bochner integral equation in $\mathbf{V}'$. i.e. for each $v \in \mathbf{V}$:

$$(u(t), v) - (u_0, v) = \int_0^t [-\nu((u(s), v)) - b(u(s), u(s), v) + (f(s), v)] \mathrm{d}s \qquad (2)$$

where u_0 is the given initial condition. The pressure vanishes in this weak formulation because $\langle \nabla p, v \rangle = -\langle p, \operatorname{div} v \rangle = 0$, but of course p can be recovered from a solution to the equation (2). The trilinear form b is the nonlinear term in (1), so that we have:

$$b(u, v, w) = \sum_{i,j=1}^n \int_D u^j(\xi) \frac{\partial v^i}{\partial \xi_j}(\xi) w^i(\xi) \mathrm{d}\xi = ((u, \nabla)v, w).$$

Note that for $u, v, z \in \mathbf{V}$ we have $b(u, v, z) = -b(u, z, v)$ so that $b(u, v, v) = 0$. There are a number of well known inequalities giving continuity of b in various topologies (see [14] and [15] p.12 for example) and we list here those (for $n \leq 4$) that we shall need.

$$|b(u, v, z)| \leq c\|u\| \|v\| \|z\| \qquad (3)$$

$$|b(u, v, z)| \leq c|u| \|v\| |Az| \qquad (4)$$

$$|b(u, v, z)| \leq c|u| |Av| \|z\| \qquad (5)$$

It is customary to write A for the operator $-\Pi \circ \Delta$ on $\mathbf{H}$ (see [15] p.15 for example), and to write $B(u) = b(u, u, \cdot) \in \mathbf{V}'$ for appropriate u. Then it is an easy consequence of (4) and (5) that:

Proposition 3.1 *For all m, $B(\cdot) \in C(K_m, \mathbf{V}'_{weak})$, where K_m is the compact subset of $\mathbf{H}$ given by $K_m = \{u \in \mathbf{H} : \|u\| \leq m\}$ with the $\mathbf{H}$-topology.* $\square$

Proof See Proposition 3.4 of [3].

We can now make precise what is taken to be a weak solution to (1)

Definition 3.2 *Given $u_0 \in \mathbf{H}$ and $f \in L^2(0, T; \mathbf{V}')$ we say that the function $u : [0, T) \to \mathbf{H}$ is a **weak solution** of the Navier-Stokes equations if*

    (i) $u \in L^2(0, T; \mathbf{V}) \cap L^\infty(0, T; \mathbf{H})$,
    (ii)   *for all $v \in \mathbf{V}$, for all $t \geq 0$, u satisfies equation (2)*

The regularity condition (i) emerges naturally by consideration of the time evolution of the energy $\frac{1}{2}|u|^2$.

The spaces $\mathbf{H}$ and $\mathbf{V}$ involved in the formulation of the Navier-Stokes equations are two from the spectrum of Hilbert spaces $\mathbf{H}^\alpha$ given by

$$\mathbf{H}^\alpha = \{u \in \mathbf{H} : \sum_{k=1}^{\infty} \lambda_k^\alpha u_k^2 < \infty\}$$

for $\alpha \geq 0$, with norm $|u|_\alpha = (\sum_{k=1}^{\infty} \lambda_k^\alpha u_k^2)^{\frac{1}{2}}$. The dual space of $\mathbf{H}^\alpha$ is written $\mathbf{H}^{-\alpha}$. For $u \in \mathbf{H}^{-\alpha}$ the norm $|u|_{-\alpha}$ is given by

$$|u|_{-\alpha} = \left(\sum_{k=1}^{\infty} \lambda_k^{-\alpha} u_k^2 \right)^{\frac{1}{2}}$$

where $u_k = u[e_k]$, and we have $u = \sum u_k e_k$ with the limit being taken in $\mathbf{H}^{-\alpha}$. Thus the elements of $\mathbf{H}^{-\alpha}$ can be represented by sequences (u_k) with

$$\sum_{k=1}^{\infty} \lambda_k^{-\alpha} u_k^2 < \infty.$$

It is easily checked that $\mathbf{H} = \mathbf{H}^0$, $\mathbf{V} = \mathbf{H}^1$ and $\mathbf{V}' = \mathbf{H}^{-1}$.

We write $\mathbf{H}_n = \text{span}\{e_1, \ldots, e_n\}$ and denote the projection from $\mathbf{H}$ onto $\mathbf{H}_n$ by Π_n. For $u \in \mathbf{H}$, $u^{(n)}$ denotes $\Pi_n u$.

4 Existence of weak solutions

In this section we prove

Theorem 4.1 *For every $u_0 \in \mathbf{H}$ there is a weak solution u to the Navier-Stokes equations with $u(0) = u_0$ and with u in the space M_0 defined below.*

This is, of course, a classical result. We shall present a proof using the classical analogue of the Approximation Theorem (Theorem 2.10) and compare it with a classical proof. The point of doing this is to show how the neocompactness machinery allows a direct generalization of our proof to give the existence result for the stochastic Navier-Stokes equations [3], for which there is no classical proof using compactness.

For the development here and in subsequent sections we define some spaces and auxiliary functions that will be important for the construction of solutions. We put

$$M = C([0, T], \mathbf{H}^{-2}) \cap \{y : y(0) \in \mathbf{H}\}$$

which is a complete separable metric space with the metric $|\cdot|_M$ given by

$$|y_1 - y_2|_M = \sup_{t \leq T} |y_1(t) - y_2(t)|_{-2} + |y_1(0) - y_2(0)|.$$

For $y \in M$ define $\theta(y)$ by

$$\theta(y) = \nu \int_0^T \|y(t)\|^2 \mathrm{d}t + \sup_{t \leq T} |y(t)|^2.$$

Let M_0 be the subset of M given by

$$M_0 = M \cap \{y : \theta(y) \leq K + |y(0)|^2\}$$

where $K = \nu^{-1} \int_0^T |f(t)|^2_{\mathbf{V}'} \mathrm{d}t$.

Proposition 4.2 *M_0 is a closed subspace of M.*

Proof   This follows easily from the fact that if u, $u^n \in \mathbf{H}^{-2}$ with $u^n \to u$ in $\mathbf{H}^{-2}$, then $|u|_\alpha \leq \underline{\lim}|u^n|_\alpha$ for any α. Apply this to $|y^n(t)| = |y^n(t)|_0$ and $\|y^n(t)\| = |y^n(t)|_1$ for any sequence $y^n \in M_0$ with $y^n \to y \in M$, and use Fatou's lemma $(\int_0^T \underline{\lim}\|y^n(t)\|^2 dt \leq \underline{\lim} \int_0^T \|y^n(t)\|^2 dt)$ and the fact that $\sup_{t \leq T} \underline{\lim}|y^n(t)|^2 \leq \underline{\lim}\sup_{t \leq T} |y^n(t)|^2$. $\square$

Remark It is easy to check that $M_0 \subseteq C([0, T], \mathbf{H}^{-\alpha})$ for all $\alpha > 0$, although we will not need this. It is a consequence of the fact that for $y \in M_0$ each $y_k(t)$ is continuous, and $|y(t)|$ is bounded.

Now for each $k > 0$ define

$$M_k = M \cap \{y : \theta(y) \leq k\}$$

Clearly each M_k is closed in M and $M_0 \subseteq \bigcup_{k>0} M_k = M_\infty$, say.

We now define a function $\gamma : M_\infty \to M$ by

$$\gamma(y)(s) = \int_0^s [-\nu Ay(t) - B(y(t)) + f(t)]\mathrm{d}t.$$

To see that γ takes its values in M, use the following key facts about A and B:

$$\begin{array}{ll} \text{(i)} & |Au|_{-1} = \|u\| \\ \text{(ii)} & |B(u)|_{-1} \le c\|u\|^2 \end{array} \qquad (6)$$

with (ii) holding by (3). These properties of A and B ensure that for $y \in M_\infty$ we have

$$\begin{aligned} \gamma(y) & \in C([0,T], \mathbf{H}_{weak}^{-1}) \cap L^\infty(0,T; \mathbf{H}^{-1}) \\ & \subseteq C([0,T], \mathbf{H}^{-\alpha}) \end{aligned}$$

for any $\alpha > 1$ and in particular for $\alpha = 2$.

Remark This shows that the choice of $\mathbf{H}^{-2}$ in the definition of M was somewhat arbitrary. We could have taken $\mathbf{H}^{-\alpha}$ for any $\alpha > 1$.

We observe that $u = y$ is a weak solution to the Navier-Stokes equations with initial condition $u_0 = x$ if and only if $y = x + \gamma(y)$.

The next two facts are the key to the construction of a solution.

Proposition 4.3 *The function γ is continuous on each set M_k (in the topology of M).*

Proof This is routine Bochner integration theory using the facts (6) above. $\square$

Proposition 4.4 *For each $k > 0$ the set $\gamma(M_k)$ is relatively compact in M.*

Proof The facts (6) can be used to show that there are constants d and $(d_m)_{m \ge 1}$ such that

$$\begin{aligned} \gamma(M_k) \subseteq \{z \in M : z(0) = 0 \ \& \ \sup_{t \le T} |z(t)|_{-1} \le d \\ \& \ |z_m(s) - z_m(t)|^2 \le d_m|s-t| \text{ for all } s,t \le T \text{ and } m \in \mathbb{N}\}. \end{aligned}$$

It is then routine to check that the set on the right is compact in M. $\square$

As with classical proofs, we now consider the finite dimensional approximations to (2) - known as the *Galerkin approximations*. For this purpose we define a sequence of functions $\gamma_n : M_\infty \to M$ by

$$\gamma_n(y)(s) = \int_0^s [-\nu Ay^{(n)}(t) - B^{(n)}(y(t)) + f^{(n)}(t)]\mathrm{d}t$$

$$= \Pi_n \gamma(y)(s).$$

It is clear that $\gamma_n(y) \in C([0,T], \mathbf{H}_n)$.

Classical ODE theory (see [14] for example) shows:

Theorem 4.5 *For each $n > 0$ and $x \in \mathbf{H}$ there is a unique $y \in M_0$ such that*
$y = x^{(n)} + \gamma_n(y)$ *(so $y \in M_{K+|x|^2}$).*

Proof   The equation $y = x^{(n)} + \gamma_n(y)$ for $x \in \mathbf{H}$ is simply the Galerkin approximation to the Navier-Stokes equations in dimension n, with initial condition $x^{(n)}$, which has a unique solution $y \in C([0,T], \mathbf{H}_n)$; consideration of the time evolution of the energy $|y(t)|^2$ shows that

$$|y(s)|^2 + \nu \int_0^s \|y(t)\|^2 \mathrm{d}t \le |y(0)|^2 + \nu^{-1} \int_0^s |f(t)|^2_{\mathbf{V}'} \mathrm{d}t$$

so that $y \in M_0$. Clearly, $y \in M_{K+|x|^2}$. $\square$

The final result in preparation for the fundamental existence result Theorem 4.1 is:

Proposition 4.6 *For $k > 0$, $\gamma_n(y) \to \gamma(y)$ uniformly on M_k.*

Proof   By Proposition 4.4, $\overline{\gamma(M_k)}$ is a compact subset of M. Dini's theorem tells us that $\mathrm{II}_n z \to z$ uniformly on any any compact set. $\square$

We now complete the proof of 4.1 by using the classical analogue of the Approximation Theorem (2.10).

Proof of Theorem 4.1   Using the classical analogue of the Approximation Theorem, it is sufficient to prove that there is a compact $D \subseteq M$ such that

$$(\forall \varepsilon > 0)(\exists y \in D^\varepsilon \cap M_0)(|x + \gamma(y) - y|_M < \varepsilon).$$

For this we take $D = x + \overline{\gamma(M_k)}$ where $k = K + |x|^2$, which is compact by Proposition 4.4. Now, given $\varepsilon > 0$ take n such that $|\gamma_n(y) - \gamma(y)| < \frac{1}{2}\varepsilon$ for all $y \in M_k$ (using Proposition 4.6), and $|x - x^{(n)}| < \frac{1}{2}\varepsilon$. Take the unique $y \in M_0$ with $y = x^{(n)} + \gamma_n(y) \in M_k$, given by Proposition 4.5. Then

$$|x + \gamma(y) - y| \le |\gamma(y) - \gamma_n(y)| + |x - x^{(n)}| < \varepsilon$$

and since $x + \gamma(y) \in D$ we have $y \in D^\varepsilon$ and the proof is complete. $\square$

This proof is really not very different from a classical proof using the preliminary results Propositions 4.4, 4.5 and 4.6 above, which runs as follows:

For $x \in \mathbf{H}$, take $y_n = x^{(n)} + \gamma_n(y_n) \in M_0 \cap M_k$ as given by Proposition 4.5, and let $z_n = x + \gamma(y_n) \in D$. By Proposition 4.4, there is a convergent subsequence $z_{n_i} \to z$, say with $z \in D$. Then

$$|y_{n_i} - z| \le |x^{(n_i)} + \gamma_{n_i}(y_{n_i}) - (x + \gamma(y_{n_i}))| + |z_{n_i} - z| \to 0$$

using Proposition 4.6. So z belongs to the closed set M_0. By Proposition 4.3 we have

$$x + \gamma(z) = x + \lim \gamma(y_{n_i}) = \lim z_{n_i} = z,$$

so that z is the required solution.

5 Existence of statistical solutions

The idea of a *statistical solution* to the Navier-Stokes equations was developed by Foias [8], and the idea is as follows. Suppose that in the Navier-Stokes equations (1) or (2) the initial condition is given by a probability measure μ_0 on **H**, with the informal idea that there is some underlying probability P such that for $A \subseteq \mathbf{H}$

$$\mu_0(A) = P(u_0 \in A).$$

Then, informally, as time evolves we can think of measures μ_t given by

$$\mu_t(A) = P(u(t) \in A). \tag{7}$$

However, this idea is only heuristic, because we do not have any meaning for the value $u(t)$ for a random initial condition u_0 - since in dimension $n \geq 3$ the uniqueness problem for equation (2) is still open. If we did have uniqueness of solutions then equation (7) could be made precise by writing $S_t(u)$ for the value at time t of the solution to (2) with initial condition u, and then the family of measures μ_t would be given by

$$\mu_t(A) = \mu_0(S_t^{-1}(A)).$$

Even though there is no such function S_t to make this precise, by arguing informally Foias [8] derived the following equation that would be satisfied by the family μ_t if S_t did exist:

$$\int_{\mathbf{H}} \varphi(u) d\mu_t(u) - \int_{\mathbf{H}} \varphi(u) d\mu(u) =$$

$$\int_0^t \int_{\mathbf{H}} [-\nu((u, \varphi'(u))) - b(u, u, \varphi'(u)) + (f(s), \varphi'(u))] d\mu_s(u) ds \tag{8}$$

where φ is any test function of the form $\varphi(u) = \exp^{i(u,v)}$ with $v \in \mathbf{V}$. This is called the *Foias equation,* and it makes sense because there is no reference to the possibly non-existent function S_t that was used informally in Foias' derivation. Solutions to the Foias equation (see below) are called *statistical solutions* to the Navier-Stokes equations. In dimensions $n \geq 3$ Foias' derivation of equation (8) is only heuristic so it does *not* guarantee the existence of statistical solutions, and these must be constructed by some other means, as Foias did in [8].

Here is the precise definition of a statistical solution.

Definition 5.1 *Suppose that a Borel probability measure μ on **H** is given, with $\int_{\mathbf{H}} |u|^2 d\mu < \infty$. Then a family of probability measures $(\mu_t)_{t\geq 0}$ is a* **statistical solution** *of the Navier-Stokes equations with initial condition μ if $\mu_0 = \mu$ and*

(i)   the function $t \mapsto \int_{\mathcal{H}} |u|^2 d\mu_t(u)$ is $L^\infty(0, T)$ for all $T < \infty$,

(ii) $\int_0^T \int_{\mathcal{H}} \|u\|^2 d\mu_t(u)dt < \infty$ for all $T < \infty$,

(iii)   for all $t \geq 0$ and test functions φ as above equation (8) holds.

Foias' proof [8] of existence of statistical solutions to the Navier-Stokes equations in dimensions $n \leq 4$ is long and complicated. A new and very short proof in [2] was based on the uniqueness of solutions to the Galerkin approximation in an infinite hyperfinite dimension N, which made Foias' heuristic derivation completely rigorous in this setting. Essentially what was proved there was the existence of a random solution to (2) for a given random initial condition. This is easily proved using the neocompactness methods from [6], together with the results of the previous section, as follows.

Theorem 5.2 *Let $x(\omega) \in \mathbf{H}$ be a random variable defined on a rich probability space (Ω, P). Then there is a random variable $y : \Omega \to M_0$ such that for almost all $\omega \in \Omega$, $y(\omega)$ is a (weak) solution to the Navier-Stokes equations with initial condition $x(\omega)$.*

Proof   Since $\mathbf{H}$ is separable there is an increasing sequence of compact sets C_n with $x(\omega) \in \cup_{n \geq 1} C_n$ a.s. Let $D_n = \left(C_n + \overline{\gamma(M_{k_n})} \right) \cap M_0$, where $k_n = K + \sup_{z \in C_n} |z|^2$. Then by Proposition 4.4, D_n is compact, and by Theorem 4.1 we have

$$(\forall z \in C_n)(\exists y \in D_n)[y = z + \gamma(y)]. \tag{9}$$

By [6], the function

$$y(\cdot) \mapsto \gamma(y(\cdot))$$

is neocontinuous and the set

$$D = \{y(\cdot) : (\forall n)(P[y(\omega) \in D_n] \geq P[x(\omega) \in C_n])\}$$

is neocompact. It follows that there are sequences $x_n(\cdot)$ and $y_n(\cdot)$ of simple random variables (i.e. random variables with finite ranges) such that x_n takes its values in C_n, $x_n(\omega) \to x(\omega)$ almost everywhere, $y_n(\cdot) \in D$, and

$$y_n(\omega) = x_n(\omega) + \gamma(y_n(\omega))$$

for almost all $\omega \in \Omega$.

Now let $C = \{x(\cdot)\}$, which is neocompact. Then we have proved

$$(\forall \varepsilon > 0)(\exists y \in D^\varepsilon)(\exists z \in C^\varepsilon)(z + \gamma(y) = y)$$

(simply take $z = x_n$ within ε of x and $y = y_n$).

The Approximation Theorem gives a random variable $y(\cdot) \in D$ with

$$y(\omega) = x(\omega) + \gamma(y(\omega))$$

for almost all ω, which is as required. $\square$

To see that this is sufficient to give statistical solutions, we have

Theorem 5.3 *Suppose that y is an M_0-valued random variable on a probability space (not necessarily a rich space) with*

$$y(\omega) = x(\omega) + \gamma(y(\omega))$$

for a.a. ω, and such that

$$E(|x(\omega)|^2) < \infty.$$

Then the family of measures μ_t on $\mathbf{H}$ given by $\mu_t(A) = P(y(\omega, t) \in A)$ is a statistical solution to the Navier-Stokes equations.

Proof   Simply follow through Foias' heuristic, as in [2].   □

6   Stochastic Navier-Stokes equations

We show here how the existence proof for stochastic Navier-Stokes equations in dimensions $n \leq 4$ can be presented using the machinery of [6]. The pattern of the proof is similar to the corresponding existence result for the deterministic case (Sec. 4), that used the classical analogue of this machinery. In that case we showed how the essential facts that made this work also led to a simple proof that avoids that machinery. In the stochastic case however, it seems unlikely that the proof below can be recast in a way that avoids the use of some form of the neocompactness machinery, because of the need for some kind of enriched probability space to carry the solution in general.

We begin by reviewing from [3] the formulation of the stochastic Navier-Stokes equations. Suppose that $Q : \mathbf{H} \to \mathbf{H}$ is a linear nonnegative trace class operator and that $w(t)$, $t \geq 0$ is an $\mathbf{H}$-valued Wiener process with covariance Q, defined on an adapted probability space $\Omega = (\Omega, \mathcal{F}, (\mathcal{F}_t)_{t \geq 0}, P)$. If we have $f : [0, T] \times \mathbf{V} \to \mathbf{V}'$ and $g : [0, T] \times \mathbf{V} \to \mathcal{L}(\mathbf{H}, \mathbf{V}')$ then the stochastic Navier-Stokes equations with full feed-back take the form

$$du(t) = [-\nu Au(t) - B(u(t)) + f(t, u(t))]dt + g(t, u(t))dw(t) \qquad (10)$$

which is to be understood as an integral equation, using the Bochner integral for the drift term and the stochastic integral of Ichikawa [10] for the noise term. The initial condition u_0 can be random in $\mathbf{H}$, although to begin with we consider only fixed initial conditions $u_0 \in \mathbf{H}$. Thus we have the following definition:

Definition 6.1 *A stochastic process $u(t, \omega)$ on Ω is a **solution to the stochastic Navier-Stokes equations** (10) if it is adapted and has almost all paths in the space*

$$Z = C(0, T; \mathbf{H}_{weak}) \cap L^2(0, T; \mathbf{V}) \cap L^\infty(0, T; \mathbf{H})$$

and the integral equation

$$u(t) = u_0 + \int_0^t [-\nu Au(s) - B(u(s)) + f(s, u(s))]ds + \int_0^t g(s, u(s))dw(s) \quad (11)$$

holds as an identity in $\mathbf{V}'$.

Now take a rich adapted probability space $\Omega = (\Omega, P, \mathcal{G}, (\mathcal{G}_t)_{t\geq 0})$ carrying an $\mathbf{H}$-valued Wiener process w with covariance Q. A process is said to be *adapted for* Ω if it is adapted with respect to the filtration $(\mathcal{F}_t)_{t\geq 0}$ where $\mathcal{F}_t$ is the P-completion of $\bigcap_{s>t} \mathcal{G}_s$.

Recall that K_m is the subset of $\mathbf{H}$ given by $K_m = \{u \in \mathbf{H} : \|u\| \leq m\}$, which is compact in the $\mathbf{H}$-topology.

The existence theorem proved in [3], in the special case that Ω is an adapted Loeb space, is

Theorem 6.2 *Let Ω be a rich adapted space. Suppose that $u_0 \in \mathbf{H}$ and that the functions $f : [0, T] \times \mathbf{V} \to \mathbf{V}'$ and $g : [0, T] \times \mathbf{V} \to \mathcal{L}(\mathbf{H}, \mathbf{H})$ are jointly measurable with the following properties:*

(i) $f(t, \cdot) \in C(K_m, \mathbf{V}'_{weak})$ for all m,

(ii) $g(t, \cdot) \in C(K_m, \mathcal{L}(\mathbf{H}, \mathbf{H})_{weak})$ for all m,

(iii) $|f(t, u)|_{\mathbf{V}'} + |g(t, u)|_{\mathbf{H}, \mathbf{H}} \leq a(t)(1 + |u|)$, for some $a \in L^2(0, T)$.

Then for each fixed $u_0 \in \mathbf{H}$ the stochastic Navier-Stokes equation (11) has a solution u on Ω, with u independent of $\mathcal{G}_0$, and such that

$$E\left(\sup_{t\leq T} |u(t)|^2 + \int_0^T \|u(t)\|^2 dt\right) < \infty.$$

(In fact u is in the space N_0 defined below).

We present a proof of this using the neocompactness machinery; first we set up some notation.

Let N be the set

$$N = \{y \in L^0(\Omega, M) : y \text{ is adapted}\},$$

and let $G^\perp$ be the set

$$G^\perp = \{y \in L^0(\Omega, M) : y \text{ is independent of } \mathcal{G}_0\}.$$

Let

$$\mathcal{E}(y) = \int_0^T \|y(t)\|^2 dt + \sup_{t\leq T} |y(t)|^2,$$

and define

$$N_0 = N \cap \{y : y(0) \in \mathbf{H} \ \& \ E(\mathcal{E}(y)) \leq K_1 + K_2|y(0)|^2 \}$$

where K_1 and K_2 are constants to be specified below. (In this definition by "$y(0) \in \mathbf{H}$" we mean that $y(0)$ is non-random and in $\mathbf{H}$.)

Proposition 6.3 N, $G^{\perp}$, and N_0 are neoclosed subsets of $L^0(\Omega, M)$.

Proof   It is shown in [6] that the set of adapted processes is neoclosed; i.e. N is neoclosed.

To show that $G^{\perp}$ is neoclosed, first show that the set

$$G_0^{\perp} = \{x \in L^0(\Omega, \mathbb{R}) : |x| \le \pi/2 \text{ and } x \text{ is independent of } \mathcal{G}_0\}$$

is neoclosed. We have

$$G_0^{\perp} = \{x : (\forall z \in C)(E(xz) = E(x)E(z) \text{ and } |x| \le \pi/2)\}$$

where C is the set of $\mathcal{G}_0$-measurable indicator functions, which is a basic neocompact set. Then use the universal quantifier rule (f) and the fact that the expectation E is neocontinuous on uniformly bounded sets of random variables. For $G^{\perp}$ itself, note that

$$y \in G^{\perp} \quad \Leftrightarrow \quad (\forall k)(\forall q \in [0, T])(\arctan(y_k(q, \cdot)) \in G_0^{\perp})$$

(q ranging over rationals). The function $y(\cdot) \mapsto \arctan(y_k(q, \cdot))$ is neocontinuous. Now use the facts that neoclosed sets are closed under countable intersections and preimages by neocontinuous functions.

The set of y with $y(0) \in \mathbf{H}$, constant, is neoclosed because the function $y \mapsto y(0)$ is neocontinuous and the space $\mathbf{H}$ is closed and separable (so the set of constant random variables in $L^0(\Omega, \mathbf{H})$ is neoclosed).

The set of y with $E(\mathcal{E}(y)) \le K_1 + K_2|y(0)|^2$ is neoclosed since the function $\mathcal{E} : M \to \mathbb{R}$ is lsc and so $E(\mathcal{E}(y(\cdot))$ is neo-lsc, and the function $y \mapsto y(0)$ is neocontinuous.

N_0 is the intersection of these two sets with N and thus is also neoclosed. $\square$

Now for each $k > 0$ define

$$N_k = N \cap \{y : E(\mathcal{E}(y)) \le k\}.$$

Clearly each N_k is neoclosed in N, and $N_0 \subseteq \cup_{k>0} N_k = N_\infty$ say.

We now define a function $\tilde{\gamma} : N_\infty \to N$ by

$$\tilde{\gamma}(y)(s) = \int_0^s [-\nu A y(t) - B(y(t)) + f(t, y(t))] dt + \int_0^s g(t, y(t)) dw(t) \quad (12)$$

Note that $u = y$ is a solution to the stochastic Navier-Stokes equations with initial condition $u_0 = x$ if and only if $y = x + \tilde{\gamma}(y)$. Moreover, we have

$$\tilde{\gamma} : G^{\perp} \cap N_\infty \to G^{\perp} \cap N,$$

because the Wiener process w has increments that are independent of $\mathcal{F}_0$ (by definition).

We wish to prove neocompact analogues of Propositions 4.4 and 4.3 for the sets N_k. For this we must deal with the two integrals in $\tilde{\gamma}$ separately. We may write

$$\tilde{\gamma}(y) = h(y) + I(y)$$

where $I(y)$ denotes the infinite dimensional stochastic integral

$$I(y) = \int_0^{\cdot} g(s, y(s)) \mathrm{d}w(s).$$

and $h(y)$ is the Bochner integral term

$$h(y)(s, \omega) = \int_0^s [-\nu Ay(t, \omega) - B(y(t, \omega)) + f(t, y(t, \omega))] \mathrm{d}t = \gamma(y(\omega))(s).$$

Here $\gamma : M_\infty \to M$ is slightly more general than the function γ in Section 4 because of the feedback in f, but Propositions 4.3 and 4.4 are still valid; i.e. γ here is also continuous on M_k and $\gamma(M_k)$ is relatively compact for each $k > 0$ (with the obvious modifications to the definition of M_k).

First we deal with h. Define the set

$$\hat{N}_k = \{y \in L^0(\Omega, M) : E(\mathcal{E}(y)) \leq k\}$$

which is neoclosed, and let

$$\hat{M}_m = \{y \in L^0(\Omega, M) : \mathcal{E}(y(\cdot)) \leq m \text{ a.s. }\} = L^0(\Omega, M_m).$$

First we note that:

Lemma 6.4 *Assume the hypotheses of Theorem 6.2. Then for each $m > 0$ the function h is neocontinuous from $\hat{M}_m$ into $L^0(\Omega, M)$ and $h(\hat{M}_m)$ is contained in a neocompact subset of $L^0(\Omega, M)$.*

Proof Since $h(y)(\omega) = \gamma(y(\omega))$ and γ is continuous, it follows that h is neocontinuous on each $\hat{M}_m$. Since $\gamma(M_m)$ is relatively compact, it follows that $h(\hat{M}_m)$ is contained in a neocompact set. $\square$

Lemma 6.5 *Assume the hypotheses of Theorem 6.2. Then for each $k > 0$ the function h is neocontinuous from $\hat{N}_k$ into $L^0(\Omega, M)$ and $h(\hat{N}_k)$ is contained in a neocompact subset of $L^0(\Omega, M)$.*

Proof For each m let D_m be a neocompact set with $h(\hat{M}_m) \subseteq D_m$. Fix k, and take $y \in \hat{N}_k$. By Chebyshev's inequality, $P(\mathcal{E}(y) \leq m) \geq 1 - \frac{k}{m}$, for each m. Define y^m by

$$y^m = \begin{cases} y & \text{on the set } \{\mathcal{E}(y) \leq m\} \\ 0 & \text{otherwise} \end{cases}$$

Then $y^m \in \hat{M}_m$ and $h(y) = h(y^m)$ on $\{\mathcal{E}(y) \leq m\}$. So $\rho(h(y), h(y^m)) \leq \frac{k}{m}$ and so $h(y) \in (D_m)^{k/m}$. Thus $h(\hat{N}_k) \subseteq D$ where

$$D = \bigcap_m (D_m)^{\frac{k}{m}}$$

which is neocompact by closure under diagonal intersections.

To see that h is neocontinuous on $\hat{N}_k$, take a neocompact set $C \subseteq \hat{N}_k$ and for each m let $C^m = \{y^m : y \in C\} \subseteq \hat{M}_m$. The function $y \mapsto y^m$ is neocontinuous, and so C^m is neocompact. The graph of h restricted to C is the neocompact set

$$\bigcap_m \{(y,v) \in C \times D : (\exists z \in C^m)\big(\rho(y,z) + \rho(v, h(z)) \leq \frac{2k}{m}\big)\}.$$

□

Corollary 6.6 *Assume the hypotheses of Theorem 6.2. Then for each k the function h is neocontinuous from N_k into N, and $h(N_k)$ is contained in a neocompact subset of N with respect to $L^0(\Omega, M)$.*

Proof This follows immediately from Lemmas 6.4 and 6.5, since $N_k \subseteq \hat{N}_k$.
□

To obtain the same result for I we must first prepare the way by dealing with a class of bounded integrands. For each $m > 0$ define

$$J_m = N \cap \{y : \sup_{t \leq T} |y(t)|^2 \leq m \text{ a.s.}\};$$

clearly J_m is neoclosed.

Lemma 6.7 *Assume the hypotheses of Theorem 6.2. Then for each m the function I is neocontinuous from J_m into N, and $I(J_m)$ is contained in a neocompact subset of N with respect to $L^0(\Omega, M)$.*

Proof Let $I_n(y)$ denote the finite dimensional stochastic integral

$$I_n(y) = \big(\textstyle\int_0^\cdot g(s, y(s)) \mathrm{d}w(s)\big)^{(n)}.$$

The results of [6] show that for each n and m, the function I_n is neocontinuous from J_m into $L^0(\Omega, M)$. The bound (iii) on g insures that for fixed m the set $\cup_n \mathrm{law}(I_n(J_m))$ is tight, so the function I maps J_m into a neocompact set D in $L^0(\Omega, M)$.

For each neocompact set $C \subseteq J_m$ in $L^0(\Omega, M)$, the graph of $I|C$ is the neocompact set

$$\bigcap_n \{(y,z) \in C \times D : \Pi_n(z) = I_n(y)\}.$$

Therefore I is neocontinuous on J_m. □

Now we must extend Lemma 6.7 to N_k.

Let ST be the set of all stopping times $\tau \in L^0(\Omega, [0, T])$ and for any process y let y^τ denote the stopped process $y(\omega)(t \wedge \tau)$. It is shown in [FK] that ST is neocompact, and that the function $(y, \tau) \mapsto y^\tau$ is neocontinuous from $N \times ST$ into N.

Proposition 6.8 *For each $k > 0$ the set $I(N_k)$ is contained in a neocompact subset of N with respect to $L^0(\Omega, M)$.*

Proof   By Lemma 6.7, for each $m \in \mathbb{N}$ there is a neocompact set $D_m \subseteq N$ such that

$$I(J_m) \subseteq D_m.$$

If $y \in N_k$ then $E(\sup_{t \leq T} |y(t)|^2) \leq k$, and by Chebyshev's inequality we have

$$P(\sup_{t \leq T} |y(t)|^2 \leq m) \geq 1 - \tfrac{k}{m}.$$

For $y \in N$ let $\tau_{m,y}(\omega)$ be the least t such that either $t = T$ or $\sup_{s \leq t} |y(s)|^2 \geq m$. Then $\tau = \tau_{m,y}$ is a stopping time; if $y \in N_k$, then y^τ belongs to J_m and we have $P(y = y^\tau) \geq 1 - \tfrac{k}{m}$; i.e. $\rho(y, y^\tau) \leq \tfrac{k}{m}$. Moreover $I(y) = I(y^\tau)$ on the set $\{y = y^\tau\}$ and so $\rho(I(y), I(y^\tau)) \leq \tfrac{k}{m}$. But $I(y^\tau) \in D_m$ and hence $I(y) \in (D_m)^{k/m}$ in the space $L^0(\Omega, M)$. It follows that $I(N_k)$ is contained in the set

$$D = N \cap \bigcap_m (D_m)^{\tfrac{k}{m}}.$$

By closure under diagonal intersections, D is neocompact in $L^0(\Omega, M)$. □

Proposition 6.9 *Assume the hypotheses of Theorem 6.2. Then for each k the function I is neocontinuous from N_k into N.*

Proof   Let C be a neocompact subset of N_k. By the preceding proposition, there is neocompact $D \supseteq I(N_k)$. Since C and ST are neocompact, the set $\tilde{C} = \{y^\tau : y \in C \wedge \tau \in ST\}$ is neocompact. Moreover, the argument in the proof of Proposition 6.8 shows that for each $y \in C$, for each m there is a stopping time τ such that $y^\tau \in \tilde{C} \cap J_m$ and $\rho(y, y^\tau) \leq \tfrac{k}{m}$ and $\rho(I(y), I(y^\tau)) \leq \tfrac{k}{m}$. It follows that the graph of $I|C$ is the neocompact set

$$\bigcap_m \{(y, v) \in C \times D : (\exists z \in \tilde{C} \cap J_m)\big(\rho(y, z) + \rho(v, I(z)) \leq \tfrac{2k}{m}\big)\},$$

so I is neocontinuous on N_k. □

The following proposition is the key to the construction of a solution. It is the neocompact analogue of propositions 4.4 and 4.3, and follow immediately from Corollary 6.6 and Propositions 6.8, 6.9.

Proposition 6.10 *Assume the hypotheses of Theorem 6.2. Then for each $k > 0$ the function $\tilde{\gamma}$ is neocontinuous from N_k into N and the set $\tilde{\gamma}(N_k)$ is contained in a neocompact subset of N with respect to $L^0(\Omega, M)$.*

Next we define the Galerkin approximation to equation (11) and show that the stochastic version of Theorem 4.5 holds. We set

$$\tilde{\gamma}_n(y) = \Pi_n \tilde{\gamma}(y).$$

Then we have

Theorem 6.11 *There are constants K_1 and K_2 such that for each $n > 0$ and $x \in \mathbf{H}$ there is $y \in N_0 \cap G^\perp$ such that $y = x^{(n)} + \tilde{\gamma}_n(y)$ (so $y \in G^\perp \cap N_{K_1 + K_2 |x|^2}$)*

Proof The equation $y = x^{(n)} + \tilde{\gamma}_n(y)$ is a finite dimensional SDE which can be solved by routine standard methods and has a unique solution. It was shown in [3] that there are K_1 and K_2, independent of n (and which can be given explicitly in terms of the parameters ν, T, trQ, $\int_0^T a^2$), such that $E(\mathcal{E}(y)) \leq K_1 + K_2 |y(0)|^2$ whenever $y = x^{(n)} + \tilde{\gamma}_n(y)$. The proof of this estimate uses the same ideas as are used for estimating the energy for the Galerkin approximation to the deterministic Navier-Stokes equations, coupled with the Burkholder-Davis-Gundy inequality and an application of Gronwall's lemma. (In [3] it was established for an infinite integer n, by transfer of finite dimensional methods which are used in the present situation.) $\square$

Theorem 6.12 (Neo-Dini's Theorem) *For a rich adapted space Ω, suppose that $D \subseteq L^0(\Omega, M)$ is neocompact and that $f_n : D \to \mathbb{R}$ is a sequence of neocontinuous functions with $f_n(x) \searrow 0$ monotonically as $n \to \infty$ for each fixed $x \in D$. Then $f_n \to 0$ uniformly on D.*

Proof Suppose f_n does not converge uniformly to 0 on D. Then there is an $\epsilon > 0$ such that for arbitrarily large $n \in \mathbb{N}$, the set

$$D_n = \{x \in D : |f_n(x)| \geq \epsilon\}$$

is nonempty. Since $f_n(x)$ converges monotonically for each x, the sets D_n form a decreasing chain. Each D_n is neocompact because D is neocompact and f_n is neocontinuous. By the countable compactness property, $\bigcap_n D_n$ is nonempty, contrary to hypothesis. $\square$

We now use the neo-Dini's theorem to prove a stochastic analogue of Proposition 4.6.

Proposition 6.13 *For $k > 0$, $\tilde{\gamma}_n(y) \to \tilde{\gamma}(y)$ uniformly on N_k.*

Proof   By Proposition 6.10, $\tilde{\gamma}(N_k) \subseteq D$ for some neocompact set $D \subseteq N$. The projection function Π_n is continuous on M and therefore neocontinuous on $L^0(\Omega, M)$. By the above neo-Dini's Theorem, $\Pi_n(z) \to z$ uniformly on D, and the result follows. $\square$

Proof of Theorem 6.2   We argue as in the proof of Theorem 4.1. Let $x = u_0$. It suffices to show that there exists $y \in G^\perp \cap N_k$ where $k = K_1 + K_2|x|^2$ such that $y = x + \tilde{\gamma}(y)$. By Proposition 12, there is a neocompact set $D \subseteq N$ such that $\tilde{\gamma}(N_k) \subseteq D$. By 6.11 and 6.13, for each $\epsilon > 0$ there exists $n \in \mathbb{N}$ and $y_n \in G^\perp \cap N_k$ such that

$$y_n = x^{(n)} + \tilde{\gamma}_n(y_n),$$

and in the space N, y_n is within ϵ of $x + \tilde{\gamma}(y_n)$ which belongs to D. Thus

$$(\forall \epsilon > 0)(\exists z \in D^\epsilon \cap G^\perp \cap N_k)(\rho(z, x + \tilde{\gamma}(z)) \le \epsilon).$$

By Proposition 6.10, $\tilde{\gamma}$ is neocontinuous on N_k. The Approximation Theorem now gives the required solution $y \in G^\perp \cap N_k$ such that $y = x + \tilde{\gamma}(y)$. $\square$

We conclude with an existence theorem for stochastic Navier-Stokes equations with a random initial condition.

Theorem 6.14 *Let Ω be a rich adapted space. Suppose that $u_0(\omega)$ is a $\mathcal{G}_0$-measurable random variable on Ω with values in $\mathbf{H}$, and f, g satisfy the hypotheses of Theorem 6.2. Then the stochastic Navier-Stokes equation (11) has a solution $u(t, \omega)$ on Ω with initial value $u(0, \omega) = u_0(\omega)$. Moreover, if $E\left[|u_0(\omega)|^2\right] < \infty$, then*

$$E\left(\sup_{t \le T} |u(t)|^2 + \int_0^T \|u(t)\|^2 dt\right) < \infty. \tag{13}$$

Proof   We shall use the fact that Theorem 6.2 gives us a solution with a deterministic initial value which is independent of $\mathcal{G}_0$.

Let $x(\omega) = u_0(\omega)$ be a $\mathcal{G}_0$-measurable initial value. Since $\mathbf{H}$ is separable there is an increasing sequence of compact sets $C_n \subseteq \mathbf{H}$ with $P[x(\omega) \in C_n] \ge 1 - 1/n$ for each $n \ge 1$. Let $k_n = K_1 + K_2 \sup_{z \in C_n} |z|^2$. Let D_n be a neocompact subset of N such that

$$D_n \supseteq \tilde{\gamma}(N_{k_n}).$$

We can take the sets D_n to be increasing. Let G_n be the neocompact set of all $\mathcal{G}_0$-measurable random variables in $L^0(\Omega, C_n)$. Then the set $B_n = G_n + D_n$ is a neocompact subset of N such that

$$B_n \supseteq G_n + \tilde{\gamma}(N_{k_n}).$$

By closure under diagonal intersections (Lemma 2.9), the set

$$B = \bigcap_{n \ge 1} \left((B_n)^{1/n}\right)$$

is neocompact.

For each $n \geq 1$ there is a simple function $x_n \in G_n$ such that for each $m \leq n$, whenever $x(\omega) \in C_m$ then $x_n(\omega) \in C_m$ and $x_n(\omega)$ is within $1/n$ of $x(\omega)$. Hence $x_n \in (G_m)^{1/m}$ for all $m \leq n$.

By piecing together finitely many solutions from Theorem 6.2 which are independent of $\mathcal{G}_0$, we see that for each n there is a $y_n \in N_{k_n} \cap B_n$ such that

$$y_n = x_n + \tilde{\gamma}(y_n) \tag{14}$$

So $y_n \in (B_m)^{1/m}$ for each $1 \leq m \leq n$, and it follows (since the sets B_n are increasing) that $y_n \in B$. Thus for each n,

$$(\exists x_n \in \{x\}^{1/n})(\exists y_n \in B)[y_n = x_n + \tilde{\gamma}(y_n)].$$

Since the addition function is neocontinuous on random variables, it follows that the function

$$(z(\cdot), y(\cdot)) \mapsto z(\cdot) + \tilde{\gamma}(y(\cdot))$$

is neocontinuous on $L^0(\Omega, \mathbf{H}) \times N_{k_n}$. So by the Approximation Theorem there is a stochastic process $y(\cdot) \in B$ with

$$y = x + \tilde{\gamma}(y)$$

as required.

Suppose finally that $E\left[|x(\omega)|^2\right] = m < \infty$. By Theorem 6.2, the y_n may be chosen so that $y_n \in N_k$, where $k = K_1 + K_2 m$. Since N_k is neoclosed, we may take the solution y to be in N_k, and therefore (13) holds. $\square$

In Theorem 6.14, it would be more natural to allow the initial value $u_0(\omega)$ to be $\mathcal{F}_0$-measurable rather than $\mathcal{G}_0$-measurable, where $\mathcal{F}_0$ is the completion of $\bigcap_{t>0} \mathcal{G}_t$. The following corollary shows that this can be done if the rich adapted space is good enough. The construction of a rich adapted space in [7] actually produces a rich adapted space

$$\Omega = (\Omega, P, \mathcal{G}, \mathcal{G}_t)$$

with the following additional property:

For every complete separable M and every $\mathcal{F}_0$-measurable random variable $x \in L^0(\Omega, M)$ there is a σ-algebra $\mathcal{G}_0' \subseteq \mathcal{F}_0$ such that x is $\mathcal{G}_0'$-measurable and the adapted space Ω with $\mathcal{G}_0'$ in place of $\mathcal{G}_0$ is still rich.

Corollary 6.15 *If Ω is a rich adapted space with the above additional property, then Theorem 6.14 holds with $\mathcal{F}_0$ in place of $\mathcal{G}_0$.*

7   Optimal solutions

We show in this section how the neocompact machinery provides existence of optimal solutions to the stochastic Navier-Stokes equations - where the term 'optimal' is capable of a wide variety of interpretations. We shall restrict attention to the case of a fixed initial condition $u_0 = x \in \mathbf{H}$, as in Theorem 6.2. Recall that we proved above that there is a solution y to the equation (11) in the set $N_{k(x)}$ where $k(x) = K_1 + K_2|x|^2$ (in fact the solution constructed is in $G^\perp$ also). It is natural to define the following **solution set** S_x for the initial condition $x \in \mathbf{H}$:

$$S_x = \{y \in N_{k(x)} : y \text{ solves the stochastic Navier-Stokes equation and } y(0) = x\}$$

$$= \{y \in N_{k(x)} : y = x + \tilde{\gamma}(y)\}$$

The key point now is the observation that

Theorem 7.1 *For each $x \in \mathbf{H}$ the set S_x is neocompact.*

Proof   (i) Let $f(y) = y - x - \tilde{\gamma}(y)$, which is neocontinuous on the neoclosed set $N_{k(x)}$, by Proposition 6.10. Clearly we have

$$S_x = N_{k(x)} \cap f^{-1}(\{0\})$$

which is neoclosed. Moreover, $S_x \subseteq x + \tilde{\gamma}(N_{k(x)})$ which is contained in a neocompact set (by Proposition 6.10 again). Hence S_x is neocompact. $\square$

The following is a general optimality result for solutions to the stochastic Navier-Stokes equations; it is a special case of a general optimization result found in [6].

Theorem 7.2 *Suppose that $g : N_0 \to \overline{\mathbb{R}}$ is a function that is neo-lsc on $N_{k(x)}$; then there is a solution $\hat{y} \in S_x$ with*

$$g(\hat{y}) = \inf\{g(y) : y \in S_x\}$$

i.e. $\hat{y}$ is an optimal solution for the quantity $g(y)$.

Proof   Since g is neo-lsc and S_x is neocompact, the upper graph

$$A = \{(y, r) \in S_x \times \overline{\mathbb{R}} : g(y) \le r\}$$

is neocompact; so the set

$$B = \{r \in \overline{\mathbb{R}} : (\exists y \in S_x)(y, r) \in A\}.$$

is a neocompact subset of $\overline{\mathbb{R}}$, which means that it is in fact compact. Thus B has a minimum element s, say, and there is $\hat{y} \in S_x$ with $g(\hat{y}) \le s$. On the other hand, for *any* $y \in S_x$, putting $r = g(y)$ we have $(y, r) \in A$ and so $r \in B$. Thus $s \leq r = g(y)$ and the optimality of $\hat{y}$ for g is established. $\square$

There are several natural functions g for which optimal solutions might be sought - for example the function $E(\mathcal{E}(y))$ that occurs naturally in the proofs of the previous section, and the expected energy integral

$$\mathbf{E}(y) = \tfrac{1}{2} E\left(\int_0^T |y(t)|^2 dt \right)$$

and the expected enstrophy integral

$$\mathbf{S}(y) = \tfrac{1}{2} E\left(\int_0^T \|y(t)\|^2 dt \right).$$

Each of these functions is neo-lsc.

On the face of it, the optimality theorem appears to be specific to the particular rich adapted space on which we are working. However, it has been shown that rich adapted spaces are universal (see [13]), so that if y' is a solution on some other adapted space Ω' then there is $y \in L^0(\Omega, M)$ with the same adapted distribution as y' and so for any neo-lsc function g of the form $g(y) = Eg_0(y(\cdot))$ where g_0 is lsc, we have the existence of globally optimal solutions. This applies to the natural examples $\mathcal{E}$, $\mathbf{E}$, and $\mathbf{S}$ above.

The optimality result above can be generalised to the case where the initial condition u_0 is random, or specialised to the deterministic setting (where $g = 0$).

References

[1] S.Albeverio, J.-E.Fenstad, R.Høegh-Krohn, and T.Lindstrøm, *Nonstandard Methods in Stochastic Analysis and Mathematical Physics*, Academic Press, New York 1986.

[2] M.Capiński and N.Cutland, A simple proof of existence of weak and statistical solutions of Navier–Stokes equations, *Proc. Royal. Soc. London* A **436** (1992), 1-11.

[3] M.Capiński & N.Cutland, Stochastic Navier-Stokes equations, *Applicandae Mathematicae* **25**(1991), 59-85.

[4] N.J.Cutland, Nonstandard measure theory and its applications, *Bull. Lond. Math. Soc.* **15**(1983), 529-589.

[5] N.J.Cutland (ed.), *Nonstandard Analysis and its Applications*, Cambridge University Press, Cambridge 1988.

[6] S.Fajardo and H.J.Keisler, A method for proving existence theorems in probability theory, to appear.

[7] S. Fajardo and H.J.Keisler, Neometric spaces, to appear.

[8] C.Foias, Statistical study of Navier-Stokes equations I, *Rend. Sem. Mat. Univ. Padova* **48**(1973), 219-348.

[9] A.E.Hurd and P.A.Loeb, *An Introduction to Nonstandard Real Analysis*, Academic Press, New York 1985.

[10] A.Ichikawa, Stability of semilinear stochastic evolution equations, *J. Math. Anal. Appl.* **90**(1982), 12-44.

[11] H.J.Keisler, An infinitesimal approach to stochastic analysis, *Mem. Amer. Math. Soc.* **48** No. 297 (1984).

[12] H. J. Keisler, From Discrete to Continuous Time, *Ann. Pure and Applied Logic* 52 (1991), 99-141.

[13] H. J. Keisler, Rich and saturated adapted spaces, to appear.

[14] R.Temam, *Navier-Stokes equations*, North-Holland, Amsterdam 1979.

[15] R.Temam, *Navier-Stokes Equations and Nonlinear Functional Analysis*, SIAM, 1983.

Numerical Experiments with S(P)DE's

J.G.Gaines

Dept. of Mathematics and Statistics

University of Edinburgh

JCMB, Kings Buildings

Mayfield Rd.

Edinburgh EH9 3JZ, UK

1 Introduction

For the last thirty years, there has been interest in numerical simulation of solutions of stochastic differential equations (SDE's). More recently, along with the growing general interest in stochastic partial differential equations (SPDE's), there has been a desire to produce numerical solutions to SPDE's. There are very few SDE's for which analytical solutions can be obtained and this is naturally also true for SPDE's. One hope is that using numerical methods to generate solutions to such equations will lead to better understanding of the equations. Theory and numerical work often go hand in hand: while theory can provide useful numerical methods, pictures obtained numerically can lead to conjectures that the analyst may then succeed in proving. Another hope is that numerical methods will be of use to people outside mathematics. There are already SPDE's used in practical applications, such as in finance.

Our aim is to show how simple techniques, already in use for numerical solution of SDE's, can be extended to provide numerical solutions to SPDE's. We will concentrate on strong solutions, that is solutions obtained for one particular realisation of Brownian motion or of a Brownian sheet. We describe how Brownian paths and sheets can be generated and stored in tree-like structures, allowing for subsequent refinement and therefore comparison of solutions using different meshes.

In Section 2 we will show what can be done when generating strong solutions to SDE's. This will be a short and far from exhaustive description of methods in current use, giving some idea of how the methods were developed. The idea is to provide enough background so that those who start looking at the numerical solution of stochastic differential equations, (whether partial or not), at this point, will be able to follow. More detailed surveys, that also describe numerical methods for weak solutions, can be found in [3], [14], [15], [17], [27], [30] . Anyone who is about to sit down in front of a computer and try to solve SDE's or SPDE's numerically is recommended to read some of them first.

In Section 2.1 we will define what we mean by strong solutions to SDE's and give examples of the kind of applications where strong solutions are required. Some numerical schemes for obtaining strong solutions are shown in Section 2.2, using the stochastic Taylor series approach to derive the schemes. Section 2.3 discusses various ways of generating and structuring Brownian paths, as sets of increments or as sets of both increments and stochastic integrals.

The main motivation of Section 2 is to introduce techniques that will be used in Section 3, where we will describe some experiments with numerical solutions to SPDE's. We say 'experiments' since, whereas results exist proving convergence and order of accuracy of methods for simulating solutions of SDE's, such results have not yet, to our knowledge, been published for the numerical solution of SPDE's. All that we can say is that the methods used seem to 'work', in that solutions obtained have properties predicted or subsequently obtained by analytical results. The methods we have used will be presented through concrete examples: first a stochastic generalised KPP equation and then the heat equation with noise. The two equations are similar in both being parabolic and therefore presenting the same stability problems, but differ in that only the second SPDE involves noise that varies with space as well as time.

2 Strong solutions to SDE's

2.1 Strong solutions: definition and motivation

We are interested in strong (or pathwise) solutions to an SDE where the driving noise $w(t)$ is a standard d-dimensional Brownian motion. In Itô form such an SDE is given by

$$
\begin{cases}
dx_t &= b(x,t)dt + \displaystyle\sum_{i=1}^{d} \sigma_i(x,t)\ dw_t^i \\
x(0) &= x_0
\end{cases}
\tag{1}
$$

with $x \in I\!\!R^n$ and $b, \sigma_i (i = 1, \dots, d) : I\!\!R^n \to I\!\!R^n$. The equation may also be written in Stratonovich form as

$$
\begin{cases}
dx_t &= a(x,t)dt + \displaystyle\sum_{i=1}^{d} \sigma_i(x,t) \circ dw_t^i \\
x(0) &= x_0
\end{cases}
$$

where

$$
a^i(x,t) = b^i(x,t) - \frac{1}{2}\sum_{j=1}^{d}\sum_{k=1}^{n} \frac{\partial \sigma_j^i}{\partial x^k}\sigma_j^k
$$

Superscripts, such as x^k, denote components of vectors. We will presume that the vector-valued functions b, σ_i, $(i = 1, \ldots, d)$ are sufficiently smooth to guarantee existence and uniqueness of solutions to (1).

When we choose or are given **one** particular Brownian path $w(t)$, $0 \leq t \leq T$, then by strong solution we mean the path $x(t)$, $0 \leq t \leq T$, unique solution of (1) for that realisation of $w(t)$. This definition is in opposition to that of a weak solution, which is the expected value of the solution or of a function of the solution over all possible Brownian paths. We will only be concerned with numerical methods used to obtain approximations to strong solutions.

There are several reasons why people should wish to obtain strong solutions to SDE's. Firstly, they may wish to examine the dependence of the solution on the initial condition or on the value of one or more parameters that appear in the functions defining the SDE. In this case one Brownian path is generated and then used repeatedly to obtain a set of solutions with varying initial conditions or parameter values. Secondly, there are situations, such as filtering problems, where the Brownian path or a noisy signal supposed to be a function of a Brownian path is not generated by a computer, but obtained as a reading of a measuring instrument. In this case, the recorded signal is used to solve an SDE just once, with the aim of getting more information about the process generating the signal. It is the first kind of application that will interest us the most. Although the second kind is very important, the fact that the driving noise is not generated 'at will' by the person performing the numerical solution puts limitations on the methods available.

We give illustrations of two uses of strong solutions to SDE's. In Figure 1 we show trajectories of the equation

$$dx = \cos(x) \; dw^1 + \sin(x) \; dw^2$$

for a range of starting values. The system has a negative Lyapunov exponent, causing the trajectories to converge, but trajectories from initial points separated by more than 2π converge separately. These facts can easily be proved analytically, but in a more complicated example this would not be the case.

The SDE used in Figure 2 is the logistic model with noise

$$dx = (x - x^2)dt + \sigma x \; dw$$

with the parameter σ taking the values $0, 0.1, 0.5, 1$.

2.2   Numerical methods for strong solutions

The standard method of obtaining numerical approximations to strong solutions of (1) is to discretise the equation. The approximate solution $\bar{x}$ will be calculated at a set of times $t_0 = 0, t_1 = t_0 + h_1, \ldots, t_{k+1} = t_k + h_{k+1}, \ldots, t_N = T$

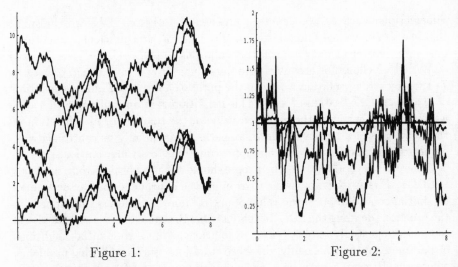

Figure 1:   Figure 2:

using a scheme of the type

$$
\begin{aligned}
\bar{x}^h(t_0) &= x(t_0) \\
\bar{x}^h(t_{k+1}) &= f\left(t_k, \bar{x}^h(t_k), h_{k+1}, \Delta w_{k+1}\right) \\
h_{k+1} &\doteq t_{k+1} - t_k \\
\Delta w_{k+1} &\doteq w(t_{k+1}) - w(t_k)
\end{aligned}
$$

for $k = 0, 1, \ldots, N - 1$. The discretisation scheme is said to converge strongly
to the solution of the SDE if

$$
\lim_{\delta \to 0} \sup_{0 \le t \le T} |x(t) - \bar{x}(t)| = 0 \ a.s.
$$

where $\delta = \max_k h_k$.

The easiest method of obtaining discretisation schemes is to truncate the
stochastic Taylor series expansion of the solution to the SDE (1), keeping only
terms up to a chosen order. This method is due to Platen and Wagner [28]
and Kloeden and Platen [16], who prove convergence of the series for small
random time and the size of the remainder term.

It is possible to expand the solution using either Itô or Stratonovich calculus. We will use the Itô stochastic Taylor series as an illustration. The
Stratonovich series is simpler and so can easily be imagined. Initially we will
take a Brownian path of dimension one.

For simplicity of notation, we write $x_k \doteq x(t_k)$ and $\bar{x}_k \doteq \bar{x}(t_k)$. The solution
to (1) at time t_{k+1} can be written as

$$
x_{k+1} = x_k + \int_{t_k}^{t_{k+1}} b(x_t) \, dt + \int_{t_k}^{t_{k+1}} \sigma(x_t) \, dw_t \tag{2}
$$

Using the Itô formula

$$
f(x_t) = f(x_k) + \int_{t_k}^{t} \mathcal{L}f(x_s)ds + \int_{t_k}^{t} f'(x_s)\sigma(x_s) \, dw_s \tag{3}
$$

where $\mathcal{L}$ is defined by

$$\mathcal{L}f(x_s) \doteq f'(x_s)b(x_s) + \frac{1}{2}f''(x_s)\sigma^2(x_s)$$

in equation (2), to expand both $b(x_t)$ and $\sigma(x_t)$ around t_k, we obtain

$$
\begin{aligned}
x_{k+1} = \;& x_k + b(x_k)h_{k+1} + \sigma(x_k)\Delta w_{k+1} \\
& + \int_{t_k}^{t_{k+1}} \int_{t_k}^{t} \mathcal{L}b(x_s)\, ds\, dt + \int_{t_k}^{t_{k+1}} \int_{t_k}^{t} b'(x_s)\sigma(x_s)\, dw_s\, dt \\
& + \int_{t_k}^{t_{k+1}} \int_{t_k}^{t} \mathcal{L}\sigma(x_s)\, ds\, dw_t + \int_{t_k}^{t_{k+1}} \int_{t_k}^{t} \sigma'(x_s)\sigma(x_s)\, dw_s\, dw_t
\end{aligned}
\tag{4}
$$

If the variance of any term in such a stochastic Taylor series is Ch^{α}, we will say the term is of order $O(h^{\alpha})$. Discarding the double integrals in (4), each of which is of order $O(h^{\alpha})$ with $\alpha \geq 1$, we obtain the Euler-Maruyama scheme

$$\bar{x}_{k+1} = \bar{x}_k + b(\bar{x}_k)h_{k+1} + \sigma(\bar{x}_k)\Delta w_{k+1}$$

This scheme, the simplest discretisation scheme for SDE's, is named after Maruyama, [19], one of the first to examine it. If we use (3) to expand $\sigma'(x_s)\sigma(x_s)$ in (4) and discard all terms of order $O(h^{\alpha})$ for $\alpha > 1$, we obtain

$$\bar{x}_{k+1} = \bar{x}_k + b(\bar{x}_k)h_{k+1} + \sigma(\bar{x}_k)\Delta w_{k+1} + \sigma'(\bar{x}_k)\sigma(\bar{x}_k) \int_{t_k}^{t_{k+1}} \int_{t_k}^{t} dw_s\, dw_t$$

or

$$\bar{x}_{k+1} = \bar{x}_k + b(\bar{x}_k)h_{k+1} + \sigma(\bar{x}_k)\Delta w_{k+1} + \frac{1}{2}\sigma'(\bar{x}_k)\sigma(\bar{x}_k)\left((\Delta w_{k+1})^2 - h_{k+1}\right)$$

which is the Milshtein scheme, first proposed by Milshtein [20] for weak solutions.

Most often we take $h_k = h = T/N$, $(k = 1, \dots, N)$ in the above schemes, but they will both converge to the true solution for any set t_k of stopping times. With a constant time step $h = T/N$, a discretisation scheme is said to have order of strong convergence α if

$$h^{-\alpha} \sup_k |x_k - \bar{x}_k^h| \to 0 \ a.s.$$

Newton ([25], [26]) and Faure ([9], [10]) proved that with a constant time step the Euler-Maruyama and the Milshtein schemes have strong order of convergence $1/2 - \epsilon$ and $1 - \epsilon$ respectively, $\forall \epsilon > 0$, in the one-dimensional case and that the Euler-Maruyama scheme has the same order of strong convergence in the multi-dimensional case.

If one tries to improve on the order of accuracy of the scheme, by keeping more terms from the Taylor series, then in general one will need to include multiple stochastic integrals in the scheme as well as increments of the Brownian path. For example, inclusion of terms of order $O(h^{3/2})$ will in general mean inclusion of the integrals

$$\int_{t_k}^{t_{k+1}} \int_{t_k}^{t} dw_s \, dt$$

These integrals have a normal distribution and it is therefore possible to generate them jointly with the increments Δw_{k+1}.

In higher dimensions the situation is even worse. In general the order of accuracy of the Euler-Maruyama scheme is the best order obtainable without the generation of multiple integrals. The stochastic Taylor series development of x_{k+1}^i in terms of x_k, for $x \in I\!R^n$, $w(t) \in I\!R^d$ is

$$x_{k+1}^i = x_k^i + a^i(x_k)h_{k+1} + \sum_{j=1}^{d} \sigma_j^i(x_k)\Delta w_{k+1}^j + \sum_{j=1}^{n} \sum_{p,q=1}^{d} \frac{\partial \sigma_p^i}{\partial x^j}\sigma_q^j(x_k)I_{pq}(k,k+1) + R$$

(5)

where

$$I_{pq}(k,k+1) = \int_{t_k}^{t_{k+1}} \int_{t_k}^{t} dw_s^p \, dw_t^q$$

and R consists of terms of order $O(h^\alpha)$ for $\alpha > 1$. If $d \le p,q \le 1$, $p \ne q$, then integration by parts gives

$$I_{pq}(k,k+1) + I_{qp}(k,k+1) = \Delta w_{k+1}^p \Delta w_{k+1}^q \doteq B_{pq}(k,k+1)$$

Define

$$A_{pq}(k,k+1) \doteq I_{pq}(k,k+1) - I_{qp}(k,k+1)$$

(6)

In (5) we can substitute for the I_{pq} in terms of B_{pq} and A_{pq}, to obtain

$$\begin{aligned}
x_{k+1}^i = \; & x_k^i + a^i(x_k)h + \sum_p \sigma_p^i(x_k)\Delta w_{k+1}^p \\
& + \frac{1}{2}\sum_{j=1}^{n}\sum_{p=1}^{d} \frac{\partial \sigma_p^i}{\partial x^j}\sigma_p^j(x_k)\left((\Delta w_{k+1}^p)^2 - h_{k+1}\right) \\
& + \sum_{j=1}^{n} \sum_{0<p<q\le d} \frac{1}{2}\left(\frac{\partial \sigma_q^i}{\partial x^j}\sigma_p^j + \frac{\partial \sigma_p^i}{\partial x^j}\sigma_q^j\right)(x_k)B_{pq}(k,k+1) \\
& + \sum_{j=1}^{n} \sum_{0<p<q\le d} \frac{1}{2}\left(\frac{\partial \sigma_q^i}{\partial x^j}\sigma_p^j - \frac{\partial \sigma_p^i}{\partial x^j}\sigma_q^j\right)(x_k)A_{pq}(k,k+1) + R
\end{aligned}$$

(7)

If $\forall i,p,q$

$$\sum_{j=1}^{n} \left(\frac{\partial \sigma_q^i}{\partial x^j}\sigma_p^j - \frac{\partial \sigma_p^i}{\partial x^j}\sigma_q^j\right) = 0$$

(8)

then the A_{pq} terms drop out of equation (7). Condition (8) is called the commutativity condition, since it can also be written as

$$[\sigma_p, \sigma_q] = 0$$

and is equivalent to saying that the vector fields σ_p commute. When the commutativity condition is not satisfied, then the terms A_{pq}, known as the Lévy areas, need to be included in any discretisation scheme that is to attain an order of convergence greater than $1/2$. For the proof see [4]. Clearly, by dropping the remainder term in (7), we obtain a discretisation scheme for general SDE's, of higher order than the Euler-Maruyama scheme, that involves generation of the Lévy areas.

2.3 Representation of Brownian paths

The simple discretisation schemes shown in Section 2.2 suggest at least three possible ways that one might wish to represent a Brownian path when using it to approximate a strong solution to an SDE. The first and simplest representation is as a set of increments

$$\{\Delta w_{k+1} = w(t_{k+1}) - w(t_k), \ k = 1, \ldots, N\}$$

Such a set can be used with the Euler-Maruyama scheme, say, in any number of dimensions.

More complicated ways of viewing and generating Brownian paths consist of generating not only increments but also sets of integrals of the path. When the Brownian path is of dimension one, or when the vector fields defining the diffusion of the SDE commute, then it makes sense to generate the Brownian path as a set of increments and integrals against time,

$$\{\Delta w_{k+1}, A_{i0}(k, k+1), \ k = 1, \ldots, N\}$$

where

$$A_{i0}(k, k+1) = I_{i0}(k, k+1) - I_{0i}(k, k+1) = \int_{t_k}^{t_{k+1}} \int_{t_k}^{t} dw_s^i \, dt - \int_{t_k}^{t_{k+1}} \int_{t_k}^{t} ds \, dw_t^i$$

Since Δw_{k+1}^i and $A_{i,0}(k, k+1)$ have a joint normal distribution, this is perfectly feasible and practical.

A third possibility, which one would like to realise in the case of a multi-dimensional Brownian path, when the commutativity condition is not satisfied, is to generate at the same time increments of the Brownian path and Lévy area integrals, $A_{ij}(k, k+1)$, as defined in (6) above. This is a much harder task, because the joint distribution is complicated. So far, Lévy areas have been generated jointly with increments of the Brownian path in the case $d = 2$ (see [13]) and work is in progress to try and generalise to higher dimensions. It does however make sense in some cases to approximate Lévy areas by adding up increments generated over smaller time steps. To get the same order of accuracy over bigger steps using approximate Lévy areas as over smaller steps using only increments of the Brownian path, one has to generate all the increments over the smaller steps any way, but the saving is in function evaluations, since the various functions of x needed in the discretisation scheme need only be evaluated at the beginning of each large time step.

In a filtering problem, it might sometimes be possible to measure integrals of the observation as well as increments. However, when only increments are available, it would always be possible to approximate integrals by appropriate summation and therefore save function evaluations.

Having represented a Brownian path as a set of increments, or increments and integrals, it is a good idea to go one step further and view the path as a tree. Since it is strong solutions that we are interested in, it makes good sense to generate the same path over different time steps. We have several applications in mind. Firstly, when choosing a suitable time step for a numerical solution, it is helpful to be able to repeat the simulation with smaller or larger time steps and then compare the trajectories obtained. This is only possible when we use the same noise each time but sample it more or less frequently. Secondly, there are SDE's where it is useful to be able to vary the size of the time step, taking smaller steps in "difficult" areas of the solution and larger steps elsewhere.

For these purposes, it is possible to use Lévy's construction, which consists in generating the Brownian path at points halfway between those points at which it has already been generated, using

$$w(t + h/2) = \frac{1}{2}(w(t) + w(t + h)) + \mathcal{N}(0, h/4)$$

Using this recipe, we generate increments of the Brownian path first over steps of length 1, then over steps of size 1/2, and so on, down to the smallest time step we wish to take. Better still, after generating the Brownian path over large time steps, we can refine pieces of it later, as required, when approximating the solution to an SDE. As long as the refinements are stored each time, the process can be repeated consistently as often as required. The set of increments obtained is best stored as a binary tree. The increments of the Brownian path over steps of length h are the parents of the increments over steps of length $h/2$. When a piece of path is refined, each increment gives birth to two children. In order to minimise time spent reading from storage, we can choose to store several consecutive increments, rather than just one, at each node of the tree.

In Figures 3 and 4 we illustrate Lévy's construction used to refine a Brownian path and a Brownian path stored as a tree. More details of the use of 'Brownian trees' and of numerical solution of SDE's using variable time steps can be found in [11].

Lévy's construction can easily be generalised for use with the second representation of a Brownian path described above, where the Brownian path consists of integrals against time as well as increments: given Δw_{k+1} and $A_{i0}(k, k + 1)$, it is easy to generate correctly distributed Δw and A_{i0} elements over the two half intervals $[t_k, t_k + h/2]$, $[t_k + h/2, t_{k+1}]$. However we have not yet succeeded in obtaining such an algorithm for the third representation, where Lévy areas are involved.

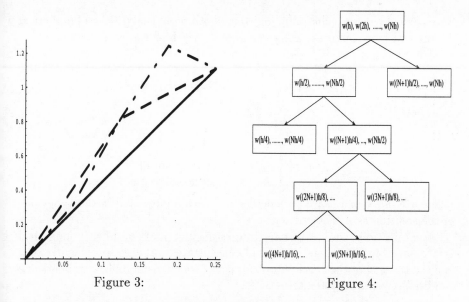

Figure 3:   Figure 4:

3   Experiments with SPDE's

Having shown briefly how strong solutions to SDE's can be obtained, it is
now time to show how these methods can be applied to numerical solution of
SPDE's. The main idea is to discretise the SPDE spatially, obtaining a system
of SDE's, that can then be solved using schemes such as those in Section 2.
However, as with PDE's, it is not possible to describe methods that work well
for all possible SPDE's. Each SPDE has to be examined when it arises and
appropriate methods found for it. That is why we have chosen two particular
SPDE's that we have worked with as illustration and will simply recount which
methods were found to work for them. Both of the examples are parabolic and
have only one space and one time dimension. The first example, a stochastic
KPP equation, involves noise that is dependent only on time, whereas the
second SPDE is driven by noise that is dependent on both time and space and
is therefore more general.

Previous work on numerical solution of SPDE's has mainly been concerned
with the Zakai equation of nonlinear filtering ([1], [2], [5], [6], [12], [18]) and
since that equation is also parabolic the methods used are essentially the same
as those described below. The noise in the Zakai equation is time dependent
(and not space dependent).

3.1   The generalised stochastic KPP equation

The equation in this section has been studied by Elworthy and Zhao [8], following on from the analysis of the deterministic generalised KPP equation by
Elworthy, Truman and Zhao [7]. The numerical simulations described below

were carried out for inclusion in [8], where both more analytical and numerical details and more pictures can be found than will be given here.

The SPDE is

$$\begin{cases} du_t^\mu(x) &= \left[\frac{1}{2}\mu^2 \Delta u_t^\mu(x) + \frac{1}{\mu^2}(1 - u_t^\mu(x))u_t^\mu(x) \right] dt + F_\mu(w_t)u_t^\mu(x) \\ u_0^\mu(x) &= \phi(x) \end{cases} \tag{9}$$

with

$$F_\mu(w_t) = \begin{cases} k/\mu \; dw_t \; \text{(mild noise)} \\ k/\mu^2 \; dw_t \; \text{(strong noise)} \\ k \; dw_t \; \text{(weak noise)} \end{cases}$$

where $x \in I\!R$, k and μ are parameters and we are interested in the behaviour of the solution as μ tends to zero.

The mild noise case is the most interesting and does not differ in numerical treatment from the others, so we will concentrate on that case. The first step is discretisation of the space interval $[x_1, x_n]$, which, using reflective boundary conditions, yields a system of n SDE's:

$$\begin{cases} d\tilde{u}_t^1 &= \left[\frac{\mu^2}{2(\Delta x)^2}(\tilde{u}_t^2 - \tilde{u}_t^1) + \frac{1}{\mu^2}(1 - \tilde{u}_t^1)\tilde{u}_t^1 \right] dt + \frac{k}{\mu}\tilde{u}_t^1 \; dw_t \\ d\tilde{u}_t^i &= \left[\frac{\mu^2}{2(\Delta x)^2}(\tilde{u}_t^{i-1} + \tilde{u}_t^{i+1} - 2\tilde{u}_t^i) + \frac{1}{\mu^2}(1 - \tilde{u}_t^i)\tilde{u}_t^i \right] dt + \frac{k}{\mu}\tilde{u}_t^i \; dw_t \\ & \qquad\qquad (i = 2, \ldots, n-1) \\ d\tilde{u}_t^n &= \left[\frac{\mu^2}{2(\Delta x)^2}(\tilde{u}_t^{n-1} - \tilde{u}_t^n) + \frac{1}{\mu^2}(1 - \tilde{u}_t^n)\tilde{u}_t^n \right] dt + \frac{k}{\mu}\tilde{u}_t^n \; dw_t \end{cases} \tag{10}$$

where $\Delta x = (x_n - x_1)/n$. Here $\tilde{u}_t$ denotes the approximate solution to (9) given by the solution of (10).

Since the dimension of the Brownian path is one, the Milshtein scheme described in Section 2 should yield first order approximations to the solution of (10). However, as with parabolic PDE's, there is a stability problem. To ensure stability, we have the choice of using an explicit discretisation scheme, such as the Milshtein scheme, with a time step h dependent on the space step Δx and satisfying

$$\mu^2 h < 1/2(\Delta x)^2 \tag{11}$$

or else of using an implicit method. Since it is the drift of the SDE and not the diffusion that causes the instability problem, it makes sense in fact to use a semi-implicit method, implicit in the drift terms and explicit in the diffusion terms. A semi-implicit version of the Milshtein scheme, first suggested in [29], is as follows:

$$\bar{x}_{k+1} = \bar{x}_k + \alpha b(\bar{x}_k, t)h + (1-\alpha)b(\bar{x}_{k+1}, t)h + \sigma(\bar{x}_k, t)\Delta w + \frac{1}{2}\sigma\sigma'(\bar{x}_k, t)(\Delta w^2 - h) \tag{12}$$

The semi-implicit version of the Euler-Maruyama scheme is the same without the last term.

Equation (10) can be written as

$$d\tilde{u}_t = A\tilde{u}_t\, dt + v_t$$

with the tridiagonal matrix

$$A = \frac{\mu^2}{2(\Delta x)^2} \begin{pmatrix} -1 & 1 & 0 & \cdot & \cdot & \cdot & 0 & 0 \\ 1 & -2 & 1 & 0 & \cdot & \cdot & \cdot & 0 \\ \cdot & \cdot & \cdot & \cdot & \cdot & \cdot & \cdot & \cdot \\ 0 & \cdot & \cdot & \cdot & 0 & 1 & -2 & 1 \\ 0 & 0 & \cdot & \cdot & \cdot & 0 & 1 & -1 \end{pmatrix}$$

and

$$v_t^i = \frac{1}{\mu^2}(1 - \tilde{u}_t^i)\tilde{u}_t^i\, dt + \frac{k}{\mu}\tilde{u}_t^i\, dw_t$$

The semi-implicit Milshtein scheme, with $\alpha = 1/2$, then gives

$$\left(I - \frac{\Delta t}{2}A\right)\bar{u}(t_{j+1}) = \left(I + \frac{\Delta t}{2}A\right)\bar{u}(t_j) + r(t_j) \tag{13}$$

where

$$r^i(t_j) = \frac{1}{\mu^2}(1 - \bar{u}^i(t_j))\bar{u}^i(t_j)\Delta t + \frac{k}{\mu}\bar{u}^i(t_j)\Delta w(j) + \frac{k^2}{\mu^2}\bar{u}^i(t_j)(\Delta w(j)^2 - \Delta t)$$

Here $\bar{u}(t)$ denotes the approximate solution to (10). See [8] for details of how (13) may be solved efficiently.

Using initial conditions such as a step function or a point mass, it turns out that small space steps are needed to obtain the correct speed of propagation of the wave (even with $k = 0$). The speed of propagation, when starting from a δ-function, is proved by Elworthy and Zhao in [8] to be $\sqrt{2 - k^2}$.

In Figure 5, we show pictures of approximate solutions to (9) in the mild noise case, starting from

$$\phi(x) = \begin{cases} 1; & x = 0 \\ 0; & \text{otherwise} \end{cases}$$

with $\mu = 0.1$ and k taking the two values 0.1 and 1. The space step used was $\Delta x = 0.01$. Note that, just as in Figure 2 in the previous section, where solutions were obtained for different parameter values, the same Brownian path has been used for the two pictures.

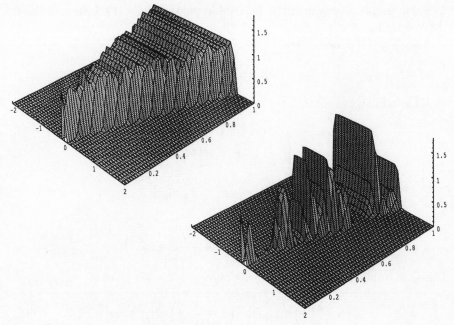

Figure 5:

3.2   The heat equation with noise

We now consider an SPDE with space-dependent noise, i.e. where the driving noise is a Brownian sheet. The SPDE is

$$\begin{cases} \dfrac{\partial u(x,t)}{\partial t} - \dfrac{\partial^2 u(x,t)}{\partial x^2} = u(x,t)^\gamma \dot{w}(x,t) \\ u(.,0) = u_0 \\ u(0,t) = u(1,t) \end{cases} \tag{14}$$

where $\gamma \in [1/2, 2]$ is a parameter. Much theoretical work has been done on this equation (see [21], [23], [22], [24]).

The spatial discretisation, similar to that of the generalised KPP equation, but with a difference in boundary condition, corresponding to solving the SPDE on a circle, yields

$$\begin{cases} du^1(t) = \dfrac{1}{(\Delta x)^2}(\tilde{u}^n + \tilde{u}^2 - 2\tilde{u}^1)\, dt + \dfrac{1}{\sqrt{\Delta x}}(\tilde{u}^1)^\gamma\, dw^1(t) \\ d\tilde{u}^i(t) = \dfrac{1}{(\Delta x)^2}(\tilde{u}^{i-1} + \tilde{u}^{i+1} - 2\tilde{u}^i)\, dt + \dfrac{1}{\sqrt{\Delta x}}(\tilde{u}^i)^\gamma\, dw^i(t), \\ \hspace{6cm} i = 2, \dots, n-1 \\ d\tilde{u}^n(t) = \dfrac{1}{(\Delta x)^2}(\tilde{u}^{n-1} + \tilde{u}^1 - 2\tilde{u}^n)\, dt + \dfrac{1}{\sqrt{\Delta x}}(\tilde{u}^n)^\gamma\, dw^n(t) \end{cases} \tag{15}$$

Since the equation is again parabolic, an explicit time discretisation scheme will only be stable for time steps satisfying $h < 1/2(\Delta x)^2$, so we again chose to use the semi-implicit Milshtein scheme (12). Although the Brownian path is multi-dimensional, the SDE (15) satisfies the commutativity condition (8), so it is still possible to obtain first order strong approximations to solutions.

The matrix corresponding to the matrix A in the previous example is, due to the boundary condition, no longer purely tridiagonal. However it is possible to modify the traditional method used for inversion of a tridiagonal matrix to accommodate the 2 off-diagonal elements.

What really makes this example more challenging than the previous one, is the handling of the Brownian sheet $w(x, t)$, which needs to be discretised both in time and in space. Spatial discretisation yields an independent Brownian path $w^i(t)$ on each subinterval $[x_i, x_{i+1}]$. We would like to be able to subdivide the noise at will, both in time, as described in Section 2.3, and also spatially, so as to be able to compare approximate solutions obtained with different time steps and with different space steps. In other words, having initially chosen a value for n, and generated n independent Brownian paths, we would like to see how the results change when the circle is divided into more intervals, ie when n is increased. Just as the easiest thing to do when decreasing time steps is to halve them, the easiest method here is to halve the length of the space intervals, so doubling n.

There are two possibilities, either subdivide in time first and then in space or vice-versa. We have tried both. In early experiments we generated sets of Brownian paths for increasing values of n, starting with $n = 1$ and doubling n at each stage. Each path $w^i(t)$ was split into two independent paths $\tilde{w}^{2i}(t)$ and $\tilde{w}^{2i+1}(t)$, by setting

$$
\begin{aligned}
\Delta \tilde{w}^{2i}(j) &= c(\Delta w^i(j) + z^i(j)) \\
\Delta \tilde{w}^{2i+1}(j) &= c(\Delta w^i(j) - z^i(j))
\end{aligned}
$$

$\forall i = 1, \ldots, n$, $\forall j = 1, \ldots, N$, where the $z^i(j)$ are all independent normally distributed numbers with the same variance as $\Delta w^i(j)$. However, with this method it was not easy to subdivide the time steps consistently.

In subsequent experiments, the time discretisation took precedence over the space discretisation. We started by fixing a maximum value for $n = 2^m$. (We used a maximum m of 6,7 or 8.) Then we generated a set of n independent Brownian paths (or rather one Brownian path of dimension n) stored in a file using the tree structure described in Section 2.3. Once the complete tree is stored, it can be used to generate a similar tree over half as many space steps, that is a Brownian path of dimension $n/2$. This is done simply by adding together pairs of increments that are adjacent in space and scaling to ensure that the new path has the standard variance, i.e.

$$
\tilde{\Delta} w^i(j) = \frac{1}{\sqrt{2}}(\Delta w^{2i}(j) + \Delta w^{2i+1}(j))
$$

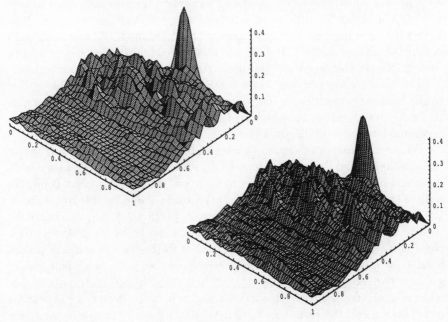

Figure 6:

$\forall i = 1, \ldots, n/2, \forall j$. The new smaller path is then stored in a file in its turn using exactly the same tree structure. This procedure can be repeated as many times as desired, creating each time a wider spatial discretisation.

With this method, it is possible to further subdivide the time steps after the Brownian paths have already been generated and used. The only constraint is that the subdivision must be performed on the initial tree, the one with the largest value of n. Then the Brownian paths corresponding to smaller values of n can be regenerated by adding up the increments in the largest tree. The resulting paths will be completely compatible with the original ones. This turns out to be very useful when working with the SPDE (14). The solution typically contains peaks and troughs, leading to large differences in value of the solution at neighbouring points. When solving numerically, this situation often leads locally to negative values of $u(x,t)$ (which is theoretically impossible). The answer is to retreat a step and take a smaller time step whenever this situation occurs, decreasing the time step as far as necessary to keep the solution positive. In practice we have often had to take time steps as small as $h = 2^{-20}$ somewhere along the solution. Taking such small steps throughout would lead to excessive amounts of both computation time and storage (for the Brownian path).

In Figure 6, we show two solutions to (14) using the same Brownian sheet, taking first 32 and then 64 space steps. The value of γ used is 0.9.

References

[1] J. F. BENNATON, *Discrete time Galerkin approximations to the nonlinear filtering solution*, J. Math. Analysis Appl., 110 (1985), pp. 364–383.

[2] A. BENSOUSSAN, R. GLOWINSKI, AND A. RASCANU, *Approximation of the Zakai equation by the splitting up method*, Siam J. Control and Optimization, 28 (1990), pp. 1420–1431.

[3] J. M. C. CLARK, *The discretization of stochastic differential equations: A primer*, in Road Vehicle Systems and Related Mathematics, Proc. Second Workshop Turino, H. Neunzert, ed., Series Mathematical Methods in Technology, 1987.

[4] J. M. C. CLARK AND R. J. CAMERON, *The maximum rate of convergence of discrete approximations for stochastic differential equations*, in Stochastic Differential Systems, B. Grigelionis, ed., no. 25 in Lecture Notes in Control and Information Sciences, Springer-Verlag, Berlin, 1980.

[5] M. H. A. DAVIS, *A pathwise solution of the equations of nonlinear filtering*, Theory Prob. Appl., 27 (1982), pp. 167–175.

[6] R. J. ELLIOT AND R. GLOWINSKI, *Approximations to solutions of the Zakai filtering equation*, Stoch. Analysis and Appl., 7 (1989), pp. 145–168.

[7] K. D. ELWORTHY, A. TRUMAN, AND H. Z. ZHAO, *Approximate travelling waves for generalized KPP equations and classical mechanics (with an appendix by J.G. Gaines)*. To appear in Proceedings of the Royal Society of London, Series A.

[8] K. D. ELWORTHY AND H. Z. ZHAO, *The propagation of travelling waves for stochastic generalized KPP equations (with an appendix by J.G. Gaines)*. To appear in Mathematical and Computer Modelling (Oxford).

[9] O. FAURE, *Numerical pathwise approximation of stochastic differential equations*, tech. report, CERMA-ENPC, 1990.

[10] ——, *Simulation du mouvement brownien et des diffusions*, PhD thesis, Ecole Nationale des Ponts et Chaussées , CERMA-ENPC, 1991.

[11] O. FAURE AND J. G. GAINES, *Simulation trajectorielle des diffusions*, in Probabilités Numériques, N. Bouleau and D. Talay, eds., INRIA, Rocquencourt, France, 1992, pp. 186–192.

[12] P. FLORCHINGER AND F. LEGLAND, *Time-discretization of the Zakai equation for diffusion-processes observed in correlated noise*, Lecture notes in control and information sciences, 144 (1990), pp. 228–237.

[13] J. G. GAINES AND T. J. LYONS, *Random generation of stochastic area integrals.* To appear in SIAM J. of Applied Math.

[14] T. C. GARD, *Introduction to Stochastic Differential Equations*, Monographs and Textbooks in Pure and Applied Mathematics, Marcel Dekker, Inc., New York and Basel, 1988.

[15] P. E. KLOEDEN AND E. PLATEN, *A survey of numerical methods for stochastic differential equations*, Journal of Stochastic Hydrology and Hydraulics, 3 (1989), pp. 155–178.

[16] ——, *Stratonovich and Itô stochastic Taylor expansions*, Math. Nachr., 151 (1991), pp. 33–50.

[17] ——, *Numerical Solution of Stochastic Differential Equations*, vol. 23 of Applications of Mathematics, Springer-Verlag, 1992.

[18] F. LEGLAND, *Splitting-up approximation for SPDEs and SDEs with application to nonlinear filtering*, Lecture notes in control and information sciences, 176 (1992), pp. 177–187.

[19] MARUYAMA, *Continuous Markov processes and stochastic equations*, Com Rend. Circ. Mat. Palermo, 4 (1955), pp. 48–90.

[20] G. N. MILSHTEIN, *Approximate integration of stochastic differential equations*, Th. Prob. Appl., 19 (1974).

[21] C. MUELLER, *Long-time existence for the heat-equation with a noise term*, Proba. Theory Rel. Fields, 90 (1991), pp. 505–517.

[22] ——, *Coupling and invariant-measures for the heat-equation with noise*, Annals of Proba., 21 (1993), pp. 2189–2199.

[23] C. MUELLER AND E. A. PERKINS, *The compact support property for solutions to the heat equation with noise*, Probab. Theory Related Fields, 93 (1992), pp. 325–358.

[24] C. MUELLER AND R. SOWERS, *Blowup for the heat-equation with a noise term*, Proba. Theory Rel. Fields, 97 (1993), pp. 287–320.

[25] N. J. NEWTON, *An asymptotically efficient difference formula for solving stochastic differential equations*, Stochastics, 19 (1986), pp. 175–206.

[26] ——, *Asymptotically efficient Runge-Kutta methods for a class of Itô and Stratonovich equations*, SIAM J. of Applied Mathematics, 51 (1991), pp. 542–567.

[27] E. PARDOUX AND D. TALAY, *Discretization and simulation of stochastic differential equations*, Acta Appl. Math., 3 (1985), pp. 23–47.

[28] E. PLATEN AND W. WAGNER, *On a taylor formula for a class of Itô processes*, Prob. and Math. Stats., 3 (1982), pp. 37–51.

[29] D. TALAY, *Analyse numérique des équations différentielles stochastiques*, PhD thesis, Univ. Provence, 1982.

[30] ——, *Simulation and numerical analysis of stochastic differential systems: A review*, Tech. Report 1313, INRIA, 1990.

Contour Processes of Random Trees

J. GEIGER

Department of Mathematics, Frankfurt University.

D-60054 Frankfurt, Germany.

Abstract. The contour process of a planar tree is the 'distance to the root'- process when the edges of the tree are traversed in a certain linear order. We study a class of random binary trees whose contour processes are Markovian. We heavily make use of the correspondence between random trees and point processes.

AMS (1991). 60J80, 60G55, 60J25.

Key words. Random tree, branching process, Poisson point process, contour process.

1 Introduction

Consider the family tree of a finite population with a single founding ancestor. The edges represent individuals and the lengths of the edges are the associated lifetimes. Suppose that births occur only at times of individuals' deaths and that offspring are distinguished as first, second, third, ... The succession of offspring induces a linear order on the edges of the tree : the order of succession to the throne under primogeniture. (The number 1 is the edge of the founding ancestor. If k edges have been labelled, we search for the one with the highest number that still has unlabelled offspring and attach number $k+1$ to the first of these unlabelled successors.)

FIG. 1

The founding ancestor at the bottom of the tree has three children. His children have 0, 2, and 0 children respectively (by convention we draw the first child left-most).

An ordered tree can be represented by its *contour*. The contour is the 'distance to the root'- process when the edges of the tree are traversed in the order of succession to the throne. The edges are traversed with constant speed and in direction leading away from the root : We begin with the edge of the founding ancestor. At its branching point we switch into the edge of the first

72

child. Coming to the end of an edge we jump back to the beginning of the edge following (in the order of succession to the throne) the one that has just been traversed (see figure 2 for the contour of the tree in figure 1).

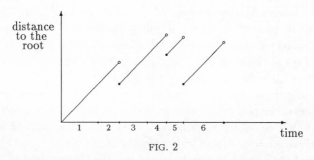

FIG. 2

The numbers below the time axis indicate the edges being traversed at that moment. By convention the speed of the traversal (= slope of the contour) is 1.

An ordered tree can be recovered from its contour providing that no individual has exactly one child. The heights of local minima correspond to times of births and the heights of local maxima correspond to times of deaths. The relationship of individuals in the tree can be seen by reading the contour from the bottom to the top. The local minima of lowest height correspond to the branching point of the founding ancestor. The number of these minima is one less than the number of children: This is because no jump occurs at time of entrance into the first child's edge (and this is also the reason that the contour of a tree consisting of a single edge cannot be distinguished from that of a tree consisting of an ancestor that has exactly one child). The part of the contour up to the time of the first minimum of lowest height describes the subtree of the first child with his possible descendants; the part between the first and the second of these minima that of the second child. Finally, the part of the contour following the last minimum of lowest height describes the subtree of the last child with his possible descendants. Now each part of the contour is again decomposed at its lowest minima. Eventually the contour falls to single peaks. The corresponding subtrees consist of single edges, the heights of the peaks are the lengths of the edges (see figure 3).

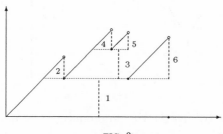

FIG. 3

The *contour process* of a random ordered tree (called *search depth process* in computer science) is the family of coordinate mappings of the random contour. An especially nice example is the critical binary Galton-Watson tree with exponential lifetime distribution: in this case the contour process is a time-homogeneous strong Markov process. A result by Le Gall ([LG], 1989) states that the contour process of a critical binary Galton-Watson tree performs an alternating exponential random walk (the times between jumps as well as the heights of jumps are independent identical exponentially distributed) up to the time of first entrance into the negative real half line.

If the contour process of a random tree is a time-homogeneous strong Markov process, the population size at time t, conditioned on non-extinction up to time t, has geometric distribution. Each individual alive at time t generates exactly one upcrossing of level t by the contour (see figure 3). Since the transition probabilities of the contour process are independent of the elapsed time (homogeneity) and of the history of the process (Markov property), the number of visits to a certain state has geometric distribution once this state has been hit.

It is easy to see that binary branching is a crucial condition that the contour processes be Markovian. If an individual has more than three offspring (as the founding ancestor of the tree in figure 1 has), minima of same height show up in the contour of the tree (compare figure 2). Thus the contour process of a random tree with general offspring distribution tends to jump to points it already jumped to in the past. This tendency is not compatible with the lack of memory of Markov processes.

The binary Galton-Watson tree is not alone: there exists a whole family of random binary trees with time-homogeneous Markovian contour processes. We call these trees *self-similar* (the terminology is not standard). To introduce them we modify our interpretation of binary branching. Branching points are no longer interpreted as one individual ceasing to exist and being replaced by two others, but as the birth of one individual during the life of its parent. (Or one might say that the parent lives on in the second child. See figure 4; the horizontal lines are merely to indicate connections).

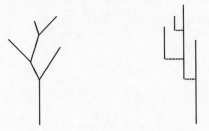

FIG. 4

The prescription for tracing the tree to produce the contour process then becomes simply 'children first'. At a branching point one switches into the edge of the child. Coming to the end of an edge one jumps back into the edge of the parent, namely to the point where the just traversed edge branches off.

Now let μ be a probability measure on $\mathbb{R}^+$. Consider the following stochastic model for the evolution of a population:

1) Individuals live, die and reproduce independently of each other.

2) The lifetime of an individual exceeds a with probability $\mu([a,\infty))$.

3) While alive an individual gives birth with rate 1 independent of age, lifetime and previous reproduction.

The random family tree of a population with a single founding ancestor arising from this model is called *binary self-similar tree with parameter* μ. Binary self-similar trees belong to the class of age-dependent branching processes: the age-dependence comes in through the survival rate. The term *self-similar* expresses the fact that the subtrees founded by individuals not in the same line of descent are independent copies of the whole tree (see the encircled descendant trees in figure 5).

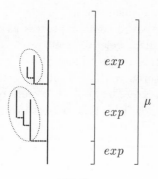

FIG. 5

One can also construct binary self-similar trees and their contour processes for certain infinite measures μ. A sufficient condition on μ is $\int_0^1 y\,\mu(dy) + \int_1^\infty \mu(dy) < \infty$, which means that individuals may have infinitely many offspring provided that only a few survive for long. In this paper we restrict ourselves for simplicity to the case where μ is a probability measure in order to avoid technical difficulties. The general case is treated in [GE].

A basic tool for constructing and studying contour processes is the correspondence between binary ordered trees and counting measures on $\mathbb{R}_0^+ \times \mathbb{R}^+$. Such a measure can be associated with any ordered binary tree. Each edge of the tree generates a point of the support of the measure. The length of the edge is chosen to be the ordinate of the point of the support. The time of first entrance into this edge during the traversal underlying the contour process becomes the abscissa. (See figure 6 for an ordered tree and the support of the associated measure.)

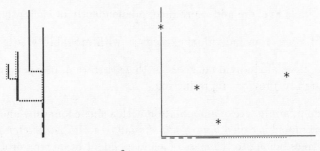

FIG. 6

In section 2 we reverse this procedure : we recover the structure of a binary ordered tree from the support of a counting measure on $\mathbb{R}_0^+ \times \mathbb{R}^+$. The suitable randomization of the measure to generate binary self-similar trees with parameter μ turns out to be the Poisson point process with intensity measure $\lambda \otimes \mu$, where λ is Lebesgue measure on $\mathbb{R}_0^+$.

In section 3 we construct the contour process as a component of a more complex process, the so-called delete & shift process. There the tree is traversed according to the prescription 'children first' and edges are cut off the moment they are passed through. The resulting pruned trees are the states of the delete & shift process. The delete & shift process is easily seen to be Markovian. We then show why this property carries over to the contour process.

2 The tree in a point process

In this section we will introduce a linear and a partial order of the support of a counting measure on $\mathbb{R}_0^+ \times \mathbb{R}^+$ that induce the structure of one or several ordered trees. Moreover there is a chronology within the associated trees, individuals give birth and die at certain ages. In the second part we will show that the trees associated with homogeneous Poisson point processes are binary self-similar trees.

Let $\quad \mathcal{M}_f := \{ \, \Phi = \sum_i \delta_{z_i} \mid z_i = (x_i, y_i) \in \mathbb{R}_0^+ \times \mathbb{R}^+, (A1), (A2) \} \, ,$

where $\quad (A1) \quad : \quad x_i \neq x_j \, , \quad \forall \, i \neq j$

$\qquad\qquad (A2) \quad : \quad \sum_i 1_{B \times \mathbb{R}^+}(z_i) \; < \infty \, , \quad \forall \, B \subset \mathbb{R}_0^+ \text{ bounded} \, .$

Let S denote the shift operator on $\mathcal{M}_f$ and R the trace operator

$$S_x \Phi(A) \ := \ \Phi((x,0)+A) \ , \quad \forall \ x \geq 0, \ A \subset \mathbb{R}_0^+ \times \mathbb{R}^+$$
$$R_D \Phi(A) \ := \ \Phi(D \cap A) \ , \quad \ \forall \ D, \ A \ \subset \mathbb{R}_0^+ \times \mathbb{R}^+ \ .$$

By means of the following functions we will order the support of a measure in $\mathcal{M}_f$. For $\Phi = \sum_i \delta_{z_i} \in \mathcal{M}_f$; $y, t \geq 0$ define

$$g_t(\Phi) \ := \ t - \sum_{0 < x_i \leq t} y_i$$

$$\overline{g}_t(\Phi) \ := \ \sup_{s \leq t} g_s(\Phi)$$

$$h_t(\Phi) \ := \ (\overline{g}_t - g_t)(\Phi)$$

$$\tau_y(\Phi) \ := \ \inf \left\{ s \geq 0 \mid \overline{g}_s(\Phi) = y \right\}$$

Note that $(g_t(\cdot))_{t \geq 0}$ does not perceive points of the support in $\{0\} \times \mathbb{R}^+$, but separates Φ and Φ' in the case where $R_{\mathbb{R}^+ \times \mathbb{R}^+} \Phi \neq R_{\mathbb{R}^+ \times \mathbb{R}^+} \Phi'$.

Example

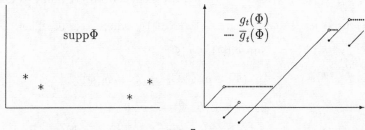

FIG. 7

Consider a system where at time x_i an item with service time y_i arrives. Let the queue discipline be 'Last Come – First Served', i.e. the items are stacked and operation occurs at the top of the pile. If all $x_i > 0$, then $h_t(\Phi)$ is the residual work in the system and $\overline{g}_t(\Phi)$ is the time up to t that no service is performed. $\tau_{y_i}(S_{x_i} \Phi)$ is the total time that item i spends in the system.

We collect some basic properties of the introduced functions. (The statements are not very surprising if one thinks in terms of the service system, see figure 7.)

Proposition 2.1 *Let* $\Phi = \sum_i \delta_{z_i} \in \mathcal{M}_f$, *then*

1) $g_\cdot(\Phi)$ *is an additive functional*

$$g_{t_1 + t_2}(\Phi) = g_{t_1}(\Phi) + g_{t_2}(S_{t_1} \Phi) \ .$$

2) $\overline{g}_\cdot(\Phi)$ *is the occupation time of 0 by $h_\cdot(\Phi)$ up to time t*

$$\overline{g}_t(\Phi) = \int_0^t 1 \{ h_s(\Phi) = 0 \} \, ds \ .$$

3) *Let* $A_i = A_i(\Phi) = [x_i, x_i + \tau_{y_i}(S_{x_i}\Phi))$, *then*

   a) $\{t > 0 \mid h_t(\Phi) > 0\} = \bigcup_{x_i > 0} A_i$

   b) $A_i \cap A_j \in \{A_i, A_j, \emptyset\}$

   c) $\lambda\left(\bigcup_{i \in I} A_i\right) = \sum_{x_j \in \bigcup_{i \in I} A_i} y_j$, λ *Lebesgue measure on* $\mathbb{R}_0^+$.

4) *The following partition equation holds*

$$t = \overline{g}_t(\Phi) + \sum_{0 < x_i < t} \min(\overline{g}_{t-x_i}(S_{x_i}\Phi), y_i) .$$

Proof

1) by definition.

2) By $(A2)$ the function $1\{h.(\Phi) = 0\}$ is piecewise constant and
$\frac{d}{ds}\overline{g}_s(\Phi) = 1\{h_s(\Phi) = 0\}$ holds except for finitely many $s \in [0, t]$.

3)   a) First let $t \in A_i = [x_i, x_i + \tau_{y_i}(S_{x_i}\Phi))$, where $x_i > 0$. Then

$$\begin{aligned}
h_t(\Phi) &= \sup_{s \leq t} g_s(\Phi) - g_t(\Phi) \\
&\geq g_{x_i^-}(\Phi) - g_{x_i}(\Phi) + g_{x_i}(\Phi) - g_t(\Phi) \\
&\overset{1)}{=} y_i - g_{t-x_i}(S_{x_i}\Phi) \quad > \quad 0 ,
\end{aligned}$$

since $t - x_i < \tau_{y_i}(S_{x_i}\Phi) = \inf\{s \geq 0 \mid \overline{g}_s(S_{x_i}\Phi) = y_i\}$.

Now let $h_t(\Phi) > 0$.
Define $x_t := \sup\{s \leq t \mid h_{s^-}(\Phi) = 0\}$ and $y_t := h_{x_t}(\Phi)$. Note
that $(x_t, y_t) \in \operatorname{supp}\Phi$ and that 2) implies $\overline{g}_{x_t}(\Phi) = \overline{g}_t(\Phi)$.
Thus

$$\begin{aligned}
\overline{g}_{t-x_t}(S_{x_t}\Phi) &\overset{1)}{=} \sup_{s \leq t-x_t}(g_{x_t+s}(\Phi) - g_{x_t}(\Phi)) \\
&= \sup_{s \leq t-x_t}(g_{x_t+s}(\Phi) - \overline{g}_{x_t+s}(\Phi)) + \overline{g}_{x_t}(\Phi) - g_{x_t}(\Phi) \\
&= h_{x_t}(\Phi) - \inf_{s \leq t-x_t} h_{x_t+s}(\Phi) \quad < \quad y_t ,
\end{aligned}$$

where the inequality holds by construction of x_t and y_t.
Consequently $t \in [x_t, x_t + \tau_{y_t}(S_{x_t}\Phi))$.

   b) Let $x_i < x_j$.
3.a) applied to $S_{x_i}\Phi$ yields $h_{s-x_i}(S_{x_i}\Phi) > 0$, $\forall\, s \in A_j$, such that
in view of 2) $\overline{g}_{x_j-x_i}(S_{x_i}\Phi) = \overline{g}_{x_j+\tau_{y_j}(S_{x_j}\Phi)-x_i}(S_{x_i}\Phi)$.

Thus

$$A_i \cap A_j = \begin{cases} A_j & , \text{ if } \overline{g}_{x_j-x_i}(S_{x_i}\Phi) < y_i \\ \emptyset & , \text{ if } \overline{g}_{x_j-x_i}(S_{x_i}\Phi) \geq y_i \end{cases} .$$

c) To begin with let $I = \{i\}$. Note that

$$\lambda(A_i) = \tau_{y_i}(S_{x_i}\Phi) = \inf\{s \geq 0 \mid g_{s-}(S_{x_i}\Phi) = y_i\} ,$$

since $g_{t-}(\Phi) \geq g_t(\Phi)$, $\forall t > 0$, $\Phi \in \mathcal{M}_f$.

If $\tau_{y_i}(S_{x_i}\Phi) = \infty$, then $\limsup_{u \to \infty} \left(u - \sum_{x_j \in (x_i, x_i+u]} y_j \right) < y_i$, and consequently $\sum_{x_j \in A_i} y_j = \sum_{x_j \in [x_i, \infty)} y_j = \infty$.

If $\tau_{y_i}(S_{x_i}\Phi) < \infty$, then the infimum is attained and consequently

$$\begin{aligned}
\lambda(A_i) &= \tau_{y_i}(S_{x_i}\Phi) - g_{(\tau_{y_i}(S_{x_i}\Phi))-}(S_{x_i}\Phi) + y_i \\
&= \sum_{x_i < x_j < x_i + \tau_{y_i}(S_{x_i}\Phi)} y_j + y_i = \sum_{x_j \in A_i} y_j .
\end{aligned}$$

Now let $|I| < \infty$. In view of 3.b) a subsequence $i_1, \ldots, i_k$ exists such that $\bigcup_{i \in I} A_i = A_{i_1} \dot{\cup} \ldots \dot{\cup} A_{i_k}$. Thus the first part yields

$$\lambda\left(\bigcup_{i \in I} A_i\right) = \sum_{j=1}^{k} \lambda(A_{i_j}) = \sum_{j=1}^{k} \sum_{x_\ell \in A_{i_j}} y_\ell = \sum_{x_\ell \in \bigcup_{i \in I} A_i} y_\ell .$$

Finally if $|I| = \infty$ chose finite I_n such that $I_n \uparrow I$ and pass to the limit

$$\lambda\left(\bigcup_{i \in I} A_i\right) = \lim_{n \to \infty} \lambda\left(\bigcup_{i \in I_n} A_i\right) = \lim_{n \to \infty} \sum_{x_j \in \bigcup_{i \in I_n} A_i} y_j = \sum_{x_j \in \bigcup_{i \in I} A_i} y_j .$$

4) First of all 2) and 3.a) imply

$$\begin{aligned}
t &= \lambda(\{s \leq t \mid h_s(\Phi) = 0\}) + \lambda(\{s \leq t \mid h_s(\Phi) > 0\}) \\
&= \overline{g}_t(\Phi) + \lambda\left(\bigcup_{x_i > 0} A_i \cap [0, t]\right) .
\end{aligned}$$

In view of 3.b) we may substitute $\bigcup_{x_i > 0} A_i$ by $\dot{\bigcup}_{x_i > 0} B_i$, where $B_i := A_i \backslash \bigcup_{A_j \subsetneq A_i} A_j$. Thus 2), 3.a) and 3.b) imply

$$\begin{aligned}
t &= \overline{g}_t(\Phi) + \sum_{x_i > 0} \lambda\left(B_i \cap [0, t]\right) \\
&= \overline{g}_t(\Phi) + \sum_{x_i > 0} \lambda(\{0 \leq s \leq \min(t - x_i, \tau_{y_i}(S_{x_i}\Phi)) \mid h_s(S_{x_i}\Phi) = 0\}) \\
&= \overline{g}_t(\Phi) + \sum_{0 < x_i < t} \min\left(\overline{g}_{t-x_i}(S_{x_i}\Phi), y_i\right) .
\end{aligned}$$

We can now define relationship of points in supp Φ .

Definition 2.2   *Let* $\Phi = \sum_i \delta_{z_i} \in \mathcal{M}_f$,

$$z_i <_\Phi z_j \quad :\Longleftrightarrow \quad x_i < x_j$$
$$z_i \prec_\Phi z_j \quad :\Longleftrightarrow \quad A_j(\Phi) \subsetneq A_i(\Phi) \quad .$$

Note that in view of proposition 2.1.3.b)

$$z_i \prec_\Phi z_k \text{ and } z_j \prec_\Phi z_k \, , \, i \neq j \quad \Longrightarrow \quad z_i \prec_\Phi z_j \text{ or } z_j \prec_\Phi z_i \, ,$$

which means that the directed graph (supp Φ, $\prec_\Phi$) contains no cycles and thus consists of one or several trees (note that the graph is possibly disconnected). The trees are ordered since the relation $<_\Phi$, which is linear by $(A1)$, induces a succession of offspring. Moreover there is a chronology within the trees. We regard y_i as lifetime of the 'individual z_i' . In case that z_i is an ancestor of z_j ($z_i \prec_\Phi z_j$) we interpret $\overline{g}_{x_j - x_i}(S_{x_i}\Phi)$ as age of z_i at which it gives birth to the child whose descendant z_j is (see figure 8).

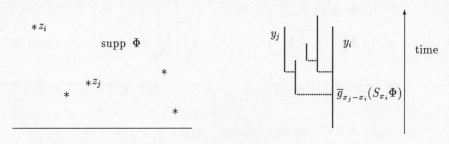

FIG. 8

We will now identify the tree associated with a homogeneous Poisson point process as binary self-similar tree: while alive and independent of age, life-time and previous reproduction an individual splits off single descendants with rate 1. We restrict ourselves to the subcritical case $\int_{\mathbb{R}+} y \, \mu(dy) \leq 1$. In the supercritical case the probability that the tree is infinite (family does not become extinct) is positive. Due to the embedding in the order of succession to the throne the point process gets caught in the first infinite subtree such that the description of the whole tree is incomplete.

Having in mind the independence properties of Poisson point processes we may condition on the lifetime and the position of the founding ancestor. For $\Phi \in \mathcal{M}_f, a > 0$ let

$$\Phi^a := \delta_{(0,a)} + R_{(0,\tau_a(\Phi)) \times \mathbb{R}+} \, \Phi \, .$$

Note that $g_s(\Phi) = g_s(\Phi^a)$, $\forall \, 0 \leq s < \tau_a(\Phi)$, such that in particular $\tau_a(\Phi) = \tau_a(\Phi^a)$.

Let $BL_1(\Phi^a)$ denote the times of birth and the lifetimes of the ancestor's children

$$BL_1(\Phi^a) := \bigcup_{(0,a) \dashv_\Phi a z_i} \{ (\overline{g}_{z_i}(\Phi^a), y_i) \} \, ,$$

where $z_i \dashv_\Phi z_j \; :\Longleftrightarrow$ 1) $z_i \prec_\Phi z_j$

2) $z_k \prec_\Phi z_j \implies z_k \prec_\Phi z_i$ or $i = k$.

Theorem 2.3 *Let N be a Poisson point process on $\mathbb{R}_0^+ \times \mathbb{R}^+$ with intensity measure $\lambda \otimes \mu$, where λ denotes Lebesgue measure on $\mathbb{R}_0^+$ and μ is a probability measure on $\mathbb{R}^+$ with $\int_{\mathbb{R}^+} y \, \mu(dy) \leq 1$. Then $BL_1(N^a)$ is a Poisson point process on $(0,a) \times \mathbb{R}^+$ with intensity measure $\lambda_{|(0,a)} \otimes \mu$.*

Remarks

– We (mis)use the term point process for set-valued as well as for measure-valued random variables (set $\hat{=}$ support of measure).

– Let T be a finite stopping time with respect to $(\mathcal{F}_{t+})_{t \geq 0}$, where $\mathcal{F}_t := \sigma\{R_{[0,t] \times \mathbb{R}^+}N\}$. Then $S_T R_{(T,\infty) \times \mathbb{R}^+}N$ is independent of $\mathcal{F}_T$ and distributed as $R_{(0,\infty) \times \mathbb{R}^+}N$. Thus the reproduction law of the ancestor carries over to descending individuals.

– The subcritical binary Galton-Watson tree with lifetime distribution $\exp(p_0/p_2)$ is generated by N^X if $\mathcal{L}(X) = \mu = \exp(p_0/p_2)$.

(We use the notation $\exp(\gamma)$ to represent the distribution of an exponential random variable with mean $1/\gamma$. Note that we consider the binary Galton-Watson tree as self–similar tree ('individuals split off single descendants') and consequently use the term lifetime distribution for μ . The time until the first birth, the quantity which is traditionally understood by lifetime in Galton–Watson process literature, has distribution $\exp(1/p_2)$ if $\mu = \exp(p_0/p_2)$.)

$$\begin{aligned} P\left(|BL_1(N^X)| = k\right) &= \int_0^\infty \frac{y^k}{k!} e^{-y} \frac{p_0}{p_2} e^{-\frac{p_0}{p_2} y} \, dy \\ &= p_0 \, p_2^k \frac{1}{k!} \int_0^\infty \left(\frac{y}{p_2}\right)^k e^{-\frac{y}{p_2}} \frac{dy}{p_2} \\ &= p_0 \, p_2^k , \qquad k = 0, 1, 2, \ldots \end{aligned}$$

Proof We want to apply Itô's theorem, which states that under slight assumptions the point process of excursions of a homogeneous Markov process is a Poisson point process, to $(h_t(N))_{t \geq 0}$ and state 0 . For this reason we have to check that $(h_t(N))_{t \geq 0}$ is a standard process and that 0 is a regular point.

First of all we show that the semigroup $(\tilde{P}_t)_{t\geq 0}$ associated with $(h_t(N))_{t\geq 0}$ has the Feller property (which is fairly obvious if one thinks of $(h_t(N))_{t\geq 0}$ as the 'residual work'–process in the service system introduced subsequent to figure 7.)

Let $\Phi \in \mathcal{M}_f$; $s, t \geq 0$ then

$$
\begin{aligned}
h_{s+t}(\Phi) &= h_t(\Phi) + \overline{g}_{s+t}(\Phi) - \overline{g}_t(\Phi) - (g_{s+t}(\Phi) - g_t(\Phi)) \\
&= h_t(\Phi) + (\sup_{u \leq s} g_{t+u}(\Phi) - \overline{g}_t(\Phi))^+ - g_s(S_t\Phi) \\
&= h_t(\Phi) + (\sup_{u \leq s} g_u(S_t\Phi) + g_t(\Phi) - \overline{g}_t(\Phi))^+ - g_s(S_t\Phi) \\
&= h_t(\Phi) + (\overline{g}_s(S_t\Phi) - h_t(\Phi))^+ - g_s(S_t\Phi) \tag{1}
\end{aligned}
$$

Let $h_t^x(\Phi) := x + (\overline{g}_t(\Phi) - x)^+ - g_t(\Phi)$

$\tilde{P}_t f(x) := \int_{\mathbb{R}_0^+} f(y)\, \tilde{P}_t(x, dy) := E\, f(h_t^x(N))\,, \ \forall\, f \in \mathcal{C}_b(\mathbb{R}_0^+)\,, \ x \geq 0.$

$(\tilde{P}_t)_{t\geq 0}$ is the semigroup associated with $(h_t(N))_{t\geq 0}$.

This is since $h_{s+t}(\Phi) = h_s^{h_t(\Phi)}(S_t\Phi)$ by (1), and since $S_t N$ is independent of $R_{[0,t]\times\mathbb{R}^+} N$ and has distribution $\mathcal{L}(N)$.

$(\tilde{P}_t)_{t\geq 0}$ is a Feller semigroup: For $x, y \geq 0$

$$
\begin{aligned}
h_t^{x+y}(\Phi) - h_t^x(\Phi) &= y + (\overline{g}_t(\Phi) - (x+y))^+ - (\overline{g}_t(\Phi) - x)^+ \\
&= \max(\overline{g}_t(\Phi) - x, y) - \max(\overline{g}_t(\Phi) - x, 0)
\end{aligned}
$$

and consequently $0 \leq h_t^{x+y}(\Phi) - h_t^x(\Phi) \leq y$, $\forall\, x, y, t \geq 0$.

Thus the Prohorov distance between $\tilde{P}_t(x+y, \cdot)$ and $\tilde{P}_t(x, \cdot)$ is bounded by y which yields

$\tilde{P}_t(x+y, \cdot) \xrightarrow[y\to 0]{} \tilde{P}_t(x, \cdot)$ weakly, $\forall\, x \in \mathbb{R}_0^+$, $t \geq 0$.

Since in addition $(h_t(N))_{t\geq 0}$ has a.s. càdlàg paths, it is a standard process.

The intensity measure of N is shift invariant, such that the holding time at 0 by $(h_t(N))_{t\geq 0}$ has exponential distribution. Thus 0 is a regular state. Since $(g_t(N))_{t\geq 0}$ is a submartingale ($\int_{\mathbb{R}^+} y\, \mu(dy) \leq 1$!) with stationary increments, $\overline{g}_t(N) \xrightarrow[t\to\infty]{} \infty$ a.s.. By proposition 2.1.2 $(\overline{g}_t(N))_{t\geq 0}$ is the local time at 0 by $(h_t(N))_{t\geq 0}$, such that 0 is recurrent.

Applying Itô's theorem (see for instance [BL], theorem (3.18)) to the standard process $(h_t(N))_{t\geq 0}$ and regular, recurrent state 0 yields

$$
\Pi(N) := \{(t, e_t) \mid \tau_t(N) \neq \tau_{t+}(N)\}, \tag{2}
$$

where $e_t(s) = h_{\min(\tau_t(N)+s,\, \tau_{t+}(N))}(N)$, $s \geq 0$,

is a Poisson point process on $\mathbb{R}^+ \times U$ with intensity measure $\lambda \otimes n$, where U is the space of excursions and n is a σ-finite measure on $(U, \mathcal{B}(U))$.

Consequently since Poisson point processes are preserved by restrictions and projections

$$\Pi_a^*(N) := \left\{ (t, h_{\tau_t(N)}(N)) \mid \tau_t(N) \neq \tau_{t+}(N),\ 0 < t < a \right\} \tag{3}$$

is a Poisson point process on $(0, a) \times \mathbb{R}_0^+$ with intensity measure $\lambda_{|(0,a)} \otimes \pi_0 n$, where $\pi_0 n(\cdot) = n(\{f \mid f(0) \in \cdot\})$.

Obviously $\pi_0 n$ is μ , since $\forall A \in \mathcal{B}(\mathbb{R}_0^+)$

$$
\begin{aligned}
\pi_0 n(A) &= \lim_{h \to 0} h^{-1} P\left(\Pi_a^*(N) \cap (0, h) \times A \neq \emptyset \right) \\
&= \lim_{h \to 0} h^{-1} P\left(N((0, h) \times A) = 1 \right) = \mu(A) \ .
\end{aligned}
\tag{4}
$$

The claim of the theorem follows from (3) and (4) if we finally show

$$BL_1(N^a) \overset{\text{a.s.}}{=} \Pi_a^*(N) \ . \tag{5}$$

Let $\Phi = \sum_i \delta_{z_i} \in \mathcal{M}_f$. Note that proposition 2.1.3 implies

$$(0, a) \dashv_\Phi z_i \iff h_{x_i^-}(\Phi) = 0,\ 0 < \bar{g}_{x_i}(\Phi) < a \ , \tag{6}$$

and that in view of $(A2)$

$$\tau_t(\Phi) \neq \tau_{t+}(\Phi) \iff 0 = h_{(\tau_t(\Phi))^-}(\Phi) < h_{\tau_t(\Phi)}(\Phi) \ . \tag{7}$$

(5) is a consequence of (6), (7) and the fact that $P(N \in \mathcal{M}_f) = 1$.

3   The delete & shift process and the contour process

In this section we will construct the contour process of a binary self-similar tree as the real component of a more complex Markov process, the so-called delete & shift process. If the tree is traversed according to the prescription 'children first' and edges are cut off the moment they are passed through, a sequence of decreasing subtrees is obtained. These pruned trees marked with the actual progress of the traversal are the states of the delete & shift process. The 'distance to the root'–process of the mark is the contour process. The term *delete & shift process* expresses the fact that cutting off edges corresponds to deleting parts of the support of the associated measure and closing up the gaps by shifting the remaining parts. We will start with constructing the state of the delete & shift process for fixed $\Phi \in \mathcal{M}_f$ and fixed $t \geq 0$. We will then show that the delete & shift procedure constitutes a semigroup.

Note that in general subsequent trees are not adjacent. We will also delete the parts of the support of the measure in between subsequent trees to prevent that a mass free area builds up at the origin. With this in mind we define the part of the support of the measure that is to be deleted $D_{t,\Phi} \times \mathbb{R}^+$, the measure resulting by closing up the gaps $F_t \Phi$, and the distance to the root of the mark $c_{t,\Phi}(t)$, as follows

Definition 3.1   *For* $\Phi = \sum_i \delta_{z_i} \in \mathcal{M}_f$, $t \geq 0$, *let*

$$D_{t,\Phi} := \bigcup_{A_i \subset [0,t)} A_i(\Phi) \ \dot{\cup} \ \{\tau_{\hat{y}(\Phi)}(\Phi) \leq s < t \,|\, h_s(\Phi) = 0\} \ ,$$

where

$$\hat{y}(\Phi) := \begin{cases} 0, & if \ \ \Phi(\{0\} \times \mathbb{R}^+) = 0 \\ y, & if \ \ \Phi(\{(0,y)\}) = 1 \ . \end{cases}$$

$$F_t\Phi(A \times B) := \Phi(d_{t,\Phi}(A) \times B) \ , \quad A \subset \mathbb{R}_0^+, \ B \subset \mathbb{R}^+ \ ;$$

where $d_{t,\Phi}(\cdot)$ *is the right–continuous inverse of* $c_{t,\Phi}(\cdot) := \lambda([0,\cdot) \cap D_{t,\Phi}^c)$,

$$d_{t,\Phi}(s) = \sup\{u \geq 0 \,|\, c_{t,\Phi}(u) = s\} \ .$$

Note that $F_t\Phi \in \mathcal{M}_f$ and that in view of proposition 2.1.3 $D_{t,\Phi}$ is indeed the disjoint union of $\bigcup_{A_i \subset [0,t)} A_i$ and $\{\tau_{\hat{y}(\Phi)}(\Phi) \leq s < t \,|\, h_s(\Phi) = 0\}$. By proposition 2.1.3.b)

$$x_i \in D_{t,\Phi} \iff A_i \subset D_{t,\Phi} \ . \tag{8}$$

Since $D_{t,\Phi}$ is the finite union of half open intervals (not containing the upper endpoint) such is $D_{t,\Phi}^c$. Consequently since $d_{t,\Phi}(\cdot)$ is right–continuous

$$d_{t,\Phi}(c_{t,\Phi}(x)) = x \ , \quad \forall \ x \in D_{t,\Phi}^c \tag{9}$$

$$d_{t,\Phi}(\mathbb{R}_0^+) \cap D_{t,\Phi} = \emptyset \ . \tag{10}$$

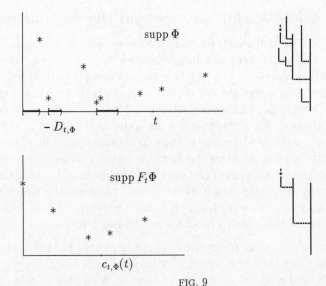

FIG. 9

supp $F_t\Phi$ *is obtained by deleting the parts of* supp Φ *above the thick lines and shifting the remaining parts. The upper tree on the right–hand side is the one associated with* Φ , *the tree below is the one associated with* $R_{[0,c_{t,\Phi}(t)] \times \mathbb{R}^+}F_t\Phi$.

Here comes a collection of illustrating facts about $F_t\Phi$ and $c_{t,\Phi}(t)$.

Proposition 3.2 *Let* $\Phi = \sum_i \delta_{z_i} \in \mathcal{M}_f$, $t \geq 0$, *then*

1) $\displaystyle F_t\Phi = \sum_{x_i \in D^c_{t,\Phi}} \delta_{(c_{t,\Phi}(x_i), y_i)}$

2) $\displaystyle c_{t,\Phi}(t) = \sum_{x_i \in D^c_{t,\Phi} \cap [0,t)} \overline{g}_{t-x_i}(S_{x_i}\Phi)$

3) *For* $x_i \in D^c_{t,\Phi}$, $u \geq 0$,

$$g_u(S_{c_{t,\Phi}(x_i)}\, F_t\Phi) = g_{d_{t,\Phi}(u+c_{t,\Phi}(x_i))-x_i}\,(S_{x_i}\Phi)$$

$$\overline{g}_u(S_{c_{t,\Phi}(x_i)}\, F_t\Phi) = \overline{g}_{d_{t,\Phi}(u+c_{t,\Phi}(x_i))-x_i}\,(S_{x_i}\Phi)$$

$$c_{t,\Phi}(x_i + \tau_{y_i}(S_{x_i}\Phi)) = c_{t,\Phi}(x_i) + \tau_{y_i}(S_{c_{t,\Phi}(x_i)}\, F_t\Phi)$$

4) $D_{c_{t,\Phi}(t), F_t\Phi} = \emptyset$.

Remarks

- By 3) the tree associated with $F_t\Phi$ inherits the relationship and the chronology from the tree associated with Φ. Thus it is a subtree.

- The tree associated with $R_{[0,c_{t,\Phi}(t)]\times\mathbb{R}^+}\, F_t\Phi$ consists of a chain (ancestor, child, grandchild, ...). This is since 3) implies for $x_i < x_j \in D^c_{t,\Phi} \cap [0,t]$

$$(c_{t,\Phi}(x_i), y_i) \prec_{F_t\Phi} (c_{t,\Phi}(x_j), y_j) \quad .$$

In view of 2) no further points exist in $\operatorname{supp} F_t\Phi \cap [0, c_{t,\Phi}(t)] \times \mathbb{R}^+$.

Proof

1) follows by (9) and (10).

2) Note that in view of proposition 2.1.2 and 2.1.3.b)

$$\lambda(\{\tau_{\hat{y}(\Phi)}(\Phi) \leq s < t \mid h_s(\Phi) = 0\}) = (\overline{g}_t(\Phi) - \hat{y}(\Phi))^+ \tag{11}$$

$$\lambda(\bigcup_{A_i \subset [0,t)} A_i) \overset{(8)}{=} \lambda(\bigcup_{x_i \in D_{t,\Phi}} A_i) = \sum_{x_i \in D_{t,\Phi}} y_i \ . \tag{12}$$

and consequently

$$c_{t,\Phi}(t) = t - \lambda(D_{t,\Phi}) = t - (\overline{g}_t(\Phi) - \hat{y}(\Phi))^+ - \sum_{x_i \in D_{t,\Phi}} y_i \ . \tag{13}$$

On the other hand the partition equation (proposition 2.1.4) implies

$$
\begin{aligned}
t &= \overline{g}_t(\Phi) + \sum_{0 < x_i < t} \min\left(\overline{g}_{t-x_i}(S_{x_i}\Phi), y_i\right) \\
&= \overline{g}_t(\Phi) - \min\left(\overline{g}_t(\Phi), \widehat{y}(\Phi)\right) + \sum_{0 \le x_i < t} \min\left(\overline{g}_{t-x_i}(S_{x_i}\Phi), y_i\right) \\
&= (\overline{g}_t(\Phi) - \widehat{y}(\Phi))^+ + \sum_{\substack{0 \le x_i < t \\ \overline{g}_{t-x_i}(S_{x_i}\Phi) \ge y_i}} y_i + \sum_{\substack{0 \le x_i < t \\ \overline{g}_{t-x_i}(S_{x_i}\Phi) < y_i}} \overline{g}_{t-x_i}(S_{x_i}\Phi) \\
&= (\overline{g}_t(\Phi) - \widehat{y}(\Phi))^+ + \sum_{x_i \in D_{t,\Phi}} y_i + \sum_{x_i \in D_{t,\Phi}^c \cap [0,t)} \overline{g}_{t-x_i}(S_{x_i}\Phi)
\end{aligned}
\tag{14}
$$

Claim 2) follows if in equation (13) t is substituted by the expression in (14).

3) Since $d_{t,\Phi}(\cdot)$ is the inverse of $c_{t,\Phi}(\cdot) = \lambda([0,\cdot) \cap D_{t,\Phi}^c)$

$$
d_{t,\Phi}(b) - d_{t,\Phi}(a) = b - a + \lambda\left(D_{t,\Phi} \cap (d_{t,\Phi}(a), d_{t,\Phi}(b)]\right) , \quad a \le b.
\tag{15}
$$

Note that proposition 2.1.3 implies

$$
\{\tau_{\widehat{y}(\Phi)}(\Phi) \le s < t \mid h_s(\Phi) = 0\} \subset [0, d_{t,\Phi}(0)) ,
\tag{16}
$$

such that

$$
D_{t,\Phi} \cap (d_{t,\Phi}(a), d_{t,\Phi}(b)] = \bigcup_{x_j \in D_{t,\Phi} \cap (d_{t,\Phi}(a), d_{t,\Phi}(b)]} A_j .
\tag{17}
$$

Let $x_i \in D_{t,\Phi}^c$, then (15), (17) and (9) imply (set $a = c_{t,\Phi}(x_i)$ and $b = c_{t,\Phi}(x_i) + u$)

$$
d_{t,\Phi}(u + c_{t,\Phi}(x_i)) - x_i = u + \sum_{x_j \in D_{t,\Phi} \cap (x_i, d_{t,\Phi}(u+c_{t,\Phi}(x_i))]} y_j .
\tag{18}
$$

Thus

$$
\begin{aligned}
g_{d_{t,\Phi}(u+c_{t,\Phi}(x_i))-x_i}(S_{x_i}\Phi) &= u + \sum_{x_j \in D_{t,\Phi} \cap (x_i, d_{t,\Phi}(u+c_{t,\Phi}(x_i))]} y_j \\
&\qquad - \sum_{x_i < x_j \le d_{t,\Phi}(u+c_{t,\Phi}(x_i))} y_j \\
&\overset{(9)}{=} u - \sum_{\substack{x_j \in D_{t,\Phi}^c \\ c_{t,\Phi}(x_i) < c_{t,\Phi}(x_j) \le c_{t,\Phi}(x_i)+u}} y_j \\
&\overset{1)}{=} g_u(S_{c_{t,\Phi}(x_i)} F_t \Phi) ,
\end{aligned}
\tag{19}
$$

which is the first claim of 3) .

To conclude the second equality of 3) from the first one, note that for $x_i \in D^c_{t,\Phi}$ (16) and proposition 2.1.3.b) imply

$$D_{t,\Phi} \cap [x_i, \infty) \subset \bigcup_{x_j > x_i} A_j .$$

Since $\bar{g}.(S_{x_i}\Phi)$ is constant on $A_j - x_i$ by propositions 2.1.2 and 2.1.3

$$
\begin{aligned}
\bar{g}_{d_{t,\Phi}(u+c_{t,\Phi}(x_i))-x_i} (S_{x_i}\Phi) &= \sup_{r \in [x_i, d_{t,\Phi}(u+c_{t,\Phi}(x_i))]} g_{r-x_i}(S_{x_i}\Phi) \\
&= \sup_{r \in [x_i, d_{t,\Phi}(u+c_{t,\Phi}(x_i))] \cap D^c_{t,\Phi}} g_{r-x_i}(S_{x_i}\Phi) \\
&\overset{(9)}{=} \sup_{r' \le u} g_{d_{t,\Phi}(r'+c_{t,\Phi}(x_i))-x_i} (S_{x_i}\Phi) \\
&= \bar{g}_u(S_{c_{t,\Phi}(x_i)} F_t\Phi) , \hspace{2cm} (20)
\end{aligned}
$$

where the last equality holds by the first claim of 3) .

Finally note that $d_{t,\Phi}(\cdot)$ is continuous and strictly increasing on $[t, \infty)$ since $D_{t,\Phi} \subset [0, t)$. Thus $(x_i \in D^c_{t,\Phi} \Rightarrow x_i + \tau_{y_i}(S_{x_i}\Phi) > t$!) the third equation of 3) follows from the second one.

4) Let $c_{t,\Phi}(t) > 0$ (the statement is obvious if $c_{t,\Phi}(t) = 0$).

By 3) and since $c_{t,\Phi}(\cdot)$ is strictly increasing on $[t, \infty)$

$$
\begin{aligned}
c_{t,\Phi}(x_i) &+ \tau_{y_i}(S_{c_{t,\Phi}(x_i)} F_t\Phi) \\
&= c_{t,\Phi}(x_i + \tau_{y_i}(S_{x_i}\Phi)) > c_{t,\Phi}(t) , \quad \forall x_i \in D^c_{t,\Phi} . \hspace{1cm} (21)
\end{aligned}
$$

Claim 4) follows from (21) and part 1) of the proposition if only

$$\hat{y}(F_t\Phi) > 0 . \hspace{4cm} (22)$$

Assume $\hat{y}(F_t\Phi) = 0$. Then part 2) of the proposition implies

$$c_{t,\Phi}(t) = \sum_{x_i \in D^c_{t,\Phi} \cap (0,t)} \bar{g}_{t-x_i}(S_{x_i}\Phi) . \hspace{2cm} (23)$$

Since $\bar{g}_s(\Phi') > 0, \forall s > 0, \Phi' \in \mathcal{M}_f$, (23) contradicts the partition equation (proposition 2.1.4 applied to $c_{t,\Phi}(t)$ and $F_t\Phi$) :

$$
\begin{aligned}
c_{t,\Phi}(t) &\overset{1)}{=} \bar{g}_{c_{t,\Phi}(t)}(F_t\Phi) + \sum_{x_i \in D^c_{t,\Phi} \cap (0,t)} \bar{g}_{c_{t,\Phi}(t)-c_{t,\Phi}(x_i)}(S_{c_{t,\Phi}(x_i)} F_t\Phi) \\
&\overset{3)}{=} \bar{g}_{c_{t,\Phi}(t)}(F_t\Phi) + \sum_{x_i \in D^c_{t,\Phi} \cap (0,t)} \bar{g}_{t-x_i}(S_{x_i}\Phi) . \hspace{1cm} (24)
\end{aligned}
$$

Thus (22) holds and claim 4) follows.

We now define the delete & shift operator P_t, $t \geq 0$, on $\mathcal{M}_f \times \mathbb{R}_0^+$

$$P_t \; : \; \mathcal{M}_f \times \mathbb{R}_0^+ \; \longrightarrow \; \mathcal{M}_f \times \mathbb{R}_0^+$$

$$P_t(\Phi, u) \; := \; (F_{t+u}\Phi, \; c_{t+u,\Phi}(t+u)) \; .$$

$(P_t)_{t \geq 0}$ is well defined since $F_t\Phi \in \mathcal{M}_f$.

Theorem 3.3 $(P_t)_{t \geq 0}$ *constitutes a semigroup*

$$P_s P_t = P_{s+t} \, , \quad \forall \, s, t \geq 0 \; .$$

In terms of the tree theorem 3.3 simply states that the pruned tree is determined by specifying the edges to be cut off, whereas the succession of cuts is of no importance.

Proof   Let $\Phi = \sum_i \delta_{z_i} \in \mathcal{M}_f$; $s, t \geq 0$, and assume $u = 0$ (this is no restriction since $P_t(\Phi, u) = P_{t+u}(\Phi, 0)$).
Note that by proposition 2.1.3 $r \in D^c_{s+t,\Phi}$ is equivalent to $A_{i_r} \not\subset [0, s+t)$, where $x_{i_r} = \max\{ x_i \,|\, z_i \in [0, r] \times \mathbb{R}^+ \}$.   Thus (9), (10)   and proposition 3.2.3   imply

$$r \in D^c_{s+t,\Phi} \iff r \in D^c_{t,\Phi} \, , \; c_{t,\Phi}(r) \in D^c_{c_{t,\Phi}(s+t), F_t\Phi} \tag{25}$$

and consequently

$$
\begin{aligned}
c_{s+t,\Phi}(\cdot) &= \int_0^{\cdot} 1\{r \in D^c_{s+t,\Phi}\} \, dr \\
&\overset{(25)}{=} \int_0^{\cdot} 1\{c_{t,\Phi}(r) \in D^c_{c_{t,\Phi}(s+t), F_t\Phi}\} \, 1\{r \in D^c_{t,\Phi}\} \, dr \\
&= \int_0^{\cdot} 1\{c_{t,\Phi}(r) \in D^c_{c_{t,\Phi}(s+t), F_t\Phi}\} \, dc_{t,\Phi}(r) \\
&= \int_0^{c_{t,\Phi}(\cdot)} 1\{r' \in D^c_{c_{t,\Phi}(s+t), F_t\Phi}\} \, dr' \\
&= c_{c_{t,\Phi}(s+t), F_t\Phi}(c_{t,\Phi}(\cdot)) \; . \tag{26}
\end{aligned}
$$

From (26) follows the equality of the real components of $P_s P_t(\Phi, 0)$ and $P_{s+t}(\Phi, 0)$ (note that $c_{t,\Phi}(s+t) = s + c_{t,\Phi}(t)$ since $D_{t,\Phi} \subset [0, t)$) as well as the equality of the $\mathcal{M}_f$ - components. The latter holds since in view of proposition 3.2.1

$$\mathrm{supp}F_{s+t}\Phi \; = \; \bigcup_{x_i \in D^c_{s+t,\Phi}} \{(c_{s+t,\Phi}(x_i), y_i)\}$$

$$\mathrm{supp}F_{s+c_{t,\Phi}(t)}F_t\Phi \; = \; \bigcup_{\substack{x_i \in D^c_{t,\Phi} \\ c_{t,\Phi}(x_i) \in D^c_{s+c_{t,\Phi}(t), F_t\Phi}}} \left\{ \left(c_{s+c_{t,\Phi}(t), F_t\Phi}(c_{t,\Phi}(x_i)), y_i \right) \right\} \; ,$$

where the index sets agree by (25).

We now obtain the delete & shift process by taking the argument of $(P_t)_{t \geq 0}$ at random and covering the 'untouched' part of the support of $F_t \Phi$.

Definition 3.4   *Let N be a Poisson point process on $\mathbb{R}_0^+ \times \mathbb{R}^+$ with intensity measure $\lambda \otimes \mu$, where λ denotes Lebesgue measure on $\mathbb{R}_0^+$ and μ is a probability measure on $\mathbb{R}^+$.*
Then $(\Psi_t, C_t)_{t \geq 0} := (R_{[0, c_{t,N}(t)] \times \mathbb{R}^+} F_t N, c_{t,N}(t))_{t \geq 0}$ is called delete & shift process and $(C_t)_{t \geq 0}$ is called contour process with parameter μ.

Due to the construction via the semigroup $(P_t)_{t \geq 0}$ and the independence properties of N the delete & shift process is Markovian :

Theorem 3.5   *The delete & shift process is a time–homogeneous strong Markov process with respect to $(\mathcal{F}_{t+})_{t \geq 0}$.*

Proof   Let $\Phi = \sum_i \delta_{z_i} \in \mathcal{M}_f, t \geq 0$.
Since $g_s(\Phi) = g_s(R_{[0,t] \times \mathbb{R}^+} \Phi), \forall \, 0 \leq s \leq t$, and since $\hat{y}(\Phi) = \hat{y}(R_{[0,t] \times \mathbb{R}^+} \Phi)$

$$D_{t,\Phi} = D_{t, R_{[0,t] \times \mathbb{R}^+} \Phi} \, , \tag{27}$$

such that

$$c_{t,\Phi}(t) = c_{t, R_{[0,t] \times \mathbb{R}^+} \Phi}(t) \tag{28}$$

$$R_{[0, c_{t,\Phi}(t)] \times \mathbb{R}^+} F_t \Phi = F_t \, R_{[0,t] \times \mathbb{R}^+} \Phi \, . \tag{29}$$

By (28) and (29) $(\Psi_t, C_t)_{t \geq 0}$ is adapted to $(\mathcal{F}_{t+})_{t \geq 0}$.

In view of proposition 3.2.1 and since $c_{t,\Phi}(x_i) = c_{t,\Phi}(t) + x_i - t$ for $x_i \geq t$

$$R_{(c_{t,\Phi}(t), \infty) \times \mathbb{R}^+} F_t \Phi = S_{t - c_{t,\Phi}(t)} \, R_{(t, \infty) \times \mathbb{R}^+} \Phi \, , \tag{30}$$

such that

$$\begin{aligned}
F_t \Phi &= R_{[0, c_{t,\Phi}(t)] \times \mathbb{R}^+} F_t \Phi + R_{(c_{t,\Phi}(t), \infty) \times \mathbb{R}^+} F_t \Phi \\
&\overset{(29),(30)}{=} F_t \, R_{[0,t] \times \mathbb{R}^+} \Phi + S_{t - c_{t,\Phi}(t)} \, R_{(t, \infty) \times \mathbb{R}^+} \Phi \, .
\end{aligned} \tag{31}$$

Since $\mathcal{F}_{t+}$ and $\sigma\{R_{(t,\infty) \times \mathbb{R}^+} N\}$ are independent σ–algebras, the Markov property of $(\Psi_t, C_t)_{t \geq 0}$ follows from the representations (28), (31), and the construction via the semigroup $(P_t)_{t \geq 0}$ (note that $F_{s+t} \Phi = F_{s + c_{t,\Phi}(t)} F_t \Phi$ by theorem 3.3). If T is a $\mathcal{F}_{t+}$–stopping time, then conditioned on T, $R_{(T, \infty) \times \mathbb{R}^+} N$ is independent of $\mathcal{F}_T$, such that $(\Psi_t, C_t)_{t \geq 0}$ is even strong Markov by (28) and (31). The stationarity of the transition probabilities is a consequence of the shift invariance of the intensity measure $\lambda \otimes \mu$.

We will now show that the contour process is Markovian, too. If the Markov property of the delete & shift process is to carry over to the real component, the contour process, then the history of the contour process may not contain any additional information about the actual state of the $\mathcal{M}_f$ - component.

We will see that the path of the contour process up to time t determines the part of the support that is to be deleted and the trace of the measure on that part. However, due to the independence properties and the homogeneity of N it is of no importance where the deleted part is and what it looks like. The only information of interest is about properties of the trace of the measure on the remaining part, that are still observable having closed up the gaps. But all we learn about the measure Ψ_t from the history of the contour process is the fact that $D_{C_t, \Psi_t} = \emptyset$, which we already knew by proposition 3.2.4 .

Another possibility to prove the Markov property of the contour process would be via time reversal. If the reversal of time is translated into a change of the prescription for tracing the tree, the time reversed contour process is seen to be distributed as $(h_t(N))_{t \geq 0}$. Since $(h_t(N))_{t \geq 0}$ is Markovian (see proof of theorem 2.3), such must be $(C_t)_{t \geq 0}$, since the Markov property is preserved by time reversal. However, since binary self–similar trees other than the Galton–Watson tree are lopsided trees (the lifetime distribution of an individual and the distribution of the residual lifetime of its parent differ), the characteristics of the contour process change by time reversal. This is why we prefer the first approach which also gives insight into the reason for the Markovian behaviour of $(C_t)_{t \geq 0}$.

In terms of the tree the Markov property of the contour process states, that knowledge of the length and the height of those edges that have already been cut off does not give any information about the number of generations in the pruned family tree. This is because in case that the contour processes of trees agree up to a certain time, the ratio of the densities of the pruned trees is equal to that of the initial trees (see figure 10).

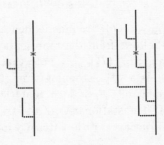

FIG. 10

The contour processes of the above trees agree up to the indicated marks. The ratio of the densities of the trees is equal to the one of the pruned trees.

Theorem 3.6 *The contour process* $(C_t)_{t\geq 0}$ *is a time–homogeneous Markov process with respect to* $(\sigma\{C_s, s \leq t\}_+)_{t\geq 0}$.

Proof To begin with we show that the path $(c_{s,\Phi}(s))_{s\leq t}$ determines the part of the support of Φ that is to be deleted as well as the trace of Φ on that part. In terms of the tree the statement is, that a binary tree is determined by its contour.

Fix $\Phi = \sum_i \delta_{z_i} \in \mathcal{M}_f$ and let $c(s) := c_{s,\Phi}(s)$, $s \geq 0$.
Proposition 2.1 implies for $s_1 \leq s_2$

$$
\begin{aligned}
c(s_2) - c(s_1) &= s_2 - s_1 - \lambda(D_{s_2,\Phi}\backslash D_{s_1,\Phi})\\
&= s_2 - s_1 - \lambda(\bigcup_{A_i\subset[0,s_2)} A_i \backslash \bigcup_{A_j\subset[0,s_1)} A_j)\\
&\quad - \lambda(\{r \in [s_1,s_2)\cap[\tau_{\hat{y}(\Phi)}(\Phi),\infty)\,|\,h_r(\Phi)=0\})\\
&= s_2 - s_1 - \sum_{x_i+\tau_{y_i}(S_{x_i}\Phi)\,\in\,(s_1,s_2]} y_i - \left(\bar{g}_{s_2}(\Phi)-\max\left(\bar{g}_{s_1}(\Phi),\hat{y}(\Phi)\right)\right)^+ \quad (32)
\end{aligned}
$$

Since $x_i + \tau_{y_i}(S_{x_i}\Phi) \neq x_j + \tau_{y_j}(S_{x_j}\Phi)$, $\forall\, i \neq j$, ($\tau.$ is left–continuous !) (32) implies

$$
c(s^-) - c(s) = \begin{cases} y_i\,, & \text{if } s = x_i + \tau_{y_i}(S_{x_i}\Phi)\\ 0\,, & \text{else}. \end{cases} \quad (33)
$$

We claim

$$
x_i = \sup\{s < x_i + \tau_{y_i}(S_{x_i}\Phi)\,|\,c(s)=c(x_i+\tau_{y_i}(S_{x_i}\Phi))\}. \quad (34)
$$

By proposition 2.1.3 $D_{x_i+\tau_{y_i}(S_{x_i}\Phi),\Phi} = D_{x_i,\Phi}\,\dot\cup\,A_i$, such that

$$
c(x_i + \tau_{y_i}(S_{x_i}\Phi)) = c(x_i). \quad (35)
$$

On the other hand if $u \in (0,\tau_{y_i}(S_{x_i}\Phi))$, then by proposition 2.1.2 $\left(\bar{g}_{x_i+u}(\Phi)-\max(\bar{g}_{x_i}(\Phi),\hat{y}(\Phi))\right)^+ = 0$, such that in view of (32)

$$
\begin{aligned}
c(x_i + u) - c(x_i) &= u - \sum_{x_j+\tau_{y_j}(S_{x_j}\Phi)\in(x_i,x_i+u]} y_j\\
&\overset{\text{prop. 2.1.3}}{=} u - \sum_{\substack{x_j\in(x_i,x_i+u]\\ \bar{g}_{x_i+u-x_j}(S_{x_j}\Phi)\geq y_j}} y_j\\
&\overset{\text{prop. 2.1.4}}{\geq} \bar{g}_u(S_{x_i}\Phi) > 0. \quad (36)
\end{aligned}
$$

(34) follows from (35) and (36).

Let $\bigcup_j\{(t_j, h_j)\}$ be the discontinuity points of $(c(s))_{s\geq 0}$ with the respective heights of jumps ($c(t_j^-)-c(t_j)=h_j$),
and define

$$
b(s) := \sup\{r < s\,|\,c(r)=c(s)\}\,, \quad s > 0,
$$

then (see figure 11)

$$R_{D_{t,\Phi} \times \mathbb{R}^+} \Phi = \sum_{t_j \le t} \delta_{(b(t_j), h_j)} \qquad (37)$$

$$\lambda \Big(D_{t,\Phi} \, \triangle \, \Big(\bigcup_{t_j \le t} [b(t_j), t_j) \, \cup \, \{ s < t \mid c(s) = 0 \} \Big) \Big) = 0 \ . \qquad (38)$$

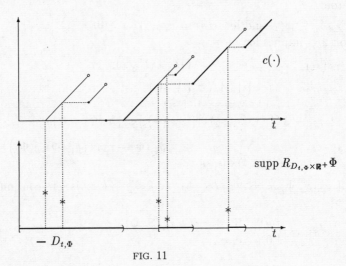

FIG. 11

(37) follows from (33) and (34) since $x_i \in D_{t,\Phi} \iff x_i + \tau_{y_i}(S_{x_i}\Phi) \le t$. To establish (38) first note that in view of proposition 2.1.2

$$h_s(\Phi) = 0 \ , \ s \ge \tau_{\hat{y}(\Phi)}(\Phi) \implies D_{s,\Phi} = [0, s) \implies c(s) = 0 \ .$$

On the other hand propositions 3.2.2 and 2.1.3 imply

$$c(s) = 0 \ , \ s > 0 \implies A_i \subset [0, s), \ \forall \, 0 \le x_i < s$$
$$\implies h_{s-}(\Phi) = 0 \ , \ s \ge \tau_{\hat{y}(\Phi)}(\Phi) \ ,$$

such that $\{ s > 0 \mid c(s) = 0, \ h_s(\Phi) > 0 \} \subset \{ x_i \mid h_{x_i^-}(\Phi) = 0 \}$. Since $x_i \in D_{t,\Phi} \implies x_i \in \bigcup_{t_j \le t} \{ b(t_j) \}$, and since

$$x_i < x_j \in D^c_{t,\Phi} \overset{\text{prop. 2.1.3}}{\implies} c(x_j) \ge \overline{g}_{x_j - x_i}(S_{x_i}\Phi) > 0,$$

the symmetric difference in (38) is a singleton $(= \inf \{ x_i < t \mid x_i \in D^c_{t,\Phi} \})$ or empty.

We will now show that any measure $\Gamma \in \mathcal{M}_f$ with the property $D_{c_{t,\Phi}(t),\Gamma} = \emptyset$ can be 'built around' $R_{D_{t,\Phi} \times \mathbb{R}^+} \Phi$, such that the associated contour process and the one of Φ agree up to time t. Let

$$\Gamma = \sum_i \delta_{(a_i,b_i)} \in \mathcal{M}_f \quad \text{with} \quad D_{c_{t,\Phi}(t),\Gamma} = \emptyset ,$$

and define

$$\Gamma^* := \sum_i \delta_{(d_{t,\Phi}(a_i),b_i)} + R_{D_{t,\Phi} \times \mathbb{R}^+} \Phi .$$

We claim

$$
\begin{align}
&1) \quad \Gamma^* \in \mathcal{M}_f \notag \\
&2) \quad c_{s,\Gamma^*}(s) = c_{s,\Phi}(s) , \quad \forall s \le t \tag{39} \\
&3) \quad F_t \Gamma^* = \Gamma . \notag
\end{align}
$$

1) follows from (10).

2) First of all note that (10) implies

$$\tau_{y_i}(S_{x_i}\Phi) = \tau_{y_i}(S_{x_i}\Gamma^*) , \quad \forall x_i \in D_{t,\Phi} . \tag{40}$$

Following the same lines as in the proof of proposition 3.2.3 we will show that

$$\overline{g}_{d_{t\Phi}(a_i+u)-d_{t,\Phi}(a_i)}(S_{d_{t,\Phi}(a_i)}\Gamma^*) = \overline{g}_u(S_{a_i}\Gamma) , \tag{41}$$

such that

$$\overline{g}_{t-d_{t,\Phi}(a_i)}(S_{d_{t,\Phi}(a_i)}\Gamma^*) = \overline{g}_{c_{t,\Phi}(t)-a_i}(S_{a_i}\Gamma) < b_i , \tag{42}$$

where the inequality holds since $D_{c_{t,\Phi}(t),\Gamma} = \emptyset$.

By (40), (42) and (33) the points of discontinuity and the respective heights of jumps by $(c_{s,\Phi}(s))_{s \le t}$ and those by $(c_{s,\Gamma^*}(s))_{s \le t}$ agree, such that claim 2) follows if for $s \le t$

$$c_{s,\Phi}(s) = 0 \iff c_{s,\Gamma^*}(s) = 0 . \tag{43}$$

To establish (41) note that $\forall (a_i, b_i) \in \operatorname{supp}\Gamma$

$$
\begin{align}
g_{d_{t,\Phi}(a_i+u)-d_{t,\Phi}(a_i)}(S_{d_{t,\Phi}(a_i)}\Gamma^*) &= d_{t,\Phi}(a_i + u) - d_{t,\Phi}(a_i) \notag \\
&\quad - \sum_{d_{t,\Phi}(a_i)<d_{t,\Phi}(a_j)\le d_{t,\Phi}(a_i+u)} b_j - \sum_{x_j \in D_{t,\Phi} \cap (d_{t,\Phi}(a_i), d_{t,\Phi}(a_i+u)]} y_j \notag \\
&\overset{(15)}{=} u + \lambda\big(D_{t,\Phi} \cap (d_{t,\Phi}(a_i), d_{t,\Phi}(a_i + u)])\big) \notag \\
&\quad - \sum_{x_j \in D_{t,\Phi} \cap (d_{t,\Phi}(a_i), d_{t,\Phi}(a_i+u)]} y_j - \sum_{a_i<a_j\le a_i+u} b_j \notag \\
&\overset{(17)}{=} u - \sum_{a_i<a_j\le a_i+u} b_j = g_u(S_{a_i}\Gamma) . \tag{44}
\end{align}
$$

Since $\bar{g}.(S_{d_{t,\Phi}(a_i)}\Gamma^*)$ is constant on $A_j - d_{t,\Phi}(a_i)$, $x_j > a_i$, and since

$D_{t,\Phi} \cap [d_{t,\Phi}(a_i), \infty) \subset \bigcup\limits_{x_j > a_i} A_j$, (41) follows from (44).

To establish (43) note that $R_{[0,s)\times\mathbb{R}^+}\Gamma^* = R_{[0,s)\times\mathbb{R}^+}\Phi$, $\forall s \leq d_{t,\Phi}(0)$, and consequently $D_{s,\Gamma^*} = D_{s,\Phi}$, $\forall s \leq d_{t,\Phi}(0)$. Thus (43) obviously holds if $s \leq d_{t,\Phi}(0)$.
But if $s \leq t$, then $c_{s,\Phi}(s) = 0$ as well as $c_{s,\Gamma^*}(s) = 0$ imply $s \leq d_{t,\Phi}(0)$, since on the one hand

$$c_{s,\Phi}(s) = 0 \implies D_{s,\Phi} = [0,s)$$
$$\implies D_{t,\Phi} \cap [0,s) = [0,s) \implies d_{t,\Phi}(0) \geq s.$$

On the other hand proposition 3.2.2 and (42) imply for $s \leq t$

$$c_{s,\Gamma^*}(s) = 0 \implies s \leq d_{t,\Phi}(a_i), \ \forall\, (a_i, b_i) \in \text{supp}\Gamma \implies s \leq d_{t,\Phi}(0),$$

since the assumption $D_{c_{t,\Phi}(t),\Gamma} = \emptyset$ forces $\inf_i a_i = 0$ or $c_{t,\Phi}(t) = 0$.

3) Note that in view of (40)

$$x_i \in D_{t,\Phi} \implies x_i \in D_{t,\Gamma^*}, \tag{45}$$

and that (42) implies

$$d_{t,\Phi}(a_i) \in D^c_{t,\Gamma^*}, \ \forall\, (a_i, b_i) \in \text{supp}\Gamma. \tag{46}$$

In view of (38) claim 2) implies

$$c_{t,\Gamma^*}(d_{t,\Phi}(a_i)) = c_{t,\Phi}(d_{t,\Phi}(a_i)) = a_i. \tag{47}$$

Claim 3) follows from (45), (46), (47) and proposition 3.2.1.

We can now show that the regular versions of the conditional distributions $P(\Psi_t \in \cdot \mid C_s, \ s \leq t)$ and $P(\Psi_t \in \cdot \mid C_t)$ agree. Let $x \geq 0$, $k \in \mathbb{N}_0$, and $A \in \mathcal{B}(\mathbb{R}_0^+)$, then

$$P\left(\Psi_t([0,x] \times A) = k \mid C_s, \ s \leq t\right)$$

$$\overset{(37),(38)}{=} P\left(\Psi_t([0,x] \times A) = k \mid C_s, s \leq t; D_{t,N}; R_{D_{t,N}\times\mathbb{R}^+}N\right) \text{ a.s.}$$

$$\overset{(39)}{=} P\left(\Psi_t([0,x] \times A) = k \mid C_t; F_{C_t,\Psi_t} = \emptyset\right) \text{ a.s.}$$

$$\overset{3.2.6}{=} P\left(\Psi_t([0,x] \times A) = k \mid C_t\right) \text{ a.s.}, \tag{48}$$

where the second equation holds since $N(A_1)$ and $N(A_2)$ are independent if $A_1 \cap A_2 = \emptyset$ and since $d_{t,\Phi}(\cdot)$ is λ–preserving.

Since the composition $(\Psi_t, C_t)_{t \geq 0}$ is Markovian by theorem 3.3 , such is the component $(C_t)_{t \geq 0}$ in view of (48). (Note that $\sigma\{C_s,\ s \leq t\}$ may be substituted by $\sigma\{C_s,\ s \leq t\}_+$ since the respective conditional expectations differ only on sets of $\mathcal{L}(N)$–measure 0.)

The transition probabilities of $(C_t)_{t \geq 0}$ are stationary since those of $(\Psi_t, C_t)_{t \geq 0}$ are (theorem 3.3), and since $P(\Psi_t \in \cdot \,|\, C_t = x) = \nu_x(\cdot)$ in view of (39).

Equations (37), (38) and (39) hold $\forall\, \Phi \in \mathcal{M}_f,\ t \geq 0$, i.e. for $\mathcal{L}(N)$–almost all realizations simultaneously for all t . Thus we may condition in (48) on specific paths and in particular we may substitute t by a $\sigma\{C_s, s \leq t\}$–stopping time T :

Corollary 3.7   *The contour process $(C_t)_{t \geq 0}$ is strong Markov with respect to the filtration $(\sigma\{C_s, s \leq t\})_{t \geq 0}$.*

$(C_t)_{t \geq 0}$ is not strong Markov with respect to $(\sigma\{C_s, s \leq t\}_+)_{t \geq 0}$: Let T be the time that the first tree appears in the point process $T := \inf\{u > 0 \,|\, C_u > 0\}$. T is a $\sigma\{C_s, s \leq t\}_+$ – stopping time and $P(C_{T+t} \neq t) = O(t)$, whereas $P(C_t \neq 0) = O(t)$. But 0 is an extraordinary state : $(C_t)_{t \geq 0}$ is strong Markov at $\sigma\{C_s, s \leq t\}_+$– stopping times T with $P(C_T > 0) = 1$, since a.s.

$$\{\, t \,|\, C_t > 0 \,\} \subset \{\, t \,|\, \exists\, h(t) > 0 \,:\, C_{t+h} = C_t + h\ ,\ \forall\, 0 < h < h(t) \,\}\,.$$

By means of proposition 3.2.4 and the (strong) Markov property of $(C_t)_{t \geq 0}$ it is quite easy to obtain an explicit representation for the conditional distribution of the pruned tree. We just state the result (see [GE] for a proof).

Proposition 3.8   *Let $X_1 \overset{a.s.}{=} 0$, and let $X_i - X_{i-1},\ i = 2, 3, \ldots$, and $Y_j,\ j = 1, 2, \ldots$ be independent random variables having exponential distribution with mean 1 and distribution μ respectively. Define*

$$U_n \ :=\ \min(X_n - X_{n-1}, Y_{n-1})\ ,\quad n = 2, 3, \ldots$$

$$H_x \ :=\ \max\{ j \geq 1 \,\big|\, \sum_{n=2}^{j} U_n \leq x \}\ ,\quad x \in \mathbb{R}_0^+\ .$$

Then the following holds for $0 < x \leq t$

$$\mathcal{L}\,(\Psi_t \,|\, C_t = x) = \mathcal{L}\,\Big(\sum_{n=1}^{H_x} \delta_{(X_n, Y_n)} \,\Big|\, U_n < Y_{n-1},\ 2 \leq n \leq H_x \Big)\ .$$

In case of the Galton–Watson tree (i.e. if $\mu = \exp(p_0/p_2)$) the situation is extremely nice : The $X_i,\ i = 2, 3, \ldots, H_x$, constitute a homogeneous Poisson process on $(0, x]$. The overshots $Y_i - (X_{i+1} - X_i),\ i = 1, 2, \ldots, H_x$ where $X_{H_x+1} := x$, have exponential distribution and are independent of $(X_{i+1} - X_i)$ ([GE], theorem 3.11). Le Gall's theorem is easily derived from this description of the pruned Galton–Watson tree.

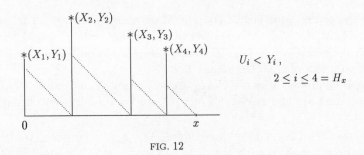

FIG. 12

References

[AL]   D.J. Aldous : The continuum random tree II: an overview. *Proc. Durham Symp. Stochastic Analysis 1990,* p. 23–70, Cambridge University Press, 1991.

[BL]   R.M. Blumenthal : *Excursions of Markov Processes,* Birkhäuser, 1992.

[GE]   J. Geiger : *Zufällige Bäume und Punktprozesse ,* unpublished manuscript, Frankfurt University, 1994.

[LG]   J.F. Le Gall :   Marches aleatoires, mouvement brownien et processus de branchement. *Seminaire de Probabilites XXIII,* p. 258–274, Springer, 1989. Lecture Notes in Math. 1372.

[NP]   J. Neveu and J.W. Pitman : The branching process in a Brownian excursion. *Seminaire de Probabilites XXIII,* p. 248–257, Springer, 1989. Lecture Notes in Math. 1372.

ON A CLASS OF QUASILINEAR STOCHASTIC DIFFERENTIAL EQUATIONS OF PARABOLIC TYPE: REGULAR DEPENDENCE OF SOLUTIONS ON INITIAL DATA

N. Yu. Goncharuk[1]

Mathematics Institute, University of Warwick, Coventry CV4 7AL, U.K.

and

Kiev Polytechnical Institute
Department of Mathematical Methods in System Analysis
37 Pobedy Pr., 252056 Kiev, Ukraine

1. INTRODUCTION

Regular dependence of solutions to nonlinear SDE's with unbounded operator – valued coefficients in Hilbert spaces on initial data was studied by several authors ([1–4]) in different contexts.

Let $H, \mathcal{H}$ be real separable Hilbert spaces, $w(t)$ a cylindrical Wiener process in $\mathcal{H}$, $\mathcal{A}$ an unbounded infinitesimal generator of a strongly continuous semigroup in H. In the case when $F(t, \cdot)$ and $B(t, \cdot)$ satisfy properly formulated differentiability and linear growth conditions in H and the space of Hilbert-Schmidt or even linear operators acting from $\mathcal{H}$ to H (under some additional assumptions about $\mathcal{A}$), respectively, smooth dependence of solutions to semilinear stochastic differential equation (SDE)

$$du(t) = (\mathcal{A}u(t) + F(t, u(t)))\, dt + B(t, u(t))\, dw(t), \quad u(t_0) = u_0 \qquad (1.1)$$

in H on u_0 was established in [1],[2] , and applied to the investigation of the properties of the transition semigroup corresponding to SDE (1.1) ([1], see also [5],[6]), and the properties of the evolution operator family, defined by the solution to (1.1) and the operator multiplicative functional generated by the solution to some linear SDE with coefficients depending on the solution to (1.1) ([2]).

---

[1] Research supported by a Royal Society Fellowship

Existence, uniqueness , and continuous dependence of solutions on initial data results were obtained in [3],[4] for an abstract stochastic Navier–Stokes equation with multiplicative noise. It allowed the authors to prove existence of stochastic flows associated with this type of nonlinear SDE's in [3].

In this paper, which is a continuation of [7], we are concerned with the smooth dependence of solutions to the abstract parabolic SDE

$$du(t) = A(t, u(t))u(t) \ dt + B(t, u(t)) \ dw(t), \quad u(t_0) = u_0 \qquad (1.2)$$

in H on initial data. Here $A(t, v)$ is a linear operator in H depending on parameters $t \in [t_0, \tau]$ and $v \in H$ with constant dense domain $\mathcal{D}(A)$ and bounded from the right in H. $B(t, v)$ is a Hilbert-Schmidt operator acting from $\mathcal{H}$ to H.

Sufficient conditions that guarantee existence of the unique (a.s.) solution to (1.2) in H on $[t_0, \tau]$ were found in [7]. It was shown there that this solution takes values in $\mathcal{D}(A)$, being continuous with respect to t on $[t_0, \tau]$, and continuous with respect to initial value u_0 in mean square (see [7] and section 3 for details).

Assuming additional smoothness of coefficients and initial data in SDE (1.2) and using an approximation procedure which is similar to the one worked out in [7], we prove that the solution to (1.2) is Gateaux differentiable along smooth directions in the sense of convergence in probability. Its directional derivative at u_0 in the direction h is bounded (a.s.), and is a solution to the equation obtained by formal termwise differentiation of (1.2) with respect to u_0 in the direction h in H on every interval $[t_0, \tau_1] \subset [t_0, \tau)$ (see section 3 for precise formulations).

In the same way one can establish that the solution to (1.2) is twice Gateaux differentiable along smooth directions in the sense of convergence in probability, and its directional derivative is bounded (a.s.).

An example of quasilinear parabolic SPDE that satisfies assumptions of theorems 3.1 and 3.2 for equation in abstract Hilbert space is given in section 4.

2. NOTATION, NOTIONS AND ASSUMPTIONS

Introduce the following notation.

1. Let X, Y be Banach spaces, $\mathcal{L}(X, Y)$ the Banach space of linear bounded operators acting from X to Y with the norm $\| \cdot \|_{\mathcal{L}(X,Y)}$, and $\mathcal{L}(X) = \mathcal{L}(X, X)$.

2. Suppose $H, \mathcal{H}, \mathcal{H}_-$ are real separable Hilbert spaces with $\mathcal{H}_- \supset \mathcal{H}$, where the embedding operator $j : \mathcal{H} \to \mathcal{H}_-$ is of Hilbert-Schmidt class, $\| \cdot \|_H$ and $(\cdot, \cdot)_H$ are the norm and the inner product in H, and $\mathcal{L}_2(\mathcal{H}, H)$ is a space of Hilbert-Schmidt operators acting from $\mathcal{H}$ to H with the norm

$$\sigma_2(B) = (\sum_{j=1}^{\infty} \|Be_j\|_H^2)^{\frac{1}{2}}$$

(here $\{e_j\}_{e_j=1}^{\infty}$ is an arbitrary orthonormal basis in $\mathcal{H}$).

3. Let $(\Omega, \mathcal{F}, P)$ be a complete probability space, $L_p(\Omega, H, P)$ the Banach space of measurable p-th integrable functions $u(\omega)(\omega \in \Omega)$ with values in H and norm

$$\ll u \gg_{H,p} = (E\|u\|_H^p)^{\frac{1}{p}} = (\int_\Omega \|u(\omega)\|_H^p \, P(d\omega))^{\frac{1}{p}},$$

$L_p([t_0, \tau] \times \Omega, H, l \times P)$ the Banach space of measurable,p-th integrable with respect to measure $l \times P$ functions $u(t, \omega)$ $(t \in [t_0, \tau], \omega \in \Omega)$ with values in H, and norm

$$\lll u \ggg_{H,p} = (\int_{t_0}^\tau dt \int_\Omega \|u(t,\omega)\|_H^p \, P(d\omega))^{\frac{1}{p}},$$

$w(t)$ a standard Wiener process in $\mathcal{H}_-$ with correlation operator the identity of $\mathcal{H}$, $w(t_0) = 0$, $\{\mathcal{F}_t\}_{t \geq t_0}$ a flow of σ-subalgebras of $\mathcal{F}$ adapted to the Wiener process $w(t)$, $S_p^t(H)$ a set of $\mathcal{F}_t$ -measurable elements from $L_p(\Omega, H, P)$ $(t \in [t_0, \tau])$, $S_p([t_0, \tau], H)$ the set of $\{\mathcal{F}_t\}_{t \geq t_0}$ adapted elements from $L_p([t_0, \tau] \times \Omega, H, l \times P)$.

4. Suppose T is a closed positive definite linear selfadjoint operator on H with dense domain $\mathcal{D}(T)$ such that $\|Tu\|_H \geq \|u\|_H (u \in \mathcal{D}(T))$. Let $\int_1^\infty \lambda \, E(d\lambda)$ be the spectral decomposition for T in H. Then set $P_n = \int_1^n E(d\lambda)$.

5. Let $H_0 \supset H_1 \supset ... \supset H_k \supset ...$ be a scale of densely embedded Hilbert spaces $H_k = \mathcal{D}(T^k)$ with the norm $\|u\|_k = \|T^k u\|_H$ $(u \in \mathcal{D}(T^k))(k \in \{0\} \cup \mathbf{N})$. By $(\cdot, \cdot)_k$ denote the inner product in H_k.

6. $a \wedge b$ denotes the minimum of a and b.

7. Use the notation $w - \lim_{n \to \infty} x_n$ for a weak limit of the sequence $\{x_n\}_{n=1}^\infty$ in X, and $P - \lim_{n \to \infty} f_n$ for a limit in probability of the sequence of $P-$ measurable functions $f_n : \Omega \to X$ $(n \in \mathbf{N})$.

Consider Itô's equation

$$u(t) = u_0 + \int_{t_0}^t \mathcal{A}(s, u(s)) \, ds + \int_{t_0}^t \mathcal{B}(s, u(s)) \, dw(s) \qquad (2.1)$$

in a real separable Hilbert space X with $\mathcal{A}(t, v)$ and $\mathcal{B}(t, v)$ being random $\{\mathcal{F}_t\}_{t \geq t_0}$ adapted functions, $\mathcal{A}(t, \cdot)$ a nonlinear operator into X with constant dense domain $D(\mathcal{A}) \subset X$, P–(a.s.), and $\mathcal{B}(t, v) \in \mathcal{L}_2(\mathcal{H}, X)$, P–(a.s.).

We shall say that $u(t)$ $(t \in [t_0, \tau])$ is a solution to equation (2.1) in X on $[t_0, \tau]$ if $u(t)$ is an $\{\mathcal{F}_t\}_{t \geq t_0}$ adapted , P–(a.s.) bounded stochastic process continuous with respect to t on $[t_0, \tau]$ with values in $D(\mathcal{A})$, and for any $t \in [t_0, \tau]$ satisfying (a.s.) (2.1) in X.

Introduce the following notation for assumptions about $A(t, u)$ and $B(t, u)$ in (1.1) :

($A1$) for every $t \in [t_0, \tau], u \in H_0$, $A(t, u)$ is a closed linear operator in H_0 with domain $\mathcal{D}(A(t, u)) = \mathcal{D}(T)$ independent of t and u,

for fixed $l \in \mathbf{N}$:

($A2_l$) for every $k \in \{0, 1, ..., l\}$,

$$(t, u) \mapsto T^k A(t, T^{-k} u) T^{-(k+1)}$$

is a continuous function from $[t_0, \tau] \times H_0$ into $\mathcal{L}(H_0)$, and

$$T^k A(t, T^{-k} u) T^{-(k+1)}$$

are bounded in $\mathcal{L}(H_0)$ uniformly with respect to (t, u) on $[t_0, \tau] \times H_0$,

($A3_l$) for every $k \in \{0, 1, ..., l\}$ there is a constant d_k such that for all $t \in [t_0, \tau]$, $u \in H_0$, $v \in H_1$,

$$(T^k A(t, T^{-k} u) T^{-k} v, v)_0 \le d_k \|v\|_0^2,$$

($A4_l$) for every $k \in \{0, 1, ..., l\}$ and every fixed $t \in [t_0, \tau]$,

$$H_0 \ni u \mapsto T^k A(t, T^{-k} u) T^{-(k+1)} \in \mathcal{L}(H_0)$$

is continuously Fréchet differentiable , and for every $k \in \{0, 1, ..., l\}$ there exists a constant $c_k > 0$ such that for all $t \in [t_0, \tau]$, $u \in H_0$,

$$\|T^k A'_u(t, T^{-k} u) T^{-(k+1)}\|_{\mathcal{L}(H_0, \mathcal{L}(H_0))} \le c_k,$$

($A5_l$) assumption ($A4_l$) holds, and for every $k \in \{0, 1, ..., l\}$ and all $t \in [t_0, \tau]$, $u, v \in H_0$,

$$\|T^k (A'_u(t, T^{-k} u) - A'_u(t, T^{-k}(u + v))) T^{-(k+1)}\|_{\mathcal{L}(H_0, \mathcal{L}(H_0))} \le c_k \|v\|_0,$$

($B1_l$) for every $k \in \{0, 1, ..., l\}$,

$$(t, u) \mapsto T^k B(t, T^{-k} u)$$

is a continuous function from $[t_0, \tau] \times H_0$ into $\mathcal{L}_2(\mathcal{H}, H_0)$,

($B2_l$) for every $k \in \{0, 1, ..., l\}$ there is a constant $b_k > 0$ such that for all $t \in [t_0, \tau]$, $u \in H_0$,

$$\sigma_2^2(T^k B(t, T^{-k} u)) \le b_k(1 + \|u\|_0^2),$$

($B3_l$) for every $k \in \{0, 1, ..., l\}$ and every fixed $t \in [t_0, \tau]$,

$$H_0 \ni u \mapsto T^k B(t, T^{-k} u) \in \mathcal{L}_2(\mathcal{H}, H_0)$$

is continuously Fréchet differentiable, and for every $k \in \{0, 1, ..., l\}$ and all $t \in [t_0, \tau]$, $u \in H_0$,

$$\sigma_2^2(T^k B'_u(t, T^{-k} u)) \le b_k,$$

($B4_l$) assumption ($B3_l$) holds, and for every $k \in \{0, 1, ..., l\}$ and all $t \in [t_0, \tau]$, $u, v \in H_0$,

$$\sigma_2^2(T^k (B'_u(t, T^{-k} u) - B'_u(t, T^{-k}(u + v)))) \le b_k \|v\|_0^2.$$

3. REGULAR DEPENDENCE OF SOLUTIONS ON INITIAL DATA

We are concerned with Itô's equation

$$\xi^x(t) = x + \int_{t_0}^t A(s, \xi^x(s))\xi^x(s)\, ds + \int_{t_0}^t B(s, \xi^x(s))\, dw(s) \qquad (3.1)$$

in H_0 on $[t_0, \tau]$. Sufficient conditions that guarantee existence, uniqueness, and continuous dependence of solution to (3.1) on initial data are given in

Theorem 3.0 ([7]). *Suppose assumptions* $(A1), (A2_l)-(A4_l), (B1_l)-(B3_l)$ *are valid, and* $x \in S_{2q+\alpha}^{t_0}(H_l)$ $(q \geq 2, \alpha > 0, l \in \{2, 3, ...\})$.
Then
1) there exists a unique (a.s.) solution $\xi^x = \xi^x(t)$ $(t \in [t_0, \tau])$ *to equation (3.1) in* H_{l-2} *on* $[t_0, \tau]$ *such that*
(i) $\xi^x(t) \in S_{2q+\beta}^t(H_l)$ $(t \in [t_0, \tau], \alpha > \beta \geq 0)$,
(ii) ξ^x *is (a.s.) continuous into* H_{l-1} ,
2) for every $t \in [t_0, \tau]$, $\xi^x(t)$ *defines a random operator family*

$$V(t, t_0) : S_{2q+\alpha}^{t_0}(H_l) \to S_{2q+\beta}^t(H_l)$$

by the formula $\xi^x(t) = V(t, t_0) \circ x$, *and* $V(t, t_0)$ *has (a.s.) the evolution property in* H_l, *i.e. for any* $u \in S_{2q+\alpha}^\lambda(H_l)$, *and for each* λ, s, t *such that* $t_0 \leq \lambda \leq s \leq t \leq \tau$,

$$V(\lambda, \lambda) \circ u = u, \quad V(t, s) \circ V(s, \lambda) \circ u = V(t, \lambda) \circ u \quad (a.s.),$$

3) for $x, x_n \in S_{2q+\alpha}^{t_0}(H_l)$ $(n \in \mathbf{N})$ *such that*

$$x = \lim_{n \to \infty} x_n$$

in $L_{2q+\beta}(\Omega, H_{l-1}, P)$ $(\alpha > \beta > 0)$, *and for every* $t \in [t_0, \tau]$,

$$V(t, t_0) \circ x = \lim_{n \to \infty} V(t, t_0) \circ x_n$$

in $L_{2q+\gamma}(\Omega, H_{l-1}, P)$ $(\alpha > \beta > \gamma \geq 0)$.

Suppose the functions $A(s, u)T^{-1}$ and $B(s, u)$ in (3.1) are continuously Fréchet differentiable with respect to u in $\mathcal{L}(H_0)$ and $\mathcal{L}_2(\mathcal{H}, H_0)$, respectively. Then by formal termwise differentiation of (3.1) with respect to x in the direction h we obtain the equation

$$\eta^{x,h}(t) = h + \int_{t_0}^t A'_u(s, \xi^x(s))(\eta^{x,h}(s), \xi^x(s))\, ds +$$

$$\int_{t_0}^t A(s, \xi^x(s))\eta^{x,h}(s)\, ds + \int_{t_0}^t B'_u(s, \xi^x(s))\, (\eta^{x,h}(s), dw(s)).$$
$$(3.2)$$

Here we give some sufficient conditions that guarantee existence of the unique (a.s.) solution to (3.2).

Theorem 3.1. *Suppose assumptions* $(A1), (A2_3)-(A4_3)$, *and* $(B1_3)-(B3_3)$ *about* $A(t, u)$ *and* $B(t, u)$ *in* (3.1) *are valid, and that* $x \in S_{4+\alpha}^{t_0}(H_3)$ $(\alpha > 0)$, *and* $h \in S_4^{t_0}(H_2)$.

Then there exists a unique (a.s.) solution to equation (3.2) *in* H_0 *on every interval* $[t_0, \tau_1] \subset [t_0, \tau)$ *such that*

(i) *for every* $t \in [t_0, \tau_1]$, $\|\eta^{x,h}(t)\|_2 < \infty$ (*a.s.*),

(ii) $\sup_{t_0 \leq s \leq \tau_1} \|\eta^{x,h}(s)\|_1 < \infty$ (*a.s.*).

We reduce the proof of theorem 3.1 to the proof of a sequence of propositions.

For $z(t) \in S_2^t(H_3)$ $(t \in [t_0, \tau])$ introduce

$$\phi_R(z)(t) = z(t) \, j\{ \int_{t_0}^t \|z(\theta)\|_3^2 \, d\theta \leq R^2 \} \quad , t \in [t_0, \tau], R > 0,$$

where $j\{\Delta\}$ is an indicator of set $\Delta \subset \Omega$. Due to theorem 3.0 there exists a unique (a.s.) solution $\xi^x = \xi^x(t)$ $(t \in [t_0, \tau])$ to equation (3.1) in H_1 on $[t_0, \tau]$ such that $\xi^x(t) \in S_4^t(H_3)$ $(t \in [t_0, \tau])$, and ξ^x is an (a.s.) continuous in H_2 stochastic process.

Then along with (3.2) consider equations

$$\eta_R^{x,h}(t) = h + \int_{t_0}^t A_u'(s, \xi^x(s))(\eta_R^{x,h}(s), \phi_R(\xi^x)(s)) \, ds +$$

$$\int_{t_0}^t A(s, \xi^x(s))\eta_R^{x,h}(s) \, ds + \int_{t_0}^t B_u'(s, \xi^x(s)) \, (\eta_R^{x,h}(s), dw(s)), \ R > 0, n \in \mathbf{N},$$
$$(3.3)$$

and

$$\eta_{R,n}^{x,h}(t) = h + \int_{t_0}^t A_u'(s, \xi^x(s))(\eta_{R,n}^{x,h}(s), \phi_R(\xi^x)(s)) \, ds +$$

$$\int_{t_0}^t A_n(s, \xi^x(s))\eta_{R,n}^{x,h}(s) \, ds + \int_{t_0}^t B_u'(s, \xi^x(s)) \, (\eta_{R,n}^{x,h}(s), dw(s)), \ R > 0, n \in \mathbf{N},$$
$$(3.4)$$

where $A_n(s, u) = P_n A(s, u) P_n$.

Proposition 3.1. *For fixed* $h \in S_{2q}^{t_0}(H_2), x \in S_{4+\alpha}^{t_0}(H_3)$ $(q \geq 2, \alpha > 0)$, $R > 0$, $n \in \mathbf{N}$, $k \in \{0, 1, 2\}$ *there exists a unique solution to the linear equation* (3.4) *in* $S_{2q}([t_0, \tau], H_k)$ *such that*

$$E\|\eta_{R,n}^{x,h}(t)\|_k^{2q} \leq 4e^{(D_1 R^2 + D_2)(t - t_0)} E\|h\|_k^{2q}.$$
$$(3.5)$$

This solution has a continuous (a.s.) modification in H_k, *for which the following estimate holds:*

$$E(\sup_{t_0 \leq s \leq t} \|\eta_{R,n}^{x,h}(s)\|_k^{2q}) \leq 4e^{(D_1 R^2 + D_3)(t - t_0)} E\|h\|_k^{2q}$$
$$(3.6)$$

with D_1, D_2, D_3 being positive constants depending on $q, \tau - t_0$, and the constants from assumptions $(A3_3), (A4_3),$ and $(B3_3)$ only .

Proof. Consider the function

$$\nu(\eta_{R,n}^{x,h})(t) = \int_{t_0}^t A_u'(s, \xi^x(s))(\eta_{R,n}^{x,h}(s), \phi_R(\xi^x)(s)) \, ds +$$

$$\int_{t_0}^t A_n(s, \xi^x(s))\eta_{R,n}^{x,h}(s) \, ds + \int_{t_0}^t B_u'(s, \xi^x(s)) \, (\eta_{R,n}^{x,h}(s), dw(s)).$$

There is a constant $\tilde{c}_{2q} > 0$ depending on q only, such that for every $t \in [t_0, \tau]$

$$\|\nu(\eta_{R,n}^{x,h})(t)\|_k^{2q} \le \tilde{c}_{2q}[(\int_{t_0}^t \|A_u'(s, \xi^x(s))(\eta_{R,n}^{x,h}(s), \phi_R(\xi^x)(s))\|_k \, ds)^{2q} +$$

$$(\int_{t_0}^t \|A_n(s, \xi^x(s))\eta_{R,n}^{x,h}(s)\|_k \, ds)^{2q} + \|\int_{t_0}^t B_u'(s, \xi^x(s)) \, (\eta_{R,n}^{x,h}(s), dw(s))\|_k^{2q}],$$

and

$$E\|\nu(\eta_{R,n}^{x,h})(t)\|_k^{2q} \le c_{2q}[(t - t_0)^{q-1} E(\int_{t_0}^t \|\eta_{R,n}^{x,h}(s)\|_k^{2q} \, ds) \times$$

$$(\int_{t_0}^t \|\phi_R(\xi^x)(s)\|_{k+1}^2 \, ds)^q + (t - t_0)^{2q-1} n^{2q} E(\int_{t_0}^t \|\eta_{R,n}^{x,h}(s)\|_k^{2q} \, ds) +$$

$$q(2q - 1)^q (t - t_0)^{q-1} E(\int_{t_0}^t \|\eta_{R,n}^{x,h}(s)\|_k^{2q} \, ds)] \le$$

$$c_{2q}(t - t_0)^{q-1}[R^{2q} + (t - t_0)^q n^{2q} + q(2q - 1)^q] \int_{t_0}^t E\|\eta_{R,n}^{x,h}(s)\|_k^{2q} \, ds,$$

where $c_{2q} > 0$ is a constant depending on q and constants from assumptions $(A3_3), (A4_3),$ and $(B3_3)$ only.

It follows from this last inequality that ν is a continuous linear mapping in $S_{2q}([t_0, \tau], H_k)$.

Consider the mapping $\tilde{\nu}$ in $S_{2q}([t_0, \tau], H_k)$ defined by

$$\tilde{\nu}(\eta)(t) = \nu(\eta)(t) + h, \quad t \in [t_0, \tau].$$

One can easily show that there is $l \ge 1$ such that $\tilde{\nu}^l$ is a contraction in $S_{2q}([t_0, \tau], H_k)$. Hence, the mapping $\tilde{\nu}$ has a unique fixed point in $S_{2q}([t_0, \tau], H_k)$, which is a solution to equation (3.4).

By Itô's formula for every $t \in [t_0, \tau]$

$$\|\eta_{R,n}^{x,h}(t)\|_k^q \le \|h\|_k^q +$$

$$q \int_{t_0}^t \|\eta_{R,n}^{x,h}(s)\|_k^{q-2} (A_u'(s,\xi^x(s))(\eta_{R,n}^{x,h}(s), \phi_R(\xi^x)(s)), \eta_{R,n}^{x,h}(s))_k \ ds +$$

$$q \int_{t_0}^t \|\eta_{R,n}^{x,h}(s)\|_k^{q-2} (A_n(s,\xi^x(s))\eta_{R,n}^{x,h}(s), \eta_{R,n}^{x,h}(s))_k \ ds +$$

$$\frac{q}{2}(q-1) \int_{t_0}^t \|\eta_{R,n}^{x,h}(s)\|_k^{q-2} \sigma_2^2 (T^k B_u'(s,\xi^x(s))\eta_{R,n}^{x,h}(s)) \ ds +$$

$$q \int_{t_0}^t \|\eta_{R,n}^{x,h}(s)\|_k^{q-2} (\eta_{R,n}^{x,h}(s), B_u'(s,\xi^x(s))(\eta_{R,n}^{x,h}(s), dw(s)))_k \le$$

$$\|h\|_k^q + c_k q \int_{t_0}^t \|\eta_{R,n}^{x,h}(s)\|_k^q \ \|\phi_R(\xi^x)(s)\|_{k+1} \ ds +$$

$$(d_k q + \frac{1}{2} b_k q(q-1)) \int_{t_0}^t \|\eta_{R,n}^{x,h}(s)\|_k^q \ ds +$$

$$q \int_{t_0}^t \|\eta_{R,n}^{x,h}(s)\|_k^{q-2} (\eta_{R,n}^{x,h}(s), B_u'(s,\xi^x(s))(\eta_{R,n}^{x,h}(s), dw(s)))_k \quad (a.s.),$$

where d_k, c_k, and b_k are the constants from assumptions $(A3_3), (A4_3)$, and $(B3_3)$, respectively. Hence,

$$\|\eta_{R,n}^{x,h}(t)\|_k^{2q} \le 4\|h\|_k^{2q} +$$

$$4[c_k^2 q^2 R^2 + (d_k q + \frac{1}{2} b_k q(q-1))^2 (\tau - t_0)] \int_{t_0}^t \|\eta_{R,n}^{x,h}(s)\|_k^{2q} \ ds +$$

$$4q^2 (\int_{t_0}^t \|\eta_{R,n}^{x,h}(s)\|_k^{q-2} (\eta_{R,n}^{x,h}(s), B_u'(s,\xi^x(s))(\eta_{R,n}^{x,h}(s), dw(s)))_k)^2 \quad (a.s.), \tag{3.7}$$

which implies (3.5). It follows from the inequality

$$E\|\eta_{R,n}^{x,h}(t) - \eta_{R,n}^{x,h}(s)\|_k^4 \le c_4(t-s)[R^4 + (t-s)n^4 + 18] \int_s^t E\|\eta_{R,n}^{x,h}(\theta)\|_k^4 \ d\theta$$

and (3.5) for $q = 2$ that $\eta_{R,n}^{x,h}(t)$ ($t \in [t_0, \tau]$) satisfies Kolmogorov's test in H_k, and therefore has a continuous modification in H_k. Estimate (3.6) for a continuous modification of $\eta_{R,n}^{x,h}$ can be easily obtained from (3.7) now.

Due to the embedding $H_2 \subset H_1 \subset H_0$, solutions to (3.4) in H_0, H_1, and H_2 coincide in H_0, and below we denote continuous modifications of all of them by $\eta_{R,n}^{x,h}$.

Proposition 3.2. For fixed $h \in S_4^{t_0}(H_2)$, $x \in S_{4+\alpha}^{t_0}(H_3)(\alpha > 0)$, $R > 0$, for every $t \in [t_0, \tau]$

(i) there exists $\eta_R^{x,h}(t) = \lim_{n \to \infty} \eta_{R,n}^{x,h}(t)$ in $S_4^t(H_1)$, and

(ii) $\eta_R^{x,h}(t) = w - \lim_{n \to \infty} \eta_{R,n}^{x,h}(t)$ in $S_4^t(H_2)$.

Proof. For every $t \in [t_0, \tau]$, $n, m \in \mathbf{N}$, $\eta_{R,n}^{x,h}(t) - \eta_{R,m}^{x,h}(t)$ satisfies Itô's equation

$$\eta_{R,n}^{x,h}(t) - \eta_{R,m}^{x,h}(t) = \int_{t_0}^t A_u'(s, \xi^x(s))(\eta_{R,n}^{x,h}(s) - \eta_{R,m}^{x,h}(s), \phi_R(\xi^x)(s)) \, ds +$$

$$\int_{t_0}^t (A_n(s, \xi^x(s))\eta_{R,n}^{x,h}(s) - A_m(s, \xi^x(s))\eta_{R,m}^{x,h}(s)) \, ds +$$

$$\int_{t_0}^t B_u'(s, \xi^x(s)) \, (\eta_{R,n}^{x,h}(s) - \eta_{R,m}^{x,h}(s), dw(s)) \quad (a.s.)$$

in H_0. Then by Itô's formula

$$\|\eta_{R,n}^{x,h}(t) - \eta_{R,m}^{x,h}(t)\|_0^2 \leq$$

$$2 \int_{t_0}^t (A_u'(s, \xi^x(s))(\eta_{R,n}^{x,h}(s) - \eta_{R,m}^{x,h}(s), \phi_R(\xi^x)(s)), \eta_{R,n}^{x,h}(s) - \eta_{R,m}^{x,h}(s))_0 \, ds +$$

$$2 \int_{t_0}^t (A_n(s, \xi^x(s))(\eta_{R,n}^{x,h}(s) - \eta_{R,m}^{x,h}(s)), \eta_{R,n}^{x,h}(s) - \eta_{R,m}^{x,h}(s))_0 \, ds +$$

$$2 \int_{t_0}^t ((A_n(s, \xi^x(s)) - A_m(s, \xi^x(s)))\eta_{R,m}^{x,h}(s), \eta_{R,n}^{x,h}(s) - \eta_{R,m}^{x,h}(s))_0 \, ds +$$

$$\int_{t_0}^t \sigma_2^2(B_u'(s, \xi^x(s)) \, (\eta_{R,n}^{x,h}(s) - \eta_{R,m}^{x,h}(s))) \, ds +$$

$$2 \int_{t_0}^t (\eta_{R,n}^{x,h}(s) - \eta_{R,m}^{x,h}(s), B_u'(s, \xi^x(s)) \, (\eta_{R,n}^{x,h}(s) - \eta_{R,m}^{x,h}(s), dw(s)))_0 \quad (a.s.).$$

Note that for all $\theta \in [t_0, \tau], u \in H_2$,

$$\|(A_n(\theta, u) - A_m(\theta, u))T^{-2}\|_{\mathcal{L}(H_0)} \leq \|(P_n - P_m)T^{-1}\| \times$$

$$(\|TA(\theta, u)T^{-2}\|_{\mathcal{L}(H_0)} + \|A(\theta, u)T^{-1}\|_{\mathcal{L}(H_0)}) \leq \tilde{c}(n \wedge m)^{-1},$$

where $\tilde{c}$ is a constant depending on the constants from assumptions $(A2_3)$

only. Whence, taking into account $(A4_3), (A3_3)$, and $(B3_3)$, we find

$$\|\eta_{R,n}^{x,h}(t) - \eta_{R,m}^{x,h}(t)\|_0^2 \le 2c_0 \int_{t_0}^t \|\eta_{R,n}^{x,h}(s) - \eta_{R,m}^{x,h}(s)\|_0^2 \|\phi_R(\xi^x)(s))\|_1 \, ds +$$

$$(2d_0 + b_0) \int_{t_0}^t \|\eta_{R,n}^{x,h}(s) - \eta_{R,m}^{x,h}(s)\|_0^2 \, ds +$$

$$2\tilde{c}(n \wedge m)^{-1} \int_{t_0}^t \|\eta_{R,m}^{x,h}(s)\|_2 \|\eta_{R,n}^{x,h}(s) - \eta_{R,m}^{x,h}(s)\|_0 \, ds +$$

$$2 \int_{t_0}^t (\eta_{R,n}^{x,h}(s) - \eta_{R,m}^{x,h}(s), B_u'(s, \xi^x(s)) \, (\eta_{R,n}^{x,h}(s) - \eta_{R,m}^{x,h}(s), dw(s)))_0 \quad (a.s.),$$

and

$$E\|\eta_{R,n}^{x,h}(t) - \eta_{R,m}^{x,h}(t)\|_0^4 \le 4[4c_0^2 R^2 + (2d_0 + b_0)^2(\tau - t_0) + 4b_0] \times$$

$$\int_{t_0}^t E\|\eta_{R,n}^{x,h}(s) - \eta_{R,m}^{x,h}(s)\|_0^4 \, ds + 4\tilde{c}^4(n \wedge m)^{-4}(\tau - t_0) \int_{t_0}^t E\|\eta_{R,m}^{x,h}(s)\|_2^4 \, ds.$$

Applying (3.5) for $k = 2$ and then Gronwall's lemma to the last inequality we obtain for every $t \in [t_0, \tau]$

$$E\|\eta_{R,n}^{x,h}(t) - \eta_{R,m}^{x,h}(t)\|_0^4 \le C_1 e^{C_2(t-t_0)}(n \wedge m)^{-4} E\|h\|_2^4, \qquad (3.8)$$

where C_1 and C_2 are positive constants depending on $\tau - t_0$, R, and constants from assumptions $(A2_3), (A3_3), (A4_3)$, and $(B3_3)$ only.

Therefore, for every $t \in [t_0, \tau]$ there exists $\eta_R^{x,h}(t) = \lim_{n \to \infty} \eta_{R,n}^{x,h}(t)$ in $S_4^t(H_0)$. Boundedness of $E\|\eta_{R,n}^{x,h}(t)\|_2^2$ uniformly with respect to n implies now the statements of the proposition (see [7], lemma 3.3).

Remark 3.1. Under the same conditions one can easily show that
  (i) $\eta_R^{x,h} = \lim_{n \to \infty} \eta_{R,n}^{x,h}$ in $S_4([t_0, \tau], H_1)$, and
  (ii) $\eta_R^{x,h} = w - \lim_{n \to \infty} \eta_{R,n}^{x,h}$ in $S_4([t_0, \tau], H_2)$,
where $\eta_R^{x,h} = \eta_R^{x,h}(t)$ $(t \in [t_0, \tau])$, $\eta_{R,n}^{x,h} = \eta_{R,n}^{x,h}(t)$ $(t \in [t_0, \tau])$.

Corollary 3.1. *The limit stochastic process $\eta_R^{x,h}(t)$ $(t \in [t_0, \tau])$ has a modification which is continuous (a.s.) in H_1.*

Really, it follows from the inequalities

$$E\|\eta_{R,n}^{x,h}(t) - \eta_{R,n}^{x,h}(s)\|_1^4 \le c_4(t - s)[R^4 + (t - s) + 18] \int_s^t E\|\eta_{R,n}^{x,h}(\theta)\|_2^4 \, d\theta,$$

$$E\|\eta_R^{x,h}(t) - \eta_R^{x,h}(s)\|_1^4 \le 27[E\|\eta_R^{x,h}(t) - \eta_{R,n}^{x,h}(t)\|_1^4 +$$

$$E\|\eta_{R,n}^{x,h}(t) - \eta_{R,n}^{x,h}(s)\|_1^4 + E\|\eta_{R,n}^{x,h}(s) - \eta_R^{x,h}(s)\|_1^4],$$

estimate (3.5) for $k = 2$, and statement (i) of proposition 3.2.

We denote a continuous modification of the limit process by $\eta_R^{x,h}(t)$ $(t \in [t_0, \tau])$.

Corollary 3.2. *For fixed $h \in S_4^{t_0}(H_2)$, $x \in S_{4+\alpha}^{t_0}(H_3)(\alpha > 0)$, $R > 0$ there are estimates*

$$E\|\eta_R^{x,h}(t)\|_k^4 \leq 4e^{(D_1 R^2 + D_2)(t-t_0)} E\|h\|_k^4, \quad t \in [t_0, \tau], k \in \{0, 1\}, \tag{3.9}$$

$$E\|\eta_R^{x,h}(t)\|_2^2 \leq 2e^{(D_1 R^2 + D_2)\frac{(t-t_0)}{2}} (E\|h\|_2^4)^{\frac{1}{2}}. \tag{3.10}$$

Here D_1, D_2 are the constants from estimate (3.5).

Remark 3.2. Note that for fixed x, h, and $t \in [t_0, \tau]$, estimates (3.9),(3.10) do not imply boundedness of the moments $E\|\eta_R^{x,h}(t)\|_k^2$ ($k \in \{0, 1, 2\}$) uniformly with respect to $R > 0$.

Proposition 3.3. *For fixed $R > 0$ the limit stochastic process from proposition 3.2 is a unique (a.s.) solution to equation (3.3) in H_0 on $[t_0, \tau]$ such that*

(i) $\eta_R^{x,h}(t) \in S_4^t(H_2)$ ($t \in [t_0, \tau]$),

(ii) $\eta_R^{x,h}$ is an (a.s.) continuous in H_1 stochastic process .

Proof. 1. $\eta_R^{x,h}(t)$ ($t \in [t_0, \tau]$) is an $\{\mathcal{F}_t\}_{t \geq t_0}$ adapted stochastic process, continuous with respect to t on $[t_0, \tau]$ with values in H_1, satisfying estimates (3.9),(3.10). Therefore to prove that $\eta_R^{x,h}$ is a solution to (3.3), it is sufficient to show that for every $t \in [t_0, \tau]$, $\eta_R^{x,h}(t)$ satisfies (3.3) (a.s.) in H_0.

2. For every $t \in [t_0, \tau]$ set

$$\mu_R(t) = \eta_R^{x,h}(t) - h - \int_{t_0}^t A_u'(s, \xi^x(s))(\eta_R^{x,h}(s), \phi_R(\xi^x)(s)) \, ds -$$
$$\int_{t_0}^t A(s, \xi^x(s))\eta_R^{x,h}(s) \, ds - \int_{t_0}^t B_u'(s, \xi^x(s)) (\eta_R^{x,h}(s), dw(s))$$

in H_0. It suffices to prove that for every $t \in [t_0, \tau]$, $\mu(t) = 0$ (a.s.) in H_0. For every $t \in [t_0, \tau]$, $n \in \mathbf{N}$, $\eta_{R,n}^{x,h}(t)$ satisfies (2.4) (a.s.) in H_0. Adding and subtracting we find

$$\mu_R(t) = \eta_R^{x,h}(t) - \eta_{R,n}^{x,h}(t)+$$

$$\int_{t_0}^t A_u'(s,\xi^x(s))(\eta_{R,n}^{x,h}(s) - \eta_R^{x,h}(s), \phi_R(\xi^x)(s))\, ds+$$

$$\int_{t_0}^t (A_n(s,\xi^x(s)) - A(s,\xi^x(s)))\eta_{R,n}^{x,h}(s)\, ds+$$

$$\int_{t_0}^t A(s,\xi^x(s))(\eta_{R,n}^{x,h}(s) - \eta_R^{x,h}(s))\, ds+$$

$$\int_{t_0}^t B_u'(s,\xi^x(s))\,(\eta_{R,n}^{x,h}(s) - \eta_R^{x,h}(s), dw(s)) \quad (a.s.),$$

$$\|\mu_R(t)\|_0^2 \leq 5[\|\eta_R^{x,h}(t) - \eta_{R,n}^{x,h}(t)\|_0^2 + c_0^2 \int_{t_0}^t \|\phi_R(\xi^x)(s)\|_1^2\, ds\times$$

$$\int_{t_0}^t \|\eta_{R,n}^{x,h}(s) - \eta_R^{x,h}(s)\|_0^2\, ds + \tilde{c}^2 n^{-2}(\tau - t_0)\int_{t_0}^t \|\eta_{R,n}^{x,h}(s)\|_2^2\, ds+$$

$$(\tau - t_0)\int_{t_0}^t \|A(s,\xi^x(s))T^{-1}\|_{\mathcal{L}(H_0)}^2 \|\eta_R^{x,h}(s) - \eta_{R,n}^{x,h}(s))\|_1^2\, ds+$$

$$\|\int_{t_0}^t B_u'(s,\xi^x(s))\,(\eta_{R,n}^{x,h}(s) - \eta_R^{x,h}(s), dw(s))\|_0^2] \quad (a.s.),$$

which implies

$$\|\mu_R(t)\|_0^2 \leq D_{1,R}(\|\eta_R^{x,h}(t) - \eta_{R,n}^{x,h}(t)\|_0^2+$$

$$\int_{t_0}^t \|\eta_R^{x,h}(s) - \eta_{R,n}^{x,h}(s))\|_1^2\, ds) + D_{2,R} n^{-2}(E\|h\|_2^4)^{\frac{1}{2}}, \qquad (3.11)$$

where $D_{1,R}$ and $D_{2,R}$ are positive constants depending on $R, \tau - t_0$, and the constants from assumptions $(A2_3), (A3_3), (A4_3)$, and $(B3_3)$ only. Passing to the limit as $n \to \infty$ in the right hand side of (3.11), we find that for every $t \in [t_0,\tau]$, $E\|\mu_R(t)\|_0^2 = 0$, and hence, $\mu_R(t) = 0$ (a.s.) in H_0.

3. Note that as conditions (i),(ii) hold, then $\mu_R(t) = 0$ (a.s.) for all $t \in [t_0,\tau]$ in H_0, and for every $t \in [t_0,\tau]$ in H_1, and one can obtain estimate (2.9) directly from Itô's formula for the norm of the solution to equation (2.3). Then uniqueness of the solution satisfying conditions (i) and (ii) follows immediately from the linearity of (2.3) and estimate (2.9) for $k = 0$. Moreover, for $k \in \{0,1\}$ and every $t \in [t_0,\tau]$ the estimate

$$E(\sup_{t_0 \leq s \leq t} \|\eta_R^{x,h}(s)\|_k^4) \leq 4e^{(D_1 R^2 + D_3)(t - t_0)} E\|h\|_k^4 \qquad (3.12)$$

holds, where D_1, D_3 are the constants from estimate (2.6).

Proposition 3.4. *There exists a unique (a.s.) solution to equation (3.2) in H_0 on every interval $[t_0, \tau_1] \subset [t_0, \tau)$ such that*
(i) *for every $t \in [t_0, \tau_1]$, $\|\eta^{x,h}(t)\|_2 < \infty$ (a.s.),*
(ii) $\sup_{t_0 \le s \le \tau_1} \|\eta^{x,h}(s)\|_1 < \infty$ *(a.s.).*

Proof. 1. First we prove existence of the solution. Let

$$\zeta_R = \inf\{\lambda \ge t_0 : \int_{t_0}^{\lambda} \|\xi^x(s)\|_3^2 \, ds > R^2\} \wedge \tau; \qquad (3.13)$$

ζ_R is a stopping time, and for every $t \in [t_0, \tau]$,

$$P\{\zeta_R < t\} \le P\{\int_{t_0}^t \|\xi^x(s)\|_3^2 \, ds > R^2\} < \frac{1}{R^2} \int_{t_0}^t E\|\xi^x(s)\|_3^2 \, ds$$

$$\le \frac{t - t_0}{R^2} [e^{d_{11}(t-t_0)} E\|x\|_3^2 + d_{21}(e^{d_{11}(t-t_0)} - 1)], \qquad (3.14)$$

where d_{11}, d_{21} are constants, depending on assumptions $(A3_3)$ and $(B2_3)$ only. Let $R_1 < R_2$, then $t_0 \le \zeta_{R_1} \le \zeta_{R_2} \le \tau$. Thus, if $\{R_n\}_{n=1}^{\infty} \nearrow +\infty$ then $\{\zeta_{R_n}\}_{n=1}^{\infty}$ is a monotone non–decreasing random sequence. Choosing $R_n = n$, $n \in \mathbf{N}$, due to (3.14) we have

$$\forall \epsilon > 0, \ P\{|\tau - \zeta_{R_n}| > \epsilon\} = P\{\zeta_{R_n} < \tau - \epsilon\} \le$$

$$\frac{\tau - t_0}{n^2} [e^{d_{11}(\tau-t_0)} E\|x\|_3^2 + d_{21}(e^{d_{11}(\tau-t_0)} - 1)],$$

and $\sum_{n=1}^{\infty} P\{|\tau - \zeta_{R_n}| > \epsilon\} < \infty$. Therefore (see, e.g., [8]), ζ_{R_n} converges to τ (a.s.) as $n \to \infty$, or

$$P\{\omega : \forall \delta > 0, \exists N = N(\delta, \omega) : \forall n \ge N, \ \zeta_n(\omega) > \tau - \delta\} = 1.$$

Note now that for all $R_1 \ge R$, $t \in [t_0, \zeta_R]$ coefficients of equations (3.3) and

$$\eta_{R_1}^{x,h}(t) = h + \int_{t_0}^t A'_u(s, \xi^x(s))(\eta_{R_1}^{x,h}(s), \phi_{R_1}(\xi^x)(s)) \, ds +$$

$$\int_{t_0}^t A(s, \xi^x(s)) \eta_{R_1}^{x,h}(s) \, ds + \int_{t_0}^t B'_u(s, \xi^x(s)) (\eta_{R_1}^{x,h}(s), dw(s))$$

coincide. Hence, because of uniqueness of the solution to (3.3), satisfying conditions (i),(ii) of proposition 3.3, for every $t \in [t_0, \zeta_R]$, $\eta_R^{x,h}(t) = \eta_{R_1}^{x,h}(t)$ (a.s.) in H_2.

Fix $\delta > 0$ and set $\tau_1 = \tau - \delta$. Then

$$1 = P(\{\omega : \exists N = N(\delta, \omega) : \forall n \ge N, \ \zeta_n(\omega) \ge \zeta_N(\omega) > \tau_1\} \cap$$

$$\{\omega : \forall N \in \mathbf{N}, \ \forall n \ge N, \ for \ every \ t \in [t_0, \zeta_N], \ \eta_n^{x,h}(t) = \eta_N^{x,h}(t) \ in \ H_2\}) \le$$

$$P\{\omega : \exists N = N(\delta, \omega) : for \ every \ t \in [t_0, \tau_1], \forall n \ge N,$$

$$\eta_n^{x,h}(t) = \eta_N^{x,h}(t) \ in \ H_2\},$$

or $\forall t \in [t_0, \tau)$ the sequence $\{\eta_n^{x,h}(t)\}$ converges a.s. in H_2 uniformly with respect to t on every interval $[t_0, \tau_1] \subset [t_0, \tau)$. Let $\eta^{x,h}(t)$ be the limit process. Then $\eta^{x,h}(t) = \eta_n^{x,h}(t)$ (a.s.) in H_2 for all $n \geq N(\omega, \delta)$, $t \in [t_0, \tau_1]$, and $\eta^{x,h}(t)$ $([t_0, \tau_1])$ is a.s. continuous into H_1 .

For every $n \in \mathbf{N}$ and all $t \in [t_0, \zeta_n]$ the coefficients of equation (3.3) for $R = n$ and equation (3.2) coincide. Therefore, for every $n \in \mathbf{N}$ and all $t \in [t_0, \zeta_n]$, $\eta^{x,h}(t)$ satisfies (3.2) (a.s.) in H_0. But for $\forall n \geq N(\delta, \omega)$, $\zeta_n(\omega) > \tau_1$ (a.s.). Hence, for all $t \in [t_0, \tau_1]$, $\eta^{x,h}(t)$ satisfies (3.2) (a.s.) in H_0.

2. To prove (i) note that for every $t \in [t_0, \tau)$, $L > 0$,

$$P\{\|\eta^{x,h}(t)\|_2 > L\} \leq P\{\|\eta^{x,h}(t)\|_2 > L \ \& \ \zeta_R \geq t\} + P\{\zeta_R < t\} \leq$$
$$P\{\|\eta_R^{x,h}(t)\|_2 > L\} + P\{\zeta_R < t\}, \ R > 0.$$

The last inequality together with (3.10) and (3.13) implies

$$\lim_{L \to +\infty} P\{\|\eta^{x,h}(t)\|_2 > L\} = 0.$$

Then

$$P\{\omega : \|\eta^{x,h}(t, \omega)\|_2 = +\infty\} = P \bigcap_{n=1}^{\infty} \bigcup_{L \geq n} \{\omega : \|\eta^{x,h}(t, \omega)\|_2 > L\} =$$
$$\lim_{L \to +\infty} P\{\omega : \|\eta^{x,h}(t, \omega)\|_2 > L\} = 0.$$

By analogy one can prove (ii).

3. A.s. uniqueness of the solution to (3.2) follows immediately from the a.s. uniqueness of the solution to equation (3.3).

The proof of theorem 3.1 is complete.

Now we show that one can differentiate solution to equation (3.1) with respect to the initial function x in the direction h not only formally.

Fix $x, h \in S_{4+\alpha}^{t_0}(H_3)$ $(\alpha > 0)$, and for every $\epsilon > 0$, $t \in [t_0, \tau]$ set

$$\eta_\epsilon(t) = \frac{1}{\epsilon}(\xi^{x+\epsilon h}(t) - \xi^x(t)), \tag{3.15}$$

where ξ^x is a solution to equation (3.1) in H_0 on $[t_0, \tau]$.

Lemma 3.1. *Assume that for every* $x \in S_{4+\alpha}^{t_0}(H_3)$ $(\alpha > 0)$, *there exists a unique solution to equation (3.1) such that for every* $t \in [t_0, \tau]$, $\xi^x(t) \in S_4^t(H_3)$, $\xi^x(t)$ $(t \in [t_0, \tau])$ *is (a.s.) continuous into* H_2 , *and conditions* $(A3_2), (A4_2)$, *and* $(B3_2)$ *hold.*

Then for every $R > 0$, $t \in [t_0, \tau]$ *the following estimates hold for the finite differences* $\eta_\epsilon(t)$ $(t \in [t_0, \tau])$ *defined by (3.15):*

$$(i) \ E(\|\eta_\epsilon(t)\|_k^4 j\{\zeta_R \geq t\}) \leq 4e^{C_1(t-t_0)} E\|h\|_k^4, \ k \in \{0, 1, 2\},$$
$$(3.16)$$

$$(ii) \ E(\sup_{t_0 \leq s \leq t} \|\eta_\epsilon(s)\|_k^4 j\{\zeta_R \geq t\}) \leq 4e^{C_2(t-t_0)} E\|h\|_k^4, \ k \in \{0, 1\},$$
$$(3.17)$$

where ζ_R *is the stopping time defined by (3.13),* C_1, C_2 *are positive constants depending only on* R, $\tau - t_0$, *and the constants from assumptions* $(A3_2), (A4_2)$, *and* $(B3_2)$.

Proof. By the assumptions of the lemma for every $t \in [t_0, \tau]$, $k\{0, 1, 2\}$, and all $t \in [t_0, \tau]$, $k\{0, 1\}$, $\eta_\epsilon(t)$ satisfies Itô's equation

$$\eta_\epsilon(t) = h + \frac{1}{\epsilon} \int_{t_0}^t [A(s, \xi^{x+\epsilon h}(s)) - A(s, \xi^x(s))] \, \xi^x(s) \, ds +$$

$$\int_{t_0}^t A(s, \xi^{x+\epsilon h}(s)) \eta_\epsilon(s) \, ds + \frac{1}{\epsilon} \int_{t_0}^t [B(s, \xi^{x+\epsilon h}(s)) - B(s, \xi^x(s))] \, dw(s) \quad (a.s.)$$

in H_k. Applying Itô's formula to $\|\eta_\epsilon(t)\|_k^2$ (for every $t \in [t_0, \tau]$, $k\{0, 1, 2\}$, and all $t \in [t_0, \tau]$, $k\{0, 1\}$) we find

$$\|\eta_\epsilon(t)\|_k^2 = \|h\|_k^2 + \frac{2}{\epsilon} \int_{t_0}^t ([A(s, \xi^{x+\epsilon h}(s)) - A(s, \xi^x(s))] \, \xi^x(s), \eta_\epsilon(s))_k \, ds +$$

$$2 \int_{t_0}^t (A(s, \xi^{x+\epsilon h}(s)) \eta_\epsilon(s), \eta_\epsilon(s))_k \, ds +$$

$$\frac{1}{\epsilon^2} \int_{t_0}^t \sigma_2^2(T^k[B(s, \xi^{x+\epsilon h}(s)) - B(s, \xi^x(s))]) \, ds +$$

$$\frac{2}{\epsilon} \int_{t_0}^t (\eta_\epsilon(s), [B(s, \xi^{x+\epsilon h}(s)) - B(s, \xi^x(s))] \, dw(s))_k \quad (a.s.).$$

Under assumptions $(A3_2), (A4_2)$, and $(B3_2)$, the last equality implies for every $t \in [t_0, \tau]$,

$$E\|\eta_\epsilon(t \wedge \zeta_R)\|_k^4 \leq 4E\|h\|_k^4 + 4[4c_k^2 R^2 + (2d_k + b_k)^2(\tau - t_0) + 4b_k] \times$$

$$E \int_{t_0}^{t \wedge \zeta_R} \|\eta_\epsilon(s)\|_k^4 \, ds, \ k \in \{0, 1, 2\},$$
$$(3.18)$$

$$E(\sup_{t_0 \leq s \leq t} \|\eta_\epsilon(s \wedge \zeta_R)\|_k^4) \leq 4E\|h\|_k^4 + 4[4c_k^2 R^2 + (2d_k + b_k)^2(\tau - t_0) + 16b_k] \times$$

$$E \int_{t_0}^{t \wedge \zeta_R} \sup_{t_0 \leq s \leq \theta} \|\eta_\epsilon(s)\|_k^4 \, d\theta, \ k \in \{0, 1\}.$$
$$(3.19)$$

Estimates (3.16) and (3.17) follow from Gronwall's lemma and estimates (3.18) and (3.19), respectively.

Theorem 3.2. *Suppose assumptions* $(A1), (A2_3) - (A4_3), (A5_1), (B1_3) -$ $(B3_3)$, *and* $(B4_1)$ *about* $A(t,u)$ *and* $(B(t,u)$ *in* (3.1) *are valid, and* $x, h \in$ $S_{4+\alpha}^{t_0}(H_3) \, (\alpha > 0)$.

Then for every $t \in [t_0, \tau_1] \subset [t_0, \tau)$ *the solution to equation* (3.1) *is Gateaux differentiable in the sense of convergence in probability in* H_1 *along directions from* $S_{4+\alpha}^{t_0}(H_3)$, *and its directional derivative at* x *in the direction* h *is a solution to equation* (3.2) *in* H_0 *on every interval* $[t_0, \tau_1] \subset [t_0, \tau)$, *satisfying conditions* (i) *and* (ii) *of theorem* 3.1.

Proof. 1. The conditions of lemma 3.1 hold. For every $t \in [t_0, \tau]$, $x, h \in$ $S_{4+\alpha}^{t_0}(H_3)$, let $\eta_\epsilon(t)$ be defined by (3.15). Introduce stopping times ζ_R and ν_r by (3.13) and

$$\nu_r = \inf\{\lambda \geq t_0 : \|\eta_\epsilon(\lambda)\|_1 > r\} \wedge \tau, \qquad (3.20)$$

respectively. For $P\{\zeta_R < t\}$ we have estimate (3.14), and for $P\{\nu_r < t \ \& \ \zeta_R \geq t\}$ using (3.17) one can find

$$P\{\nu_r < t \ \& \ \zeta_R \geq t\} = P\{ \sup_{t_0 \leq s \leq t} \|\eta_\epsilon(s)\|_1 > r \ \& \ \zeta_R \geq t\} \leq$$

$$\frac{1}{r^4} E(\sup_{t_0 \leq s \leq t} \|\eta_\epsilon(s)\|_1^4 j\{\zeta_R \geq t\}) \leq \frac{4}{r^4} e^{C_2(t-t_0)} E\|h\|_1^4, \qquad (3.21)$$

where C_2 is a positive constant, depending on $R, \tau - t_0$, and constants from assumptions $(A3_2), (A4_2)$, and $(B3_2)$ only.

2. For fixed $x, h \in S_{4+\alpha}^{t_0}(H_3)$, suppose $\eta^{x,h}(t) \ (t \in [t_0, \tau_1] \subset [t_0, \tau))$ is the unique (a.s.) solution to equation (3.2) in H_0 on every $[t_0, \tau_1] \subset [t_0, \tau)$, satisfying conditions (i),(ii) of theorem 3.1. For fixed $\tau_1 < \tau$ and every $t \in [t_0, \tau_1]$, set $\mu_\epsilon(t) = \eta_\epsilon(t) - \eta^{x,h}(t)$. Then for every $t \in [t_0, \tau_1]$, $\mu_\epsilon(t)$ satisfies Itô's equation

$$\mu_\epsilon(t) = \int_{t_0}^{t} A_u'(s, \xi^x(s)) \, (\mu_\epsilon(s), \xi^x(s)) \, ds + \int_{t_0}^{t} A(s, \xi^x(s)) \mu_\epsilon(s) \, ds +$$

$$\frac{1}{\epsilon} \int_{t_0}^{t} [A(s, \xi^{x+\epsilon h}(s)) - A(s, \xi^x(s)) -$$

$$A_u'(s, \xi^x(s)) \, (\xi^{x+\epsilon h}(s) - \xi^x(s))] \xi^x(s) \, ds +$$

$$\frac{1}{\epsilon} \int_{t_0}^{t} [A(s, \xi^{x+\epsilon h}(s)) - A(s, \xi^x(s))] \, (\xi^{x+\epsilon h}(s) - \xi^x(s)) \, ds +$$

$$\int_{t_0}^{t} B_u'(s, \xi^x(s)) \, (\mu_\epsilon(s), dw(s)) +$$

$$\frac{1}{\epsilon} \int_{t_0}^{t} [B(s, \xi^{x+\epsilon h}(s)) - B(s, \xi^x(s)) -$$

$$B_u'(s, \xi^x(s)) \, (\xi^{x+\epsilon h}(s) - \xi^x(s))] \, dw(s) \quad (a.s.)$$

in H_1. Applying Itô's formula to $\|\mu_\epsilon(t)\|_1^2$ under conditions $(A3_2), (A4_2)$, $(A5_1), (B3_2)$, and $(B4_1)$ we obtain the inequality

$$\|\mu_\epsilon(t)\|_1^2 \le 2c_1 \int_{t_0}^t \|\mu_\epsilon(s)\|_1^2 \|\xi^x(s)\|_2 \, ds + (2d_1 + b_1) \int_{t_0}^t \|\mu_\epsilon(s)\|_1^2 \, ds +$$

$$\frac{2}{\epsilon} c_1 \int_{t_0}^t \|\xi^{x+\epsilon h}(s) - \xi^x(s)\|_1^2 \|\xi^x(s)\|_2 \|\mu_\epsilon(s)\|_1 \, ds +$$

$$\frac{2}{\epsilon} c_1 \int_{t_0}^t \|\xi^{x+\epsilon h}(s) - \xi^x(s)\|_1 \|\xi^{x+\epsilon h}(s) - \xi^x(s)\|_2 \|\mu_\epsilon(s)\|_1 \, ds +$$

$$\frac{b_1}{\epsilon^2} \int_{t_0}^t \|\xi^{x+\epsilon h}(s) - \xi^x(s)\|_1^4 \, ds + 2 \int_{t_0}^t (\mu_\epsilon(s), B_u'(s, \xi^x(s)) \, (\mu_\epsilon(s), dw(s)))_1 +$$

$$\frac{2}{\epsilon} \int_{t_0}^t (\mu_\epsilon(s), [B(s, \xi^{x+\epsilon h}(s)) - B(s, \xi^x(s)) -$$

$$B_u'(s, \xi^x(s)) \, (\xi^{x+\epsilon h}(s) - \xi^x(s))] \, dw(s))_1 \quad (a.s.),$$

which implies

$$E(\|\mu_\epsilon(t)\|_1^4 j\{\zeta_R \wedge \nu_r \ge t\}) \le 7[4c_1^2 R^2 + (2d_1 + b_1)^2(\tau - t_0) +$$

$$2c_1^2 r^2 (R^2 + \tau - t_0) + 2b_1] \int_{t_0}^t E(\|\mu_\epsilon(s)\|_1^4 j\{\zeta_R \wedge \nu_r \ge s\}) \, ds +$$

$$7r^2 [2c_1^2 (R^2 + \tau - t_0) + b_1^2 r^2 (\tau - t_0) + 2b_1 r^2] \int_{t_0}^t E\|\xi^{x+\epsilon h}(s) - \xi^x(s)\|_2^4 \, ds,$$

and for any $\epsilon_1 > 0$

$$P\{\|\eta_\epsilon(t) - \eta^{x,h}(t)\|_1 > \epsilon_1 \ \& \ \zeta_R \wedge \nu_r \ge t\} =$$

$$P\{\|\mu_\epsilon(t)\|_1 > \epsilon_1 \ \& \ \zeta_R \wedge \nu_r \ge t\} \le$$

$$\frac{1}{\epsilon_1^2} E(\|\mu_\epsilon(t)\|_1^4 j\{\zeta_R \wedge \nu_r \ge t\}) \le$$

$$\frac{7r^2}{\epsilon_1^2} [2c_1^2 (R^2 + \tau - t_0) + b_1^2 r^2 (\tau - t_0) + 2b_1 r^2] \times$$

$$\exp\{7[4c_1^2 R^2 + (2d_1 + b_1)^2 (\tau - t_0) + 2c_1^2 r^2 (R^2 + \tau - t_0) + 2b_1]\}(t - t_0) \times$$

$$\int_{t_0}^t E\|\xi^{x+\epsilon h}(s) - \xi^x(s)\|_2^4 \, ds. \tag{3.22}$$

As $\xi^x = \lim_{\epsilon \to 0} \xi^{x+\epsilon h}$ in $S_4([t_0, \tau], H_2)$, it follows from $(3.14), (3.21)$, and (3.22) that for every $t \in [t_0, \tau_1]$ and any $\epsilon_1 > 0$, $\lim_{\epsilon \to 0} P\{\|\eta_\epsilon(t) - \eta^{x,h}(t)\|_1 > \epsilon_1\} = 0$, or

$$\eta^{x,h}(t) = P - \lim_{\epsilon \to 0} \frac{1}{\epsilon} (\xi^{x+\epsilon h}(t) - \xi^x(t))$$

in H_1.

It turns out that the family $\eta^{x,h}(t)$ ($t \in [t_0, \tau_1]$, $x, h \in S_{4+\alpha}^{t_0}(H_3)$) depends continuously on x and h in the sense of

Theorem 3.3. Let for every $x, h \in S_{4+\alpha}^{t_0}(H_3)$, $\eta^{x,h} = \eta^{x,h}(t)$ ($t \in [t_0, \tau_1] \subset [t_0, \tau)$) be a solution to equation (3.2) in H_0 on $[t_0, \tau_1]$ satisfying conditions (i) and (ii) of theorem 3.1, and assumptions $(A3_3), (A4_3), (A5_1), (B2_3), (B3_3)$, and $(B4_1)$ about $A(t, u)$ and $B(t, u)$ in (3.1) valid.

Then

(i) for fixed $h \in S_{4+\alpha}^{t_0}(H_3)$ and for $x, x_n \in S_{4+\alpha}^{t_0}(H_3)$ such that $x = \lim_{n\to\infty} x_n$ in $L_4(\Omega, H_2, P)$, for every $t \in [t_0, \tau_1]$

$$\eta^{x,h}(t) = P - \lim_{n\to\infty} \eta^{x_n,h}(t)$$

in H_1,

(ii) for fixed $x \in S_{4+\alpha}^{t_0}(H_3)$ and for $h, h_n \in S_{4+\alpha}^{t_0}(H_3)$ such that $h = \lim_{n\to\infty} h_n$ in $L_4(\Omega, H_1, P)$, for every $t \in [t_0, \tau_1]$

$$\eta^{x,h}(t) = P - \lim_{n\to\infty} \eta^{x,h_n}(t)$$

in H_1.

Proof. 1. Let ζ_R be defined by (3.13),

$$\zeta^2 = \inf\{\lambda \geq t_0 : \|\xi^x(\lambda) - \xi^{x_n}(\lambda)\|_2 > 1\} \wedge \tau,$$

$$\zeta_r^3 = \inf\{\lambda \geq t_0 : \int_{t_0}^{\lambda} \|\eta^{x,h}(s)\|_2^2 \, ds > r^2\} \wedge \tau,$$

$$\zeta_r^4 = \inf\{\lambda \geq t_0 : \|\eta^{x,h}(\lambda)\|_1 > r\} \wedge \tau, \quad \nu_{R,r} = \zeta_R \wedge \zeta^2 \wedge \zeta_r^3 \wedge \zeta_r^4.$$

For $P\{\zeta_R < t\}$ we have estimate (3.14). By the assumptions of the theorem it is easy to show that

$$P\{\zeta^2 < t \ \& \ \zeta_R \geq t\} = P\{\sup_{t_0 \leq s \leq t} \|\xi^x(s) - \xi^{x_n}(s)\|_2 > 1 \ \& \ \zeta_R \geq t\} \leq$$

$$E(\sup_{t_0 \leq s \leq t} \|\xi^x(s) - \xi^{x_n}(s)\|_2^4 j\{\zeta_R \geq t\}) \leq 4e^{C_R(t-t_0)} E\|x - x_n\|_2^4, \tag{3.23}$$

where C_R is the constant, depending on R, $t - t_0$, and the constants from assumptions $(A3_3), (A4_3)$, and $(B3_3)$ only. Due to (3.10),(3.12)

$$P\{\zeta_r^3 < t \ \& \ \zeta_R \geq t\} = P\{\int_{t_0}^{t} \|\eta_R^{x,h}(s)\|_2^2 \, ds > r^2 \ \& \ \zeta_R \geq t\} \leq$$

$$\frac{2}{r^2} e^{(D_1 R^2 + D_2)\frac{(t-t_0)}{2}} (E\|h\|_2^4)^{\frac{1}{2}} \tag{3.24}$$

and

$$P\{\zeta_r^4 < t \ \& \ \zeta_R \geq t\} = P\{\sup_{t_0 \leq s \leq t} \eta_R^{x,h}(s)\|_1 > r \ \& \ \zeta_R \geq t\} \leq$$

$$\frac{4}{r^4} e^{(D_1 R^2 + D_3)(t-t_0)} E\|h\|_1^4 \ , \tag{3.25}$$

where D_1, D_2, D_3 are the constants from (3.10),(3.12).

2. Fix $h \in S_{4+\alpha}^{t_0}(H_3)$. Then for every $t \in [t_0, \tau_1]$

$$\eta^{x_n,h}(t) - \eta^{x,h}(t) = \int_{t_0}^t [A(s,\xi^{x_n}(s))\eta^{x_n,h}(s) - A(s,\xi^x(s))\eta^{x,h}(s)] \, ds +$$

$$\int_{t_0}^t [A'_u(s,\xi^{x_n}(s))(\eta^{x_n,h}(s),\xi^{x_n}(s)) - A'_u(s,\xi^x(s))(\eta^{x,h}(s),\xi^x(s))] \, ds +$$

$$\int_{t_0}^t [B'_u(s,\xi^{x_n}(s))\eta^{x_n,h}(s) - B'_u(s,\xi^x(s))\eta^{x,h}(s)] \, dw(s) \quad (a.s.) \tag{3.26}$$

in H_1. Set $\mu_n(t) = \eta^{x_n,h}(t) - \eta^{x,h}(t)$. Adding and subtracting in the right hand side of (3.26) we find:

$$\mu_n(t) = \int_{t_0}^t A(s,\xi^{x_n}(s))\mu_n(s) \, ds + \int_{t_0}^t [A(s,\xi^{x_n}(s)) - A(s,\xi^x(s))]\eta^{x,h}(s) \, ds +$$

$$\int_{t_0}^t A'_u(s,\xi^{x_n}(s)) \, (\mu_n(s),\xi^{x_n}(s) - \xi^x(s)) \, ds +$$

$$\int_{t_0}^t A'_u(s,\xi^{x_n}(s)) \, (\eta^{x,h}(s),\xi^{x_n}(s) - \xi^x(s)) \, ds +$$

$$\int_{t_0}^t [A'_u(s,\xi^{x_n}(s)) - A'_u(s,\xi^x(s))] \, (\eta^{x,h}(s),\xi^x(s)) \, ds +$$

$$\int_{t_0}^t B'_u(s,\xi^x(s)) \, (\mu_n(s),dw(s)) +$$

$$\int_{t_0}^t [B'_u(s,\xi^{x_n}(s)) - B'_u(s,\xi^x(s))] \, (\eta^{x,h}(s),dw(s))$$

(a.s.) in H_1, and under assumptions $(A3_3), (A4_3), (A5_1), (B3_3), (B4_1)$

$$\|\mu_n(t \wedge \nu_{R,r})\|_1^2 \leq (2d_1 + b_1 + 2c_1) \int_{t_0}^{t \wedge \nu_{R,r}} \|\mu_n(s)\|_1^2 \, ds +$$

$$4c_1 \int_{t_0}^{t \wedge \nu_{R,r}} \|\eta_R^{x,h}(s)\|_2 \|\xi^{x_n}(s) - \xi^x(s)\|_2 \|\mu_n(s)\|_1 \, ds +$$

$$2c_1 \int_{t_0}^{t \wedge \nu_{R,r}} \|\eta_R^{x,h}(s)\|_1 \|\xi^{x_n}(s) - \xi^x(s)\|_1 \|\xi^x(s)\|_2 \|\mu_n(s)\|_1 \, ds +$$

$$b_1 \int_{t_0}^{t \wedge \nu_{R,r}} \|\eta_R^{x,h}(s)\|_1^2 \|\xi^{x_n}(s) - \xi^x(s)\|_1^2 \, ds +$$

$$2 \int_{t_0}^{t \wedge \nu_{R,r}} (\mu_n(s), B_u'(s,\xi^x(s)) \, (\mu_n(s), dw(s)))_1 +$$

$$2 \int_{t_0}^{t \wedge \nu_{R,r}} (\mu_n(s), [B_u'(s,\xi^{x_n}(s)) - B_u'(s,\xi^x(s))] \, (\eta^{x,h}(s), dw(s)))_1 \quad \text{(a.s.)},$$

which implies

$$E(\|\mu_n(t)\|_1^4 j\{\nu_{R,r} \geq t\}) \leq$$

$$6[(2d_1 + b_1 + 2c_1)^2(\tau - t_0) + 2c_1^2 r^2(4 + R^2) + 2b_1(2 + r^2)] \times$$

$$\int_{t_0}^{t} E(\|\mu_n(s)\|_1^4 j\{\nu_{R,r} \geq s\}) \, ds + 6r^2 [2c_1^2(4 + R^2) + b_1^2 r^2(\tau - t_0) + 2b_1 r^2] \times$$

$$\int_{t_0}^{t} \sup_{t_0 \leq s \leq \theta} \|\xi^{x_n}(s) - \xi^x(s)\|_2^4 j\{\zeta_R \geq \theta\}) \, d\theta.$$

It follows from Gronwall's lemma and the last inequality in (3.23) that

$$E(\|\mu_n(t)\|_1^4 j\{\nu_{R,r} \geq t\}) \leq D_{1,R,r} e^{D_{2,R,r}(t-t_0)} E\|x - x_n\|_2^4, \qquad (3.27)$$

where $D_{1,R,r}, D_{2,R,r}$ are the constants depending on $R, r, \tau - t_0$, and constants from assumptions $(A3_3), (A4_3), (A5_1), (B3_3)$, and $(B4_1)$ only. Statement (i) of the theorem follows now immediately from (3.14), (3.23)-(3.25), and (3.27).

3. Equation (3.2) is linear with respect to $\eta^{x,h}$. Then for fixed $x \in S_{4+\alpha}^{t_0}(H_3)$, and $h, h_n \in S_{4+\alpha}^{t_0}(H_3)$ such that $h = \lim_{n \to \infty} h_n$ in $L_4(\Omega, H_1, P)$ due to (3.9) for any $\epsilon > 0$ and every $t \in [t_0, \tau]$, we have

$$P\{\|\eta^{x,h}(t) - \eta^{x,h_n}(t)\|_1 > \epsilon \ \& \ \zeta_R \geq t\} \leq \frac{1}{\epsilon^4} E\|\eta_R^{x,h-h_n}(t)\|_1^4 \leq$$

$$\frac{4}{\epsilon^4} e^{(D_1 R^2 + D_3)(t-t_0)} E\|h - h_n\|_1^4.$$

Statement (ii) follows now from (3.14) and the last inequality.

4. EXAMPLE

Consider the initial boundary value problem

$$\frac{\partial u}{\partial t}(t, x) = a(t, u(t))\frac{\partial^2 u}{\partial x^2}(t, x) + b(t, u(t))u(t, x)+$$

$$\int_\alpha^\beta c(t, x, y, u(t, x))\xi(t, y)\, dy, \quad x \in (\alpha, \beta),$$

for all $t \in [0, \tau]$, $u(t, \alpha) = u(t, \beta) = 0$, $u(0, x) = u_0(x)$, $x \in [\alpha, \beta]$,

where $\xi(t, y)$ is a space-time "white noise" (see precise definition below), and coefficients a, b, c satisfy conditions:

(i) for every $u \in L_2(\alpha, \beta) \equiv L_2([\alpha, \beta])$, $a(\cdot, u), b(\cdot, u)$ are real valued continuous functions on $[0, \tau]$, and for every $t \in [0, \tau]$, $a(t, \cdot), b(t, \cdot)$ are twice continuously Fréchet differentiable functionals on $L_2(\alpha, \beta)$; denote their first and second derivatives with respect to $u \in L_2([\alpha, \beta])$ by a'_u, b'_u, a''_{uu} and b''_{uu} correspondently.

(ii) There are constants a_0, a_1 such that for all $t \in [0, \tau]$, $u \in L_2(\alpha, \beta)$,

$$0 < a_0 \le a(t, u) \le a_1,$$

and $\beta, a'_u, b'_u, a''_{uu}$ and b''_{uu} are bounded with respect to t and u uniformly on $[0, \tau] \times L_2(\alpha, \beta)$.

(iii) $c(t, x, y, v) = c_1(t, x, y) + c_2(t, x, y)v$ ($t \in [0, \tau]$, $x, y \in [\alpha, \beta]$, $v \in \mathbf{R}$); for every $x \in [\alpha, \beta]$, $c_1(\cdot, x, \cdot), c_2(\cdot, x, \cdot)$ are real valued continuous functions on $[0, \tau] \times [\alpha, \beta]$; for every $(t, y) \in [0, \tau] \times [\alpha, \beta]$, $c_1(t, \cdot, y), c_2(t, \cdot, y)$ are six times continuously differentiable functions on $[\alpha, \beta]$.

(iv) c_1, c_2, and all their derivatives with respect to x up to the 6th order are bounded uniformly with respect to t, x, and y on $[0, \tau] \times [\alpha, \beta] \times [\alpha, \beta]$.

Abstract spaces and operators for this problem are defined as follows:

1. $H = \mathcal{H} = L_2(\alpha, \beta)$, $\mathcal{H}_- = W_2^{-1}([\alpha, \beta])$ a negative Sobolev space of generalized functions corresponding to the positive Sobolev space $W_2^1([\alpha, \beta])$ of real valued functions on $[\alpha, \beta]$ with square integrable m-th derivative ($m = 0, 1$); T is a closure of the operator

$$L_2(\alpha, \beta) \supset \tilde{C}^2([\alpha, \beta]) \ni u \mapsto u - \frac{\partial^2 u}{\partial x^2} \in L_2([\alpha, \beta]),$$

where $\tilde{C}^2([\alpha, \beta]) = \{u \in C^2([\alpha, \beta]) : u(\alpha) = u(\beta) = 0\}$.

2. For every $t \in [0, \tau], u \in L_2(\alpha, \beta)$ let $A(t, u)$ be a closure of the operator

$$L_2(\alpha, \beta) \supset \tilde{C}^2([\alpha, \beta]) \ni v \mapsto a(t, u)\frac{\partial^2 v}{\partial x^2} + b(t, u)v \in L_2(\alpha, \beta).$$

3. For $t \in [0, \tau], u, v \in L_2(\alpha, \beta)$ define

$$(B(t, u)v)(x) = \int_\alpha^\beta c_1(t, x, y)v(y) \, dy + \int_\alpha^\beta c_2(t, x, y)u(x)v(y) \, dy.$$

4. Define the space–time "white noise" corresponding to Hilbert space $L_2(\alpha, \beta)$ as a generalized function $\xi \in W_2^{-1}([0, \tau]) \otimes W_2^{-1}([\alpha, \beta])$ such that for any $\phi, \psi \in L_2([0, \tau] \times [\alpha, \beta])$, $t \in [0, \tau]$,

$$E\Big(\int_0^t \int_\alpha^\beta \phi(s, x)\xi(s, x) \, dx \, ds\Big)\Big(\int_0^t \int_\alpha^\beta \psi(s, x)\xi(s, x) \, dx \, ds\Big) =$$

$$\int_0^t \int_\alpha^\beta \phi(s, x)\psi(s, x) \, dx \, ds.$$

Then the standard Wiener process $w(t)$ with values in $W_2^{-1}([\alpha, \beta])$ is defined by

$$\int_0^t (\phi(s), dw(s))_{L_2(\alpha, \beta)} = \int_0^t \int_\alpha^\beta \phi(s, x)\xi(s, x) \, dx \, ds.$$

Operators A and B introduced as above satisfy assumptions of theorems 3.1 and 3.2.

ACKNOWLEDGMENTS

The author is grateful to Professor K.D.Elworthy for helpful discussions and encouragement.

REFERENCES

[1]   G. Da Prato, J. Zabczyk, *Stochastic equations in infinite dimensions*, Cambridge University Press, 1992.

[2]   Yu. L. Dalecky, S. V. Fomin, *Measures and differential equations in infinite-dimensional space*, Kluwer Academic Publishers, 1991.

[3]   Z. Brzeźniak, M. Capiński, F. Flandoli, *Stochastic Navier–Stokes equations with multiplicative noise*, Stochastic Analysis and Applications **10** (1992), no. 5, 523–532.

[4]   Z. Brzeźniak, M. Capiński, F. Flandoli, *Stochastic partial differential equations and turbulence*, Mathematical Models and Methods in Applied Sciences **1** (1991), no. 1, 41–59.

[5]   G. Da Prato, K.D.Elworthy, J. Zabczyk, *Strong Feller property for sto-chastic semilinear equations*, Warwick preprints, 1/1992. To appear in Stochastic Analysis and Applications.

[6]   S.Peszat, J. Zabczyk, *Strong Feller property and irreducibility for diffu-sions on Hilbert spaces* (1993), Preprint of Polish Academy of Sciences.

[7]   Yu. L. Dalecky, N. Yu. Goncharuk, *On a quasilinear stochastic differ-ential equation of parabolic type*, Stochastic Analysis and Applications **12** (1994), no. 1, 103–129.

[8]   M. Loève, *Probability theory*, D. Van Nostand Company, Inc., 1963.

FLUCTUATIONS OF A TWO–LEVEL CRITICAL BRANCHING SYSTEM

Luis G. Gorostiza

Departamento de Matemáticas

Centro de Investigación y de Estudios Avanzados

México

Abstract. The critical dimension for the long–time behaviour of the high–density fluctuation limit of a system of critical branching super Brownian motions is equal to 4, the same as the critical dimension for a related two–level superprocess.

1 Introduction.

Multilevel branching particle systems were introduced recently by Dawson and Hochberg [DH] as models for a class of hierarchically structured populations. In such a model individual particles migrate and branch, and in addition collections of particles are subject to independent branching mechanisms at the different levels of the hierarchical organization. The analysis of these models is complicated by the fact that the higher level branching leads to the absence of independence in the particle dynamics. Several aspects of multilevel branching systems have been investigated by Dawson, Hochberg and Wu [DHW], Dawson, Hochberg and Vinogradov [DHV1, DHV2], Dawson and Wu [DW], Etheridge [E], Gorostiza, Hochberg and Wakolbinger [GHW], Hochberg [H], and Wu [W1,W2, W3,W4]. Some of these references, specially [DHV2], include examples of areas of application where hierarchical branching structures are present.

Wu [W1,W2] studied the long–time behaviour of a critical two–level superprocess constructed from a system of branching Brownian motions in $\mathbb{R}^d$. This process is obtained in a similar way as the usual (one–level) superprocesses, i.e., as a limit under a high–density/short–life/small–mass rescaling, which in the two–level case is applied simultaneously at the two levels. He proved that this critical two–level superprocess goes to extinction, as time tends to infinity, in dimensions $d \leq 4$, and from his results in [W1] it follows that it does not become extinct in dimensions $d > 4$. Gorostiza, Hochberg and Wakolbinger [GHW] proved the stronger result that persistence holds in dimensions $d > 4$. This means that as time grows to infinity the critical two–level superprocess converges weakly towards an equilibrium state which does not loose intensity. Hence this critical two–level superprocess obeys a persistence/extinction dichotomy with critical dimension $d = 4$. This result was originally conjectured by Dawson.

Our starting point in the present paper is the following observation. The two–level superprocess can also be obtained by first going to the limit in the first level, and then in the second level. The first level limit yields a system of branching superprocesses (super Brownian motions). One reason this two–step approach is useful is that, at each time, the system of branching superprocesses is a doubly stochastic Poisson random field with random intensity given by the two–level superprocess. This fact was exploited in [GHW] for the persistence analysis of the two–level superprocess. The two–step approach motivates the study of other second level limits, different from the superprocess type limit, which also provide insight into the high–density behaviour of the two–level branching system. In particular, it is interesting to study the high–density hydrodynamic behaviour (law of large numbers) and asymptotic fluctuations of the empirical measures of the system of branching superprocesses resulting from the first level limit. A study of these fluctuations would require as state space a nuclear space of distributions on a space of tempered measures (in order to make use of Lévy's continuity theorem). Such a space is not readily available. A more modest aim is to study the high–density behaviour of the aggregated (or superposition) process of the empirical measure process. This study can be carried out using $\mathcal{S}'(\mathbb{R}^d)$, the space of tempered distributions on $\mathbb{R}^d$, as state space. Although some information is lost in the aggregated process, it turns out that its asymptotic analysis produces significant information on the system. This analysis is the subject of [G], where the asymptotic behaviour of a system of (not necessarily critical) branching superprocesses is undertaken. In the present note we give only the main results for the critical case involving Brownian motion. In this case the long–time behaviour of the fluctuation limit process has critical dimension $d = 4$, which coincides, as it should be expected, with the critical dimension for the persistence/extinction dichotomy of the critical two–level superprocess mentioned above. We refer the reader to [G] for more general results and detailed proofs.

2 A system of branching superprocesses and high–density limits.

The critical (finite variance) super Brownian motion is an $\mathcal{M}_p(\mathbb{R}^d)$–valued homogeneous Markov process $B \equiv \{B(t), t \geq 0\}$ with conditional Laplace functional given by

$$E[exp\{-\langle B(t), \varphi \rangle\} | B(0) = \mu] = exp\{-\langle \mu, u_\varphi(t) \rangle\}, \quad \varphi \in C_p(\mathbb{R}^d), \ \mu \in \mathcal{M}_p(\mathbb{R}^d),$$

where $\{u_\varphi(x, t), x \in \mathbb{R}^d, t \geq 0\}$ is a non–linear semigroup which is the unique solution of the integral equation

$$u(t) = S_t \varphi - \frac{1}{2} \int_0^t S_{t-s}(u(s)^2) ds,$$

and (S_t) is the Brownian semigroup on $\mathbb{R}^d$. We recall that this superprocess arises as the limit as $n \to \infty$ of a system of critical (finite variance) branching Brownian motions on $\mathbb{R}^d$, where the particles have mass n^{-1}, are initially distributed with a density proportional to n, and branch at rate n. The spaces

$C_p(\mathbb{R}^d)$ and $\mathcal{M}_p(\mathbb{R}^d)$ are defined for $p > 0$, respectively, as the space of non–negative continuous functions φ on $\mathbb{R}^d$ such that $\sup_x |\varphi(x)/\varphi_p(x)| < \infty$, where $\varphi_p(x) = (1+ \parallel x \parallel^2)^{-p}$, and the space of non–negative Radon measures μ on $\mathbb{R}^d$ such that $\langle \mu, \varphi_p \rangle < \infty$, where $\langle \mu, \varphi \rangle \equiv \int \varphi d\mu$. These spaces are suitably topologized, and p is taken such that $p > d/2$. See Dawson [D] for background on superprocesses.

If $B(0) = \lambda$ (Lebesgue measure on $\mathbb{R}^d$), then $B(t)$ has the same intensity (λ) for all finite t. If $d > 2$, then $B(t)$ converges weakly, as $t \to \infty$, towards an equilibrium state $B(\infty)$ which also has intensity λ (i.e., persistence holds), and if $d \leq 2$, then $B(t)$ goes to local extinction as $t \to \infty$. In the case $d > 2, B(\infty)$ is an infinitely divisible random measure on $\mathbb{R}^d$, and we denote its canonical measure by R_∞. This measure belongs to $\mathcal{M}_{LF}^2(\mathbb{R}^d)$, which is a suitable space of locally finite measures on the space of locally finite measures on $\mathbb{R}^d$, and it has intensity λ in the sense that $\langle\!\langle R_\infty, \langle \cdot, \varphi \rangle \rangle\!\rangle = \langle \lambda, \varphi \rangle, \varphi \in C_p(\mathbb{R}^d)$, where $\langle\!\langle R, \Phi \rangle\!\rangle \equiv \int \Phi(\mu) R(d\mu)$. These results are found in [D], [DP] and [GW].

Now we consider a system of branching super Brownian motions given by a collection of $\mathcal{M}_p(\mathbb{R}^d)$–valued particles, or "superparticles", which migrate according to the critical super Brownian motion described above, and branch at rate 1 with a critical (finite variance) branching law. The offspring superparticles start off from the branching site of their parent, and the motions, lifetimes and branchings of different superparticles are independent of each other and of the initial configuration of the system. We assume that the initial configuration is given by a Poisson random measure on $\mathcal{M}_p(\mathbb{R}^d)$ with locally finite intensity.

If $B_{ij}(t)$ denotes the location in $\mathcal{M}_p(\mathbb{R}^d)$ at time t of the jth superparticle in the ith branching super Brownian motion, then the *empirical measure process* $Y \equiv \{Y(t), t \geq 0\}$ is defined by $Y(t) = \sum_i \sum_j \delta_{B_{ij}(t)}$, and its *aggregated process* $\overline{Y} \equiv \{\overline{Y}(t), t \geq 0\}$ is given by $\overline{Y}(t) = \sum_i \sum_j B_{ij}(t)$. (Note that $\langle\!\langle Y(t), \langle \cdot, \varphi \rangle \rangle\!\rangle = \langle \overline{Y}(t), \varphi \rangle$). The process $\overline{Y}$ takes values in $\mathcal{M}_p(\mathbb{R}^d)$ (while the process Y takes values in $\mathcal{M}_{LF}^2(\mathbb{R}^d)$). We consider $d > 2$ and assume that the intensity of the initial Poisson random measure is given by the equilibrium canonical measure R_∞. In this case $\overline{Y}(t)$ has intensity λ for all finite t.

For the high–density behaviour we consider a sequence of such processes. For each $n = 1, 2, \ldots$, let $\overline{Y}^n(t)$ be the aggregated process corresponding to a system of branching super Brownian motions as above, with initial configuration given by a Poisson random measure on $\mathcal{M}_p(\mathbb{R}^d)$ with intensity nR_∞. The hydrodynamic behaviour of $\overline{Y}^n(t)$ is given by the following law of large numbers.

Theorem 1. *For each $t \geq 0$ and $\varphi \in C_p(\mathbb{R}^d)$,*

$$n^{-1}\langle \overline{Y}^n(t), \varphi \rangle \to \langle \lambda, \varphi \rangle$$

in L^2 as $n \to \infty$.

Now we consider the fluctuation process $Z^n \equiv \{Z^n(t), t \geq 0\}$ of $\overline{Y}^n$, defined by

$$Z^n(t) = n^{-1/2}(\overline{Y}^n(t) - \lambda).$$

Since this process is signed (tempered) measure–valued, we may consider it as $\mathcal{S}'(\mathbb{R}^d)$–valued, and therefore we now take test functions φ in $\mathcal{S}(\mathbb{R}^d)$, the space of infinitely differentiable functions on $\mathbb{R}^d$ which are rapidly decreasing at infinity. The asymptotic behaviour of Z^n is given by the following result.

Theorem 2. *Z^n converges weakly in the Skorokhod space $D([0, \infty), \mathcal{S}'(\mathbb{R}^d))$, as $n \to \infty$, towards a Gaussian process Z which is a continuous generalized Ornstein–Uhlenbeck process governed by the generalized Langevin equation*

$$dZ(t) = \frac{1}{2} \Delta^* Z(t) dt + dW(t), \qquad t > 0,$$

$$Z(0) = \mathcal{S}'(\mathbb{R}^d) - \text{valued Gaussian random variable with}$$
$$\text{covariance functional}$$
$$Cov(\langle Z(0), \varphi \rangle, \langle Z(0), \psi \rangle) = \langle\!\langle R_\infty, \langle \cdot, \varphi \rangle \langle \cdot, \psi \rangle \rangle\!\rangle,$$

and W is an $\mathcal{S}'(\mathbb{R}^d)$–Wiener process with covariance functional

$$Cov(\langle W(s), \varphi \rangle, \langle W(t), \psi \rangle) = \int_0^{s \wedge t} Q_u(\varphi, \psi) du,$$

where

$$Q_u(\varphi, \psi) = \langle \lambda, \varphi \psi \rangle + \langle\!\langle R_\infty, \langle \cdot, S_u \varphi \rangle \langle \cdot, S_u \psi \rangle \rangle\!\rangle + \int_0^u \langle \lambda, (S_v \varphi)(S_v \psi) \rangle dv.$$

The terms of the form $\langle\!\langle R_\infty, \langle \cdot, \varphi \rangle \langle \cdot, \psi \rangle \rangle\!\rangle$ are given by

$$\langle\!\langle R_\infty, \langle \cdot, \varphi \rangle \langle \cdot, \psi \rangle \rangle\!\rangle = \frac{\Gamma(\frac{d-2}{2})}{8\pi^{d/2}} \int_{\mathbb{R}^d} \int_{\mathbb{R}^d} \frac{\varphi(x)\psi(y)}{\| x - y \|^{d-2}} dx\, dy.$$

Background on $\mathcal{S}'$–Wiener processes and stochastic evolution equations of the type above can be found in [BG] and [BJ].

The long–time behaviour of the fluctuation limit process Z is given by the following result.

Theorem 3. *If $d > 4$, then $Z(t)$ converges weakly, as $t \to \infty$, towards $Z(\infty)$, which is an $\mathcal{S}'(\mathbb{R}^d)$–valued centered Gaussian random variable with covariance functional*

$$Cov(\langle Z(\infty), \varphi \rangle, \langle Z(\infty), \psi \rangle) = \int_{\mathbb{R}^d} \int_{\mathbb{R}^d} \varphi(x)\psi(y)[k_1(x, y) + k_2(x, y)] dx\, dy,$$

where

$$k_1(x, y) = \frac{\Gamma(\frac{d-2}{2})}{8\pi^{d/2} \| x - y \|^{d-2}}$$

and

$$k_2(x,y) = \frac{\Gamma(\frac{d-4}{2})}{32\pi^{(d-1)/2} \parallel x - y \parallel^{d-4}}.$$

For $d \leq 4, Z(t)$ *does not have a finite limit as* $t \to \infty$.

If $d > 4$, then $\overline{Y}^n(t)$ has a weak limit $\overline{Y}^n(\infty)$ as $t \to \infty$ for each n, and the fluctuation limit of $\overline{Y}^n(\infty)$ as $n \to \infty$ coincides with $Z(\infty)$, i.e., the limits $t \to \infty$ and $n \to \infty$ are interchangeable. Moreover, in this case $Z(\infty)$ is distributed as the sum of the fluctuation limit as $n \to \infty$ of the equilibrium state $B^n(\infty)$ with intensity $n\lambda$ and the long–time fluctuation limit of the occupation time of an independent (critical) super Brownian motion with intensity λ. The second part of this representation follows from Theorem 5.5 of Iscoe [I].

Theorem 3 shows that $d = 4$ is the critical dimension for the long–time behaviour of the fluctuation limit process Z. We recalled in the Introduction that $d = 4$ is also the critical dimension for the persistence/extinction dichotomy of the two–level superprocess constructed from the same system of critical branching Brownian motions. This coincidence is of course not surprising. However, it should be noted that the two critical dimension results have been obtained independently of each other and by different techniques. It is natural to ask if they can be derived from each other.

Acknowledgement. The author thanks CIMAT for hospitality. This research is partially supported by CONACyT grant 2059–E9302.

References.

[BG]   T. Bojdecki and L.G. Gorostiza (1986), *Langevin equations for S'-valued Gaussian processes and fluctuation limits of infinite particle systems*, Probab. Th. Rel. Fields 73, 227–244.

[BJ]   T. Bojdecki and J. Jakubowski (1990), *Stochastic integration for inhomogeneous Wiener process in the dual of a nuclear space*, J. Multiv. Anal. 34, 185–210.

[D]   D.A. Dawson (1993), *Measure–valued Markov processes*, in " Ecole d'Eté de Probabilités de Saint–Flour XXI–1991", Lect. Notes Math. 1541, 1–260 Springer–Verlag.

[DH]   D.A. Dawson and K.J. Hochberg (1991), *A multilevel branching model*, Adv. Appl. Probab. 23, 701–715.

[DHW]   D.A. Dawson, K.J. Hochberg and Y. Wu (1990), *Multilevel branching systems*, in "White Noise Analysis: Mathematics and Applications", World Scientific, 93–107.

[DHV1]   D.A. Dawson, K.J. Hochberg and V. Vinogradov (1991), *On path properties of super-2 processes I*, in "Measure–valued Processes, Stochastic

Partial Differential Equations, and Interacting Systems", CRM Proceedings and Lecture Notes 5, 69–82, Amer. Math. Soc.

[DHV2] D.A. Dawson, K.J. Hochberg and V. Vinogradov (1994), *On path properties of super-2 processes II*, (preprint).

[DP] D.A. Dawson and E.A. Perkins (1991), *Historical processes*, Memoirs Amer. Math. Soc. 454.

[DW] D.A. Dawson and Y. Wu (1993), *Multilevel multitype branching models of dynamical information systems*, (preprint).

[E] A. Etheridge (1993), *Limiting behaviour of two–level measure branching*, Adv. Appl. Prob. 25, 773–782.

[G] L.G. Gorostiza (1994), *Asymptotic fluctuations and critical dimension for a two–level branching system*, (preprint).

[GHW] L.G. Gorostiza, K.J. Hochberg and A. Wakolbinger (1994), *Persistence of a critical super-2 process*, J. Appl. Prob. (in press).

[GW] L.G. Gorostiza and A. Wakolbinger (1991), *Persistence criteria for a class of critical branching particle systems in continuous time*, Ann. Probab. 19, 266–288.

[H] K.J. Hochberg (1993), *Hierarchically structured branching populations with spatial motion*, Rocky Mountain J. Math., (in press).

[I] I. Iscoe (1986), *A weighted occupation time for a class of measure–valued branching processes*, Probab. Th. Rel. Fields 71, 85–116.

[W1] Y. Wu (1991), *Dynamic particle systems and multilevel measure branching processes*, Ph.D. thesis, Carleton University, Ottawa.

[W2] Y. Wu (1993), *Multilevel birth and death particle system and its continuous diffusion*, Adv. Appl. Prob. 25, 549–569.

[W3] Y. Wu (1994), *Asymptotic behaviour of two level measure branching processes*, Ann. Prob. 22, 854–874.

[W4] Y. Wu (1994), *A three level particle system and existence of general multilevel measure–valued processes*, in "Measure–Valued Processes, Stochastic Partial Differential Equations, and Interacting Systems", CRM Proceedings and Lecture Notes 5, 233–241, Amer. Math. Soc.

Non-persistence of two-level branching particle systems in low dimensions

Kenneth J. Hochberg
Department of Mathematics and Computer Science
Bar-Ilan University
52900 Ramat-Gan
Israel
hochberg@bimacs.cs.biu.ac.il

Anton Wakolbinger
Fachbereich Mathematik
J.W. Goethe-Universität
D-60054 Frankfurt am Main
Germany
wakolbin@math.uni-frankfurt.de

Abstract

We show that in dimensions $d \leq 4$, no non-trivial finite intensity equilibria of a critical two-level binary branching Brownian particle system exist. Our method relies on the analysis of backward trees: in dimensions $d \leq 4$, they are shown to exhibit an infinite clumping of mass, contradicting the existence of such equilibria.

1   Introduction

Hierarchically structured multilevel branching particle systems were first introduced by Dawson, Hochberg and Wu ([3],[5]) and have since been the object of much investigation. Such a two-level system consists of individuals or "first-level particles" undergoing some spatial motion and branching, which are, on the second level, grouped into clusters. Each of these clusters constitutes a "superparticle" which is simultaneously affected by some other, independent branching mechanism. The idea is to describe not only reproduction or death of individuals, but also replication or catastrophic elimination of whole families or clans. See [4] for several examples arising in various fields of applications.

In this paper, we consider a system of Brownian particles in $\mathbb{R}^d$ with particularly simple reproduction mechanisms – namely, critical binary branching at each level – and we show that in dimensions $d \leq 4$, these systems *do not persist*, in the sense that no nontrivial equilibria with finite first moments exist. (In light of a recent result of Bramson, Cox and Greven [1] on nonexistence of equilibria of branching Brownian particle systems in dimensions one and two, the restriction "with finite first moments" might be superfluous here; the determination of this, however, is beyond the scope of the techniques used in this paper.)

In particular, in dimensions three and four, the two-level system suffers local extinction when started from an equilibrium clan structure of a single-level critical binary branching Brownian particle system. We note that a corresponding result for "super-2 Brownian motions" – these are high-density, small-mass, short-individual-lifetime limits of the systems we consider here – has been proved by Wu [15] by completely different methods. He analyses, mainly by appropriate scaling arguments, the *forward evolution* of the system, whereas we *look back into the genealogy* and show that for $d \leq 4$, the two-level branching leads to infinite clumping of the population of relatives of a sampled individual.

Part of our argument relies on a well-known persistence criterion for branching particle systems in terms of genealogical trees, an idea that goes back to Kallenberg ([13]) and that was extended to continuous-time systems in [12]. There, non-persistence of a critical binary branching Brownian particle system in dimensions $d \leq 2$ was derived from the fact that in this case

$$\int_0^\infty \pi_t(W_t, B)dt = \infty \quad \text{a.s.,} \tag{1.1}$$

where $(W_t)_{t\geq 0}$ is a Brownian path, $(\pi_t)_{t\geq 0}$ denotes the Brownian transition probability, and B is some ball in $\mathbb{R}^d$. The left-hand side of (1.1) can be interpreted as the expected number of relatives in B of an individual encountered in an infinitely old population, given the individual's ancestral path (W_t); the fact that this expected number is infinite then excludes persistence. In the present work, we apply Kallenberg's method to our system of superparticles. The core argument for non-persistence in dimensions $d \leq 4$ now is Proposition 5.1, whose proper interpretation is that in such dimensions, the expected number of second-level relatives of an individual encountered in an infinitely old population, given the individual's level-1 genealogical tree, would have to be infinite. But this again excludes persistence, as we show in Section 4.

We note that in dimensions $d \geq 5$, we could either apply "forward-type arguments" of the sort presented in Gorostiza, Hochberg and Wakolbinger [9] or a "backward-type argument" along the line of that presented here, to show that the two-level branching Brownian system starting from an equilibrium single-level clan system *persists in the large-time limit*, i.e., converges to an equilibrium state whose intensity measure equals that of the initial state.

2   Description of the model and statement of the main result

A two-level critical branching Brownian particle system in $\mathbb{R}^d$ can be obtained as follows. First, one begins with a population $\sum_{i \in I_0} \delta_{\varphi_0^i}$ of particle clusters φ_0^i in $\mathbb{R}^d$. Each of these clusters then develops, independently of all the others, as an ordinary critical binary branching Brownian particle system with branching rate V_1, thus performing what we will call the *superparticle motion* (and what is called "family dynamics" in [14]). After an exponentially distributed time, independent of everything else and with rate V_2, a whole cluster duplicates or vanishes with probability one-half each; thus, the superparticle motion is the *expectation process* of the second-level branching system.

For a system of *superparticles* $\mathbf{X} = \sum_{i \in I} \delta_{\varphi^i}$, we call $\sum_{i \in I} \varphi^i$ the *aggregation* of $\mathbf{X}$ and $\mathbb{E} \sum_{i \in I} \varphi^i$ its *aggregated intensity measure*. If the latter is locally finite, we simply say that $\mathbf{X}$ has a *finite aggregated intensity*.

We will prove the following:

Theorem 2.1 *In dimensions $d \leq 4$, there is no nontrivial equilibrium system with finite aggregated intensity for the two-level branching dynamics.*

In dimensions $d \geq 3$, the single-level branching dynamics admits a nontrivial infinitely divisible equilibrium (see [6], [12]) which is the aggregation of a

Poisson system $\mathbf{X}_0 = \sum_{i \in I_0} \delta_{\varphi_0^i}$ of single-level *equilibrium clans*. The intensity
measure of this Poisson system (which is the canonical measure of the infinitely
divisible single-level equilibrium distribution) is invariant for the superparticle
motion. It then follows from a Laplace functional argument (cf. [12], p. 269
and [14], Prop. 2.3.1) that a critical branching superparticle system $\mathbf{X}_t$ starting from $\mathbf{X}_0$ converges in distribution to some equilibrium state $\mathbf{X}_\infty$ whose
intensity measure is bounded by that of $\mathbf{X}_0$. But by Theorem 2.1, in dimensions $d \leq 4$, $\mathbf{X}_\infty$ vanishes a.s. Subsuming, we have the following result in
dimensions three and four:

Corollary 2.2 *In dimensions $d = 3$ or 4, let $\mathbf{X}_t = \sum_{i \in I_t} \delta_{\varphi_t^i}$ be the two-
level branching particle system, started at time $t = 0$ from the Poisson system
$\mathbf{X}_0 = \sum_{i \in I_0} \delta_{\varphi_0^i}$ of single-level equilibrium clans φ_0^i, $i \in I_0$, of critical binary
branching motions in $\mathbb{R}^d$. Then, $\mathbf{X}_t$ suffers local extinction, in the sense that
for all bounded Borel sets $B \subseteq \mathbb{R}^d$, we have*

$$P\{\exists i \in I_t : \varphi_t^i(B) > 0\} \to 0 \quad as \quad t \to \infty.$$

Theorem 2.1 will be proved in three steps. In Section 3, we will show that
an individual picked from a nontrivial equilibrium population has an a.s. locally finite population of second-level relatives. In Section 4, we will show that
even the expectation of this population is locally finite, given the prehistory of
the first-level relatives of the selected individual. On the other hand, we will
show in Section 5 that the same quantity is locally infinite in dimensions less
than or equal to four.

3 The tree top in equilibrium is locally finite

Let $\mathbf{X}$ be a nontrivial equilibrium system of the two-level branching dynamics
on $\mathbb{R}^d$ with a locally finite aggregated intensity measure γ. The intensity measure $Q = \mathbb{E}\mathbf{X}$ of $\mathbf{X}$ is a measure on the clusters φ whose intensity measure
$\int \varphi(.)Q(d\varphi)$ also equals γ. Since Q is invariant with respect to the superparticle motion, its intensity measure is invariant with respect to the expectation
process of the single-level branching particle system, which is Brownian motion. This implies (cf. [8]) that

$$\gamma = c\lambda,$$

where λ denotes Lebesgue measure on $\mathbb{R}^d$ and c is a strictly positive constant,
which, without loss of generality, we set equal to one.

Let $H = (H^t)_{-\infty < t \le 0}$ denote a *historical path,* i.e. H^t is a population $\sum_{i \in I_t} \delta_{Y_i^t}$ on $C((-\infty, t], \mathbb{R}^d)$ obeying the consistency condition

$$\sum_{i \in I_t} \delta_{Y_i^t(s)} \le \sum_{i \in I_s} \delta_{Y_i^s(s)} \quad \text{for all} \quad s \le t. \tag{3.1}$$

If we interpret H^t as the population of ancestral paths of all the individuals in the cluster $\mu_t^H := \sum_{i \in I_t} \delta_{Y_i^t(t)}$, condition (3.1) is just the requirement that the ancestors at time s of those alive at time t are among all the individuals alive at time s. As in Dawson and Perkins [7], Chapter 6, where similar notions are given in the superprocess setting, one then infers the existence of a unique measure $\vec{Q}$ on the historical paths H for which the measure induced by μ_t^H equals Q for all $t \in (-\infty, 0]$.

Now, for any superparticle δ_ν and any $t > 0$, denote by $\mathbf{\Phi}_t^\nu$ the outcome of a two-level branching particle system at time t started off from δ_ν at time 0, and let $\mathbf{\Psi}_t^\nu$ be the aggregated $\mathbf{\Phi}_t^\nu$. Let $-t_1 > -t_2 > \dots$ be a Poisson process on the negative axis with rate V_2, and consider a "two-level backward tree" whose branches grow out of the trunk H at the time points $-t_n$, $n \in \mathbb{N}$, leading to the "tree top"

$$\mathbf{R}(H) := \delta_{\mu_0^H} + \sum_{n \in \mathbb{N}} \mathbf{\Phi}_{t_n}^{\mu_{-t_n}^H}.$$

The continuous-time version of Kallenberg's stability criterion (cf. [11], Proposition 2.4) applied to the branching superparticle system then leads to the following result:

Lemma 3.1 *For $\vec{Q}$-almost all H, $\mathbf{R}(H)$ is a.s. locally finite, in the sense that, with probability one, only finitely many superparticles in $\mathbf{R}(H)$ populate any bounded Borel set $B \subseteq \mathbb{R}^d$.*

The next step is to pass from the measure $\vec{Q}(dH)$ to the disintegration $\vec{Q}_Y(dH)$ of the measure $\vec{Q}(dH)H^0(dY)$ with respect to its second projection onto $C((-\infty, 0], \mathbb{R}^d)$. (Thus $\vec{Q}_Y$ is a kind of "Palm distribution," and might be imagined as a conditional distribution of the historical path, given that an individual ancestral path in the population H^0 equals Y.) Since the second projection of the measure $\vec{Q}(dH)H^0(dY)$ is the stationary Wiener measure (henceforth denoted by Λ) on $C((-\infty, 0], \mathbb{R}^d)$, the distributions Q_Y are defined by

$$\vec{Q}(dH)H^0(dY) = \Lambda(dY)\vec{Q}_Y(dH).$$

Lemma 3.1 now transforms easily into the following:

Lemma 3.2 *For Λ almost all $Y \in C((-\infty, 0], \mathbb{R}^d)$ and $\vec{Q}_Y$ almost all H, both $\mathbf{R}(H)$ and the aggregation of $\mathbf{R}(H)$ are a.s. locally finite.*

The advantage of passing from $\vec{Q}$ to $\vec{Q}_Y$ is that the latter admits a "backward tree representation" that we are going to exploit. First, let us define, for each $t > 0$, the cutting operator

$$c_{-t} : C((-\infty, 0], \mathbb{R}^d) \to C((-\infty, -t], \mathbb{R}^d), \quad (X_s)_{s \leq 0} \mapsto (X_s)_{s \leq -t}$$

and the shifts

$$\theta_t : C((-\infty, 0], \mathbb{R}^d) \to C((-\infty, 0], \mathbb{R}^d), \quad (X_s)_{s \leq 0} \mapsto (X_{-t+s})_{s \leq 0},$$

$$\theta_{-t} : C((-\infty, 0], \mathbb{R}^d) \to C((-\infty, -t], \mathbb{R}^d), \quad (X_s)_{s \leq 0} \mapsto (X_{t+s})_{s \leq -t}.$$

Let $-s_1 > -s_2 > \ldots$ be Poisson time points on the negative axis with rate V_1, and, for a fixed time $t > 0$, define $G_{Y,t}$ as the superposition of independent level-1 trees that grow out of $c_{-s_n} Y$, $0 \leq s_n \leq t$, and then develop over a time s_n. Let $H_{\theta_t Y}$ be a random historical path with distribution $\vec{Q}_{\theta_t Y}$, and put $H_{\theta_t Y}^{\text{red}} := H_{\theta_t Y} - \delta_{\theta_t Y}$. Finally, let $F_{Y,t}$ be the random historical path that grows out of $\theta_{-t}(H_{\theta_t Y}^{\text{red}})$ through the level-1 branching dynamics up to time 0. Then (cf. [2] and [14], Section 1.9), the independent superposition $\delta_Y + G_{Y,t} + F_{Y,t}$ has distribution $\vec{Q}_Y$.

In particular, letting $t \to \infty$, we infer that $G_{Y,\infty}$, viewed as a random historical path, has a distribution that is stochastically smaller than $\vec{Q}_Y$ in the following sense: on a suitable common probability space, a $\vec{Q}_Y$-distributed H_Y and a copy G_Y of $G_{Y,\infty}$ can be defined such that for all $s \leq 0$, the populations G_Y^s and H_Y^s on $C((-\infty, s], \mathbb{R}^d)$ obey

$$G_Y^s \leq H_Y^s \quad \text{a.s.} \tag{3.2}$$

Now, for fixed $y \in \mathbb{R}^d$, we define the *random backward tree* G_y as follows (cf. [12], p. 271). We first take as the "backward ancestral path" a Brownian trajectory $W = (W_s)_{s \geq 0}$ starting at y and, independently, a Poisson sequence $0 < s_1 < s_2 < \ldots$ with rate V_1. We then define G_y as the superposition of independent level-1 trees that grow out from the root $(W_{-s})_{\infty > s \geq -s_n}$ and develop over a time s_n, $n = 1, 2, \ldots$

Denoting the population in G_y at time $-t$ by $\nu_t(G_y)$, or simply by ν_t if there is no ambiguity, and noting that G_y equals $G_{Y,\infty}$ with $Y_{-s} := W_s$, $s \geq 0$, we obtain the following result by combining (3.2) with Lemma 3.2:

Proposition 3.4 *If the two-level branching dynamics on $\mathbb{R}^d$ admits a nontrivial equilibrium distribution with finite aggregated intensity, then, for all*

$y \in \mathbb{R}^d$, the *"aggregated two-level tree top"*

$$R_y := \nu_0 + \sum_{n \in \mathbb{N}} \Psi_{t_n}^{\nu_{t_n}}$$

is a.s. locally finite.

This proposition already proves our theorem for dimensions $d = 1$ and 2, since in these cases even the "first-level tree top" ν_0 is locally infinite (see [12], Theorem 2.2 and Proposition 4.1). In order to prove the theorem also for dimensions $d = 3$ and 4, we need to sharpen the assertion of the proposition. This we do in the next section.

4 The intensity of the tree top in equilibrium is locally finite

In this section, we will show, under the same assumptions as in the previous section, that not only are almost all populations R_y locally finite, but that also their expectations, given the trunk G_y and the Poisson sequence $-t_1, -t_2, ...$, are locally finite as well. To this end, we first remark that Proposition 3.4 implies, via the Borel-Cantelli lemma, that for almost all realisations G_y and Poisson sequences $-t_1, -t_2, ...$, we have

$$\sum_{n \in \mathbb{N}} P\{\Psi_{t_n}^{\nu_{t_n}}(B) > 0\} < \infty,$$

for any fixed ball B in $\mathbb{R}^d$. The desired result would thus follow if we could estimate the nonzero probabilities

$$P\{\Psi_t^{\nu_t}(B) > 0\}$$

(times a constant) from below by the expectations

$$\mathbb{E}\,\Psi_t^{\nu_t}(B) = \int \nu_t(dz)\pi_t(z, B)$$

along a fixed G_y, for suitably large t, where π_t denotes the Brownian transition kernel. To obtain such estimates, tightness of the Palm distributions $(\Psi_t^{\nu_t})_x$ would be helpful. However, rather than prove this directly, we apply a technique of Kallenberg [13] that involves randomizing independently all individual positions in the aggregated equilibrium population. For this purpose, we fix a symmetric, continuous and strictly positive probability density $\rho(x)$ that is constant on the unit ball in $\mathbb{R}^d$ and decays like an inverse power of $||x||$ as $||x|| \to \infty$.

Definition 4.1 *For a random population*

$$\Psi := \sum_{i \in I} \delta_{y_i},$$

put

$$\Psi' := \sum_{i \in I} \delta_{y_i + Z_i},$$

where the $\{Z_i\}$ *are independent, identically distributed random elements of* $\mathbb{R}^d$ *with distribution density* ρ.

Writing

$$\mathbf{X} = \sum_{j \in J} \delta_{\Psi_j}$$

for our two-level equilibrium system and applying the reasoning of Section 3 to the "shuffled system"

$$\mathbf{X}' := \sum_{j \in J} \delta_{\Psi_j'},$$

we obtain:

Lemma 4.2 *For all* $x \in \mathbb{R}^d$, *the random population*

$$\overline{R}_x := \delta_x + \left(\nu_0(G_{x+Z}) - \delta_{x+Z}\right)' + \sum_{n \in \mathbb{N}} \left(\Psi_{t_n}^{\nu_{t_n}(G_{x+Z})}\right)'$$

is a.s. locally finite, where, again, Z *stands for a random element in* $\mathbb{R}^d$ *with distribution density* ρ, *and* $-t_1 > -t_2 > \dots$ *is a Poisson sequence of time points on* $(-\infty, 0]$ *with parameter* V_2.

This implies, again by a Borel-Cantelli argument, the following:

Corollary 4.3 *For all* $y \in \mathbb{R}^d$, *and almost all realisations* $G = G_y$ *of the backward tree and Poisson time sequences* (t_n), *we have*

$$\sum_{n \in \mathbb{N}} P\{(\Psi_{t_n}^{\nu_{t_n}(G)})'(B) > 0\} < \infty.$$

We now show how to detect the Palm populations $((\Psi_t^\nu)')_x$ as part of the tree top $\overline{R}_x$, conditioned on $\nu_{t_n} = \nu$. For a fixed population

$$\nu = \sum_{j \in J} \delta_{u_j}$$

and a fixed $x \in \mathbb{R}^d$, a tree representation of

$$(\Psi_t^\nu)'_x := ((\Psi_t^\nu)')_x$$

is obtained as follows (cf. [2] and [14], Section 1.9). First, we choose an ancestor δ_{u_j} of the individual δ_x according to the weights $(p_t * \rho)(x - u_j)$, $j \in J$, where p_t denotes the ordinary Gaussian density with covariance matrix $t \cdot I$. Then, we connect $x + Z$, where Z is distributed according to the probability density ρ, to u_j via a Brownian bridge $W_{[0,t]}$. We denote by S_j the single-level tree that grows out of the trunk $W_{[0,t]}$, and by S_i, $i \neq j$, the single-level trees that grow out of the individuals δ_{u_i}, $i \neq j$. We let S denote the superposition of $\{S_i : i \in J\}$ and let $\{\tau_1, \tau_2, ... \tau_m\}$ be a Poisson configuration on $[0,t]$ with intensity V_2. Writing $\nu_r(S)$ for the population of individuals in the "forest" S at time $r \in [0,t]$, we conclude that the desired tree representation of $(\Psi_t^{\nu})_x'$ is given by

$$\delta_x + (\nu_t(S) - \delta_{x+Z})' + \sum_{n=1}^{m} (\Psi_{\tau_n}^{\nu_{t-\tau_n}(S)})'.$$

Since the distribution of S obviously coincides with that of the restriction of G_{x+Z} to the time interval $[-t, 0]$ given that $\nu_t(G_{x+Z}) = \nu$, we conclude from Lemma 4.2 that $(\Psi_t^{\nu})_x'$ is stochastically smaller than $\overline{R}_x$, given that $\nu_t(G_{x+Z}) = \nu$. In particular,

$$P\{\overline{R}_x(B) \leq l \mid \nu_t(G_{x+Z})\} \leq P\{(\Psi_t^{\nu_t(G_{x+Z})})_x'(B) \leq l\} \quad \text{a.s.} \quad (4.1)$$

By the Markov property for $(W_s)_{s \geq 0}$, the left-hand side of (4.1) equals

$$P\{\overline{R}_x(B) \leq l \mid \nu_s(G_{x+Z}), s \geq t\} \quad \text{a.s.}$$

From the martingale convergence theorem and the a.s. local finiteness of $\overline{R}_x$, we infer in the same way as in [12], Lemma 5.4, that

$$\lim_{l \to \infty} \liminf_{n \to \infty} P\{(\Psi_{t_n}^{\nu_{t_n}(G_{x+Z})})_x'(B) \leq l\} = 1 \quad \text{a.s.} \quad (4.2)$$

In a manner analogous to that found in [12], Lemma 5.5, assertion (4.2) can be used to prove the following:

Lemma 4.4 *For all $y \in \mathbb{R}^d$, and almost all realisations $G = G_y$ of the backward tree and Poisson time sequences $\{t_n\}$, there exists a constant $c > 0$ such that for all n sufficiently large,*

$$c \, P\{(\Psi_{t_n}^{\nu_{t_n}(G)})'(B) > 0\} \geq \int \nu_{t_n}(G)(dz) \pi_{t_n}(z, B).$$

Finally, combining Corollary 4.3 and Lemma 4.4, we obtain the following result:

Proposition 4.5 *If the two-level branching dynamics on $\mathbb{R}^d$ admits a nontrivial equilibrium distribution with finite aggregated intensity, then for all $y \in \mathbb{R}^d$, almost all realisations $G = G_y$ of the backward tree and Poisson time sequences (t_n), we have*

$$\sum_{n \in \mathbb{N}} \int \nu_{t_n}(G)(dz)\pi_{t_n}(z, B) < \infty.$$

Thus, under the assumptions of Proposition 4.5, it follows that

$$\int\limits_0^\infty \int \nu_t(dx)\pi_t(x, B)\, dt < \infty \text{ a.s.}, \tag{4.3}$$

i.e., the intensity measure of the aggregated two-level tree top R_y (defined in Proposition 3.4) is a.s. locally finite, given a realisation of the backward tree G_y.

5   The intensity of the tree top in low dimensions is locally infinite

In this section, we show that (4.3) fails in dimensions $d \leq 4$. This excludes, in this case, the possibility of nontrivial equilibria with finite aggregated intensity and, thus, completes the proof of Theorem 2.1.

Proposition 5.1 *In dimensions $d \leq 4$, there holds*

$$\int\limits_0^\infty \int \nu_t(dx)\pi_t(x, B)\, dt = \infty \text{ a.s.},$$

where ν_t denotes the population at time $-t$ in the backward tree G_0 (as defined in Section 3) of an individual at the origin at time 0, and B is the unit ball in $\mathbb{R}^d$.

Before describing all the details of the argument, let us outline the idea of the proof. Writing

$$\eta(dt, dx) = \nu_t(dx)\, dt$$

for the occupation measure of G_0 on $(-\infty, 0] \times \mathbb{R}^d$, we have to show that the "weighted occupation time"

$$\langle \eta, f \rangle = \int\limits_{(-\infty,0]\times\mathbb{R}^d} f(t,x)\eta(dt,dx)$$

is a.s. infinite, where

$$f(t,x) := \pi_{-t}(x, B), \quad t \leq 0, \ x \in \mathbb{R}^d.$$

The decomposition of G_0 into the trunk W and the sidetrees $N^1, N^2, ...$, originating in the space-time points $(-s_n, W_{s_n})$ leads to a decomposition

$$\eta = \eta_0 + \eta_1 + \eta_2 + \cdots,$$

where $\eta_0(dt, dx) = \delta_{W_{-t}}(dx)\,dt$ is the occupation measure of the trunk W, and η_n is the occupation measure of the sidetree N^n, living on $[-s_n, 0] \times \mathbb{R}^d, n = 1, 2, ...$

We first note that $\langle \eta_0, f \rangle = \infty$ a.s. in dimensions $d = 1$ and 2 (see [12], Prop. 4.4). Here, we are going to show that also in dimensions $d = 3$ and 4, the weighted occupation time $\langle \eta, f \rangle$ is a.s. infinite, due to the contributions of the $\langle \eta_n, f \rangle$, $n = 1, 2, ...$ Specifically, Lemma 5.3 implies that given that the ancestor of the sidetree N^n is not too far from the origin, its contribution $\langle \eta_n, f \rangle$ to the total weighted occupation time can be suitably estimated from below. On the other hand, Lemma 5.2 serves to show that suitably many branching points along the trunk are indeed not too far from the origin.

Lemma 5.2 *For almost all standard Brownian paths (W_t), we have*

$$\int\limits_{1}^{\infty} \frac{1}{t} \, 1_{\{\|W_t\| \leq \sqrt{t}\}} \, dt = \infty.$$

Proof: We use the following fact: if $X(t)$ is a standard Ornstein-Uhlenbeck process, then

$$B(t) := \sqrt{t} X(\frac{1}{2} \log t)$$

is a standard Brownian motion.

Starting from independent Ornstein-Uhlenbeck processes $X_1, ..., X_d$, we put

$$W_t := \Big(B_1(t), ..., B_d(t)\Big).$$

If $B_i^2(t) \leq \frac{t}{d}$ for $i = 1, \ldots, d$, then $\|W_t\| \leq \sqrt{t}$; hence, if $\|X_i\left(\frac{1}{2}\log t\right)\| \leq \frac{1}{\sqrt{d}}$ for $i = 1, \ldots, d$, then $\|W_t\| \leq \sqrt{t}$.

For fixed paths $X_1, \ldots X_d$, let

$$S_{(t)} := \left\{ t \geq 0 \ : \ \|X_i(t)\| \leq \frac{1}{\sqrt{d}} \text{ for } i = 1, \ldots, d \right\}.$$

Then

$$\int_1^\infty \frac{1}{t} 1_{\{\|W_t\| \leq \sqrt{t}\}} \, dt \geq \int_1^\infty \frac{1}{t} 1_{S_{\left(\frac{1}{2}\log t\right)}} \, dt = 2 \int_0^\infty 1_{S_{(r)}} \, dr.$$

But the latter integral is infinite due to the ergodicity of the Ornstein-Uhlenbeck processes. $\square$

Lemma 5.3 *Let* $(N_t)_{t \geq 0}$ *be a critical binary branching Brownian population starting from one ancestor at the origin, and define*

$$Z_T := \frac{k}{T^2} \int_0^{\frac{T}{2}} N_t \left(B_{\sqrt{T-t}} \right) \, dt, \quad T \geq 1,$$

where k *is a positive constant independent of* T, *and* B_r *denotes the ball with radius* r *centered at the origin. Then there holds*

$$w(T) := -\log \mathbb{E} \, e^{-Z_T} \geq \frac{c}{T},$$

for some positive constant c *independent of* T.

Proof: It follows from known properties of the weighted occupation time of branching particle systems (cf. [10]) that

$$w(T) = -\log\left(1 - u_T(0, \frac{T}{2})\right),$$

where $u_T(x, t)$ solves the evolution equation

$$\frac{\partial u_T}{\partial t}(x, t) = \frac{1}{2}\left(\Delta u_T - u_T^2\right)(x, t) + \frac{k}{T^2} 1_{B_{\sqrt{T-\left(\frac{T}{2}-t\right)}}}(x)(1 - u_T(x, t)),$$

$$0 \leq t \leq \frac{T}{2},$$

$$u_T(x, 0) = 0.$$

Using the transformation

$$y\sqrt{T} = x, \quad sT = t, \quad v_T(y, s) := u_T(x, t),$$

we obtain

$$\frac{\partial v_T}{\partial s}(y,s) = T\frac{\partial u_T}{\partial t}(x,t)$$

$$= \frac{T}{2}(\Delta u_T - u_T^2)(x,t) + \frac{k}{T}1_{B\sqrt{\frac{T}{2}+t}}(x)\Big(1 - u_T(x,t)\Big)$$

$$= \frac{1}{2}\Big(\Delta v_T(y,s) - Tv_T^2(y,s)\Big) + \frac{k}{T}1_{B\sqrt{\frac{1}{2}+s}}(y)\Big(1 - v_T(y,s)\Big).$$

Therefore, $\tilde{v}_T(y,s) := Tv_T(y,s)$ obeys

$$\frac{\partial \tilde{v}_T}{\partial s}(y,s) = \frac{1}{2}(\Delta \tilde{v}_T - \tilde{v}_T^2)(y,s) + k1_{B\sqrt{\frac{1}{2}+s}}(y)\Big(1 - \frac{1}{T}\tilde{v}_T(y,s)\Big),$$

$$0 \le s \le \frac{1}{2},$$

$$\tilde{v}_T(y,0) = 0.$$

By a comparison argument, $\tilde{v}_T(y,s)$ is bounded from below by the solution $q(y,s)$ of

$$\frac{\partial q}{\partial s}(y,s) = \frac{1}{2}(\Delta q - q^2)(y,s) + k1_{B\sqrt{\frac{1}{2}+s}}(y)\Big(1 - q(y,s)\Big),$$

$$q(y,0) = 0.$$

Therefore,

$$u_T(0, \frac{T}{2}) = v_T(0, \frac{1}{2}) = \frac{1}{T}\tilde{v}_T(0, \frac{1}{2}) \ge \frac{1}{T}q(0, \frac{1}{2}).$$

Since

$$0 < u_T(0, \frac{T}{2}) < 1$$

and

$$-\log(1-p) \ge p \quad \text{for all } p \in (0,1),$$

we have

$$w(T) \ge \frac{1}{T}\,q(0, \frac{1}{2}) := \frac{c}{T}. \qquad \square$$

Proof of Proposition 5.1: We first note that, for fixed realisations W and (s_i) and all $n = 1, 2, ...$, the weight function $f(t,x) = \pi_{-t}(x, B)$ can be estimated from below, with a suitable positive constant k (not depending on W, (s_i) and n), by $k/s_n^{d/2} \ge k/s_n^2$ on

$$\Big\{(t,x) \ : \ -s_n \le t \le -s_n/2, \|W_{s_n} - x\| \le \sqrt{s_n + t}\Big\},$$

provided that $\|W_{s_n}\| \le \sqrt{s_n}$ and $s_n \ge 1$. Therefore, it follows from Lemma 5.3 that there exists a positive constant c, not depending on W, (s_i) and n, such that

$$\mathbb{E}\left[\exp\Big\{-\langle \eta_n, f\rangle 1_{\{\|W_{s_n}\|\le\sqrt{s_n}\}}\Big\} \mid W, (s_i)\right]$$

$$\leq \quad \exp\left\{-\frac{c}{s_n}1_{\{\|W_{s_n}\|\leq\sqrt{s_n}\}} \, 1_{\{s_n\geq 1\}}\right\}.$$

From this we conclude, recalling that (s_n) is a Poisson point sequence with intensity V_1, that

$$\mathbb{E}\,e^{-\langle\eta,f\rangle} \quad \leq \quad \mathbb{E}\,\mathbb{E}\left[\exp\left\{-\sum_{n:s_n\geq 1}\frac{c}{s_n}1_{\{\|W_s\|\leq\sqrt{s_n}\}}\right\}\,\Big|\,W\right]$$

$$= \quad \mathbb{E}\exp\left\{-\int_1^\infty V_1\left(1-e^{-\frac{c}{t}}\right)1_{\{\|W_t\|\leq\sqrt{t}\}}\,dt\right\}$$

$$\leq \quad \mathbb{E}\exp\left\{-V_1e^{-c}c\int_1^\infty\frac{1}{t}1_{\{\|W_t\|\leq\sqrt{t}\}}\,dt\right\}$$

$$= \quad 0,$$

the last equality resulting from Lemma 5.2. Hence $\langle\eta,f\rangle \equiv \infty$ a.s.   □

Remark 5.4 In the case of symmetric stable motion with parameter α, $0 < \alpha < 2$, and branching on the i-th level, $i = 1,2$, that is in the domain of normal attraction of a $(1+\beta_i)$-stable law with $0 < \beta_i < 1$, appropriate adjustments to the computations of this section lead to the conclusion that the critical dimension for persistence versus local extinction is

$$d = \alpha(1/\beta_1 + 1/\beta_2).$$

Acknowledgements We thank Alison Etheridge for an essential contribution to the proof of Proposition 5.1, and José Alfredo López-Mimbela for stimulating discussions. A.W. acknowledges the support and hospitality of Bar-Ilan University (Ramat-Gan) and of the ICMS (Edinburgh), where part of this research was done. K.J.H. acknowledges the hospitality of J. W. Goethe Universität (Frankfurt), the support of a grant from the Bar-Ilan University Research Authority, and the travel support of the Minerva Foundation in Germany through the Emmy Noether Research Institute for the Mathematical Sciences. Finally, the remarks of a careful referee helped us to improve the exposition.

References

[1] M. Bramson, J.T. Cox, A. Greven (1993), *Ergodicity of critical spatial branching processes in low dimensions*, Ann. Probab. 21, 248-289.

[2] B. Chauvin, A. Rouault and A. Wakolbinger (1991), *Growing conditioned trees*, Stochastic Processes Appl. 39, 117-130.

[3] D.A. Dawson and K.J. Hochberg (1991), *A multilevel branching model*, Adv. Appl. Probab. 23, 701-715.

[4] D.A. Dawson, K.J. Hochberg, and V. Vinogradov (1994), *On path properties of super-2 processes II*, to appear in Symposia in Pure and Applied Mathematics, American Mathematical Society.

[5] D.A. Dawson, K.J. Hochberg, and Y. Wu (1990), *Multilevel branching systems*, in "White Noise Analysis: Mathematics and Applications," World Scientific, 93-107.

[6] D.A. Dawson and G. Ivanoff (1978), *Branching diffusions and random measures*, in Branching Processes, A. Joffe and P. Ney, eds., Dekker, New York, 61-103.

[7] D.A. Dawson and E. A. Perkins (1991), *Historical Processes*, Memoirs of the American Mathematical Society, no. 454.

[8] J. Deny (1960), *Sur l'équation de convolution $\mu = \mu * \sigma$*. Séminaire Brelot-Choquet-Deny (Théorie du Potentiel), 4e année, no 5.

[9] L.G. Gorostiza, K.J. Hochberg and A. Wakolbinger (1994), *Persistence of a critical super-2 process*, to appear in J. Appl. Probab.

[10] L.G. Gorostiza and J.A. López-Mimbela (1994), *An occupation time approach for convergence of measure-valued processes and the dead mass of a branching system*, to appear in Statistics and Probability Letters.

[11] L. Gorostiza, S. Roelly and A. Wakolbinger (1992), *Persistence of critical multitype particle and measure branching processes*, Probab. Theory Relat. Fields 92, 313-335.

[12] L.G. Gorostiza and A. Wakolbinger (1991), *Persistence criteria for a class of critical branching particle systems in continuous time*, Ann. Probab. 19, 266-288.

[13] O. Kallenberg (1977), *Stability of critical cluster fields*, Math. Nachr. 77, 7-43.

[14] A. Liemant, K. Matthes and A. Wakolbinger (1988), *Equilibrium distributions of branching processes*, Akademie-Verlag, Berlin, and Academic Publishers, Dordrecht.

[15] Y. Wu (1994), *Asymptotic behavior of two level measure branching processes*, Ann. Probab., in press.

THE STOCHASTIC WICK-TYPE BURGERS EQUATION

H. Holden[1], T. Lindstrøm[2], B. Øksendal[3], J. Ubøe[4] and T.-S. Zhang[4]

Abstract

We study the multidimensional stochastic (Wick-type) Burgers equation

$$\begin{cases} \frac{\partial u_k}{\partial t} + \lambda \sum_{j=1}^{n} u_j \diamond \frac{\partial u_k}{\partial x_j} = \nu \Delta u_k + w_k(t,x) \; ; \; t > 0, x \in \mathbf{R}^n \\ u_k(0,x) = g_k(x) \; ; \; 1 \le k \le n \end{cases}$$

where $\diamond$ denotes the Wick product, λ and ν are constants ($\nu > 0$, $\lambda \ne 0$), Δ denotes the Laplacian and $\{w_k(t,x)\}_{k=1}^{n}$ are $(n+1)$-parameter stochastic processes (noise). We prove an existence and uniqueness result for the solution $u(t,x) = \{u_k(t,x)\}_{k=1}^{n}$, regarded as an $(n+1)$-parameter stochastic process with values in the Kondratiev space $(\mathcal{S})^{-1}$ of stochastic distributions.

CONTENTS

[1] Dept. of Mathematical Sciences, Norwegian Institute of Technology, University of
Trondheim, N-7034 Trondheim, Norway

[2] Dept. of Mathematics, University of Oslo, Box 1053, Blindern, N-0316 Oslo,
Norway

[3] Dept. of Mathematics/VISTA, University of Oslo, Box 1053, Blindern, N-0316 Oslo,
Norway

[4] Dept. of Mathematics, National College of Safety Engineering, Skåregt. 103,
N-5500 Haugesund, Norway.

141

§0. INTRODUCTION

The purpose of this paper is to continue the work done in [HLØUZ 2] regarding the stochastic multidimensional Burgers equation in $\{u_k(t,x)\}_{k=1}^n$:

$$(0.1) \qquad \begin{cases} \frac{\partial u_k}{\partial t} + \lambda \sum_{j=1}^{n} u_j \diamond \frac{\partial u_k}{\partial x_j} = \nu \Delta u_k + w_k(t,x); \ t > 0, \ x \in \mathbf{R}^n \\ u_k(0,x) = g_k(x); \ 1 \leq k \leq n \end{cases}$$

where $\diamond$ denotes the Wick product, λ and ν are constants ($\nu > 0$, $\lambda \neq 0$), Δ denotes the Laplacian and $w_k(t,x) = w_k(t,x)$ are $(n+1)$-parameter generalized stochastic processes; $1 \leq k \leq n$. (See details below). The use of the Wick product corresponds to an Ito/Skorohod interpretation of the equation (see e.g. [B1], [LØU 2] and also [HLØUZ 3]). We may regard $u = (u_1, \cdots, u_n)$ as the velocity field of a vorticity free fluid with viscosity ν, being exposed to the stochastic force $w = (w_1, \cdots, w_n)$.

In [HLØUZ 2] the following was proved: (For definition of functional processes etc. see §1). (Here - and in the following - all gradients are taken w.r.t. x).

A. Let $N = N(\phi, t, x, \omega)$ be a functional process and define $w = -\nabla N$. Assume that (0.1) has a solution of the form

$$u = -\nabla X$$

for some functional process $X = X(\phi, t, x, \omega)$. Moreover, assume that the functional processes

$$(0.2) \qquad Y := \mathrm{Exp}(\frac{\lambda}{2\nu} X)$$

as well as $Y \diamond X$ and $Y \diamond X^{\diamond 2}$ exist in $L^p(\mu)$ for some $p \geq 1$. Then Y solves the stochastic heat equation

$$(0.3) \qquad \begin{cases} \frac{\partial Y}{\partial t} = \nu \Delta Y + \frac{\lambda}{2\nu} Y \diamond (N + C); \quad t > 0, \ x \in \mathbf{R}^n \\ Y(0,x) = \mathrm{Exp}(\frac{\lambda}{2\nu} X(0,x)) \end{cases}$$

for some $C(t, \omega)$ not depending on x. (For simplicity we have written $Y(t,x)$ for $Y(\phi, x, x, \omega)$ etc.)

Furthermore, we proved the following:

B. Let either $N = W(\phi, t, x, \omega)$ ($(n+1)$-parameter white noise) or $N = \mathrm{Exp}W(\phi, t, x, \omega)$ ($(n+1)$-parameter positive noise), and assume that both

$f(x) := Y(0, x)$ and $C(t, \omega) = C(t)$ are bounded and deterministic (do not depend on ω).

Then (0.3) has a unique $L^2(\mu)$ (respectively $L^1(\mu)$) functional process solution $Y(t, x)$ given by

$$(0.4) \qquad Y(t, x) = \hat{E}^x[f(b_{\alpha t}) \text{Exp}(\int_0^t H(s, b_{\alpha s}) ds)],$$

where $\alpha = \sqrt{2\nu}, (b_t, \hat{P}^x)$ is standard Brownian motion in $\mathbf{R}^n$ ($\hat{E}^x$ denotes expectation w.r.t. $\hat{P}^x$) and

$$(0.5) \qquad H(t, x) = H(t, x, \omega) = \frac{\lambda}{2\nu}(N(\phi, t, x, \omega) + C(t)).$$

Thus we see that if u is a gradient solution of (0.1) then we can use it to construct an explicit solution of the stochastic heat equation (0.3). Moreover, by other methods the unique solution of (0.3) can be found explicitly. In particular, this proves that under the given assumptions there exists at most one functional process solution (of gradient form) of the stochastic Burgers equation (0.1).

The purpose of the present paper is to complete the analysis by proving that one can reverse the Wick-Cole-Hopf transformation (0.2) to construct a solution of the stochastic Burgers equation from the stochastic heat equation and hence obtain a uniqueness *and* existence result for equation (0.1). In order to accomplish this we consider processes and operations in the *Kondratiev space* $(\mathcal{S})^{-1}$ *of stochastic distributions*. This space has already found several applications in stochastic partial (and ordinary) differential equations. See e.g. [B 2], [HLØUZ 4] and [Ø]. For the stochastic Burgers equation this approach has the following advantages:

a) The transformation from the stochastic Burgers equation to the stochastic heat equation can be performed with fewer assumptions than given above if done in the space $(\mathcal{S})^{-1}$. (See Theorem 2.1).

b) The stochastic heat equation can be solved explicitly in $(\mathcal{S})^{-1}$ for a general $(\mathcal{S})^{-1}$ potential (Theorem 3.1).

c) Most importantly, in $(\mathcal{S})^{-1}$ one can also construct the *converse* transformation from the stochastic heat equation to the stochastic Burgers equation. (See Theorem 4.1).

By combining a), b) and c) we obtain a uniqueness and existence result (Theorem 5.1) for the stochastic Burgers equation (0.1). Moreover, we obtain this under weaker assumptions than what was needed for the uniqueness result in [HLØUZ 2].

One-dimensional Burgers equations with ordinary product instead of Wick product have been studied in [BCJ-L], [DDT] and [DG].

§1. WHITE NOISE, WICK PRODUCTS AND STOCHASTIC DISTRIBUTIONS

Here we briefly recall some of the basic definitions and results that we need from white noise calculus. For more information the reader is referred to [HKPS] and [KLS].

In the following we fix the parameter dimension d and let $\mathcal{S} = \mathcal{S}(\mathbf{R}^d)$ denote the Schwartz space of rapidly decreasing smooth (C^∞) functions on $\mathbf{R}^d$. The dual $\mathcal{S}' = \mathcal{S}'(\mathbf{R}^d)$ is the space of tempered distributions. By the Bochner-Minlos theorem [GV] there exists a probability measure μ on the Borel subsets $\mathcal{B}$ of $\mathcal{S}'$ with the property that

$$(1.1) \qquad \int_{\mathcal{S}'} e^{i\langle \omega, \phi \rangle} d\mu(\omega) = e^{-\frac{1}{2}\|\phi\|^2}; \forall \phi \in \mathcal{S}$$

where $\langle \omega, \phi \rangle$ denotes the action of $\omega \in \mathcal{S}'$ on $\phi \in \mathcal{S}$ and $\|\phi\|^2 = \int_{\mathbf{R}^d} |\phi(x)|^2 dx$. The triple $(\mathcal{S}', \mathcal{B}, \mu)$ is called *the white noise probability space*.

The white noise process is the map $W : \mathcal{S} \times \mathcal{S}' \to \mathbf{R}$ defined by

$$(1.2) \qquad W(\phi, \omega) = W_\phi(\omega) = \langle \omega, \phi \rangle; \quad \omega \in \mathcal{S}', \phi \in \mathcal{S}$$

Expressed in terms of Ito integrals with respect to d-parameter Brownian motion B we have

$$(1.3) \qquad W_\phi(\omega) = \int_{\mathbf{R}^d} \phi(x) dB_x(\omega) \quad ; \phi \in \mathcal{S}.$$

The *Hermite polynomials* are defined by

$$(1.4) \qquad h_n(x) = (-1)^n e^{\frac{x^2}{2}} \frac{d^n}{dx^n}(e^{-\frac{x^2}{2}}); \quad n = 0, 1, 2, \cdots$$

and the *Hermite functions* are defined by

$$(1.5) \qquad \xi_n(x) = \pi^{-\frac{1}{4}}((n-1)!)^{-\frac{1}{2}} e^{-\frac{x^2}{2}} h_{n-1}(\sqrt{2}x) \quad ; \quad n \geq 1$$

In the following we let $\{e_1, e_2, \cdots\} \subset \mathcal{S}$ denote a fixed orthonormal basis for $L^2(\mathbf{R}^d)$. For many purposes the basis can be arbitrary, but for us it is convenient to assume that the e_n's are obtained by taking tensor products of $\xi_k(x)$. Define

$$(1.6) \qquad \theta_j(\omega) := W_{e_j}(\omega) = \int_{\mathbf{R}^d} e_j(x) dB_x(\omega) \quad ; \quad j = 1, 2, \cdots$$

If $\alpha = (\alpha_1, \cdots, \alpha_m)$ is a multi-index of non-negative integers we put

$$(1.7) \qquad H_\alpha(\omega) = \prod_{j=1}^m h_{\alpha_j}(\theta_j)$$

The *Wiener-Ito chaos expansion theorem* says that any $X \in L^2(\mu)$ can be (uniquely) written

$$(1.8) \qquad X(\omega) = \sum_\alpha c_\alpha H_\alpha(\omega)$$

Moreover,

$$(1.9) \qquad \|X\|_{L^2(\mu)}^2 = \sum_\alpha \alpha! c_\alpha^2 \quad \text{where} \quad \alpha! = \alpha_1! \alpha_2! \cdots \alpha_m!$$

The Hida test function space $(\mathcal{S})$ and the Hida distribution space $(\mathcal{S})^*$ can be given the following characterization, due to T.-S. Zhang [Z]:

THEOREM 1.1 ([Z])

Part a): A function $f = \sum_\alpha c_\alpha H_\alpha \in L^2(\mu)$ belongs to $(\mathcal{S})$ if and only if

$$(1.10) \qquad \sup_\alpha c_\alpha^2 \alpha! (2\mathbf{N})^{\alpha k} < \infty \quad \forall k < \infty$$

where

$$(1.11) \qquad (2\mathbf{N})^\alpha := \prod_{j=1}^m (2^d \beta_1^{(j)} \cdots \beta_d^{(j)})^{\alpha_j} \quad \text{if } \alpha = (\alpha_1, \cdots, \alpha_m)$$

Here $\beta^{(j)} = (\beta_1^{(j)}, \cdots, \beta_d^{(j)})$ is multi-index nr. j in the fixed ordering of all d-dimensional multi-indices $\beta = (\beta_1, \cdots, \beta_d)$, related to the basis $\{e_j\}$ by

$$(1.12) \qquad e_j = \xi_{\beta_1^{(j)}} \otimes \cdots \otimes \xi_{\beta_d^{(j)}}.$$

Part b): A formal series $F = \sum_\alpha b_\alpha H_\alpha$ belongs to $(\mathcal{S})^*$ if and only if

$$(1.13) \qquad \sup_\alpha b_\alpha^2 \alpha! (2\mathbf{N})^{-\alpha q} < \infty \quad \text{for some} \quad q < \infty$$

The action of $F = \sum_\alpha b_\alpha H_\alpha \in (\mathcal{S})^*$ on $f = \sum_\alpha c_\alpha H_\alpha \in (\mathcal{S})$ is given by

$$(1.14) \qquad \langle F, f \rangle = \sum_\alpha \alpha! b_\alpha c_\alpha$$

EXAMPLE The pointwise (or singular) white noise W_x is defined by

$$(1.15) \qquad W_x(\omega) = \sum_{k=1}^\infty e_k(x) H_{\epsilon_k}(\omega) = \sum_{k=1}^\infty e_k(x) h_1(\theta_k)$$

where $\epsilon_k = (0, 0, \cdots, 0, 1)$ with 1 on k'th place.

In this case
$$b_\alpha = b_{\epsilon_k} = e_k(x) \text{ if } \alpha = \epsilon_k \text{ for some } k$$
$$b_\alpha = 0 \text{ if } \alpha \neq \epsilon_k \text{ for all } k$$

Moreover, if $\alpha = \epsilon_k$ we have

$$(2\mathbf{N})^\alpha = 2^d \beta_1^{(k)} \cdots \beta_d^{(k)}$$

So in this case we get

$$\sup_\alpha b_\alpha^2 \alpha! (2\mathbf{N})^{-\alpha q} = \sup_k e_k^2(x) (2^d \beta_1^{(k)} \cdots \beta_d^{(k)})^{-q} < \infty$$

for all $q > 0$, since

$$\sup_{t \in \mathbf{R}} |\xi_k(t)| = 0(k^{-\frac{1}{12}}) \quad ([\text{HiP}])$$

We conclude that $W_x(\omega) \in (\mathcal{S})^*$.

Note that if $1 < p < \infty$ we have

$$(1.16) \qquad (\mathcal{S}) \subset L^p(\mu) \subset (\mathcal{S})^*$$

However,

$$(1.17) \qquad L^1(\mu) \not\subset (\mathcal{S})^* \quad \text{(see e.g. [HLØUZ 1])}$$

For our purposes it turns out to be convenient to work with the *Kondratiev* spaces $(\mathcal{S})^1$ and $(\mathcal{S})^{-1}$ which are related to $(\mathcal{S})$ and $(\mathcal{S})^*$ as follows:

$$(1.18) \qquad (\mathcal{S})^1 \subset (\mathcal{S}) \subset (\mathcal{S})^* \subset (\mathcal{S})^{-1}$$

The spaces $(\mathcal{S})$ and $(\mathcal{S})^{-1}$ were originally constructed on spaces of sequences by Kondratiev [K] and later extended by him and several other authors. See [KLS] and the references there. We recall here their basic properties, stated in forms which are convenient for our purposes. For details and proofs we refer to [KLS].

DEFINITION 1.2 [KLS]

Part a): For $0 \le \rho \le 1$ let $(\mathcal{S})^\rho$ (the *Kondratiev space of stochastic test functions*) consist of all

$$f = \sum_\alpha c_\alpha H_\alpha \in L^2(\mu) \quad \text{such that}$$

(1.19) $\qquad \|f\|_{\rho,k}^2 := \sum_\alpha c_\alpha^2 (\alpha!)^{1+\rho}(2\mathbf{N})^{\alpha k} < \infty \quad \text{for all } k < \infty$

Part b): The *Kondratiev space of stochastic distributions*, $(\mathcal{S})^{-\rho}$, consists of all formal expansions
$$F = \sum_\alpha b_\alpha H_\alpha$$

such that

(1.20) $\qquad \sum_\alpha b_\alpha^2 (\alpha!)^{1-\rho}(2\mathbf{N})^{-\alpha q} < \infty \quad \text{for some } q < \infty$

The family of seminorms $\|f\|_{\rho,k}^2$; $k = 1,2,\cdots$ gives rise to a topology on $(\mathcal{S})^\rho$ and we can then regard $(\mathcal{S})^{-\rho}$ as the dual of $(\mathcal{S})^\rho$ by the action

(1.21) $\qquad \langle F, f \rangle = \sum_\alpha b_\alpha c_\alpha \alpha!$

if $F = \sum b_\alpha H_\alpha \in (\mathcal{S})^{-\rho}$ and $f = \sum c_\alpha H_\alpha \in (\mathcal{S})^\rho$.

REMARKS.

1) Regarding (1.21), note that

$$\sum_\alpha |b_\alpha c_\alpha| \alpha! = \sum_\alpha |b_\alpha c_\alpha| (\alpha!)^{\frac{1-\rho}{2}} (\alpha!)^{\frac{1+\rho}{2}} \cdot (2\mathbf{N})^{\frac{\alpha k}{2}} (2\mathbf{N})^{-\frac{\alpha k}{2}}$$
$$\le [\sum_\alpha b_\alpha^2 (\alpha!)^{1-\rho}(2\mathbf{N})^{-\alpha k}]^{\frac{1}{2}} \cdot [\sum_\alpha c_\alpha^2 (\alpha!)^{1+\rho}(2\mathbf{N})^{\alpha k}]^{\frac{1}{2}}$$
$$< \infty \quad \text{for } k \text{ large enough.}$$

2) Putting $\rho = 0$ we see by comparing (1.19), (1.20) with (1.10), (1.13) that $(\mathcal{S}) = (\mathcal{S})^0$ and $(\mathcal{S})^* = (\mathcal{S})^{-0}$. So for general $\rho \in [0,1]$ we have

$$(1.22) \quad (\mathcal{S})^1 \subset (\mathcal{S})^\rho \subset (\mathcal{S})^0 = (\mathcal{S}) \subset (\mathcal{S})^* = (\mathcal{S})^{-0} \subset (\mathcal{S})^{-\rho} \subset (\mathcal{S})^{-1}$$

(Observe that with this notation $(\mathcal{S})^0$ and $(\mathcal{S})^{-0}$ are different spaces).

DEFINITION 1.3

The Wick product $F \diamond G$ *of two elements*

$$F = \sum_\alpha a_\alpha H_\alpha, G = \sum_\beta b_\beta H_\beta \text{ in } (\mathcal{S})^{-1} \text{ is defined by}$$

(1.23)
$$F \diamond G = \sum_{\alpha,\beta} a_\alpha b_\beta H_{\alpha+\beta}$$

From Def. 1.2 we get

LEMMA 1.4

(i) $F, G \in (\mathcal{S})^{-1} \Rightarrow F \diamond G \in (\mathcal{S})^{-1}$
(ii) $f, g \in (\mathcal{S})^1 \Rightarrow f \diamond g \in (\mathcal{S})^1$

The *Hermite transform* [LØU 1-3] has a natural extension to $(\mathcal{S})^{-1}$:

DEFINITION 1.5 If $F = \sum_\alpha b_\alpha H_\alpha \in (\mathcal{S})^{-1}$ then the Hermite transform of $F, \mathcal{H}F = \tilde{F}$, is defined by

$$(1.24) \qquad \tilde{F}(z) = \mathcal{H}F(z) = \sum_\alpha b_\alpha z^\alpha \quad \text{(whenever convergent)}$$

where $z = (z_1, z_2, \cdots) \in \mathbf{C}^{\mathbf{N}}$ (the space of all sequences of complex numbers) and

$$z^\alpha = z_1^{\alpha_1} z_2^{\alpha_2} \cdots z_m^{\alpha_m} \text{ if } \alpha = (\alpha_1, \cdots, \alpha_m).$$

If $F \in (\mathcal{S})^{-\rho}$ for $\rho < 1$ then it is easy to see that $(\mathcal{H}F)(z_1, z_2, \cdots)$ converges for all finite sequences $(z_1, \cdots, z_m)$ of complex numbers.

If $F \in (\mathcal{S})^{-1}$, however, we can only obtain convergence of $\mathcal{H}F(z_1, z_2, \cdots)$ in a neighbourhood of the origin: We have

$$\sum_\alpha |b_\alpha||z^\alpha| \le [\sum_\alpha b_\alpha^2 (2\mathbf{N})^{-\alpha q}]^{\frac{1}{2}} \cdot [\sum_\alpha |z^\alpha|^2 (2\mathbf{N})^{\alpha q}]^{\frac{1}{2}},$$

where the first factor on the right hand side converges for q large enough. For such a value of q we have convergence of the second factor if

$$(1.25) \qquad z \in \mathbf{B}_q(\delta) := \{\zeta = (\zeta_1, \zeta_2, \cdots) \in \mathbf{C}^{\mathbf{N}} \; ; \; \sum_{\alpha \ne 0} |\zeta^\alpha|^2 (2\mathbf{N})^{\alpha q} < \delta^2\}$$

for some $\delta < \infty$.

The next result is an immediate consequence of Def. 1.3 and Def. 1.5:

LEMMA 1.6 If $F, G \in (\mathcal{S})^{-1}$ then

$$\mathcal{H}(F \diamond G)(z) = \mathcal{H}F(z) \cdot \mathcal{H}G(z)$$

for all z such that $\mathcal{H}F(z)$ and $\mathcal{H}G(z)$ exist.

The topology on $(\mathcal{S})^{-1}$ can conveniently be expressed in terms of Hermite transforms as follows:

LEMMA 1.7 [KLS]

The following are equivalent
(i) $X_n \to X$ in $(\mathcal{S})^{-1}$
(ii) $\exists \delta > 0, q < \infty, M < \infty$ such that

$$\mathcal{H}X_n(z) \to \mathcal{H}X(z) \quad \text{as } n \to \infty \quad \text{for } z \in \mathbf{B}_q(\delta)$$

and

$$|\mathcal{H}X_n(z)| \leq M \quad \text{for all } n = 1, 2, \cdots; z \in \mathbf{B}_q(\delta).$$

THEOREM 1.8 [KLS] (Characterization theorem for $(\mathcal{S})^{-1}$)

Suppose $g(z_1, z_2, \cdots)$ is a bounded analytic function on $\mathbf{B}_q(\delta)$ for some $\delta > 0, q < \infty$. Then there exists $X \in (\mathcal{S})^{-1}$ such that

$$\mathcal{H}X = g$$

From this we deduce the following useful result:

COROLLARY 1.9

Suppose $g = \mathcal{H}X$ for some $X \in (\mathcal{S})^{-1}$. Let f be an analytic function in a neighbourhood of $\zeta_0 = g(0)$ in $\mathbf{C}$. Then there exists $Y \in (\mathcal{S})^{-1}$ such that

$$\mathcal{H}Y = f \circ g$$

Proof. Let $r > 0$ be such that f is bounded analytic on $\{\zeta \in \mathbf{C}; |\zeta - \zeta_0| < r\}$. Then choose $\delta > 0$ and $q < \infty$ such that the function $z \to g(z)$ is bounded analytic on $\mathbf{B}_q(\delta)$ and such that $|g(z) - \zeta_0| < r$ for $z \in \mathbf{B}_q(\delta)$. Then $f \circ g$ is bounded analytic in $\mathbf{B}_q(\delta)$, so the result follows from Theorem 1.8.

EXAMPLE 1.10

a) Let $X \in (\mathcal{S})^{-1}$. Then $X \diamond X = X^{\diamond 2} \in (\mathcal{S})^{-1}$ and more generally $X^{\diamond n} \in (\mathcal{S})^{-1}$ for all natural numbers n. Define *the Wick exponential of* X, Exp X, by

$$\text{Exp } X = \sum_{n=0}^{\infty} \frac{1}{n!} X^{\diamond n}$$

Then by Corollary 1.9 applied to $f(z) = e^z$ we see that Exp $X \in (\mathcal{S})^{-1}$ also.

b) In particular, if we choose $X = W_x$ (the singular white noise) then $K_0 := \text{Exp} W_x$ is in fact in $(\mathcal{S})^*$. As suggested in [LØU 1], [LØU 3] the process $K_0(x, \omega)$ is a natural model for stochastic permeability in connection with fluid flow in porous media.

c) Other useful applications of Lemma 1.9 include the *Wick logarithm* $Y = \text{Log } X$, which is defined (in $(\mathcal{S})^{-1}$) for all $X \in (\mathcal{S})^{-1}$ with $\tilde{X}(0) \neq 0$. For such X we have

$$\text{Exp}(\text{Log } X) = X$$

and for all $Z \in (\mathcal{S})^{-1}$ we have

$$\text{Log}(\text{Exp } Z) = Z$$

d) Similarly we note that the *Wick-inverse* $X^{\diamond(-1)}$ and more generally the *Wick powers* $X^{\diamond \gamma} (\gamma \in \mathbf{R})$ exist in $(\mathcal{S})^{-1}$ for all $X \in (\mathcal{S})^{-1}$ with $\tilde{X}(0) \neq 0$.

REMARK.

The connection between the $\mathcal{H}$-transform and the $\mathcal{S}$-*transform* [HKPS] is

(1.26) $$\mathcal{H}F(z_1, z_2, \cdots, z_m) = (\mathcal{S}F)(z_1 e_1 + \cdots + z_m e_m)$$

(see e.g. [LØU 1])

REMARK. We can define what we could call the *generalized expectation* of an arbitrary $F \in (\mathcal{S})^{-1}$, in spite of the fact that such an F need not even be in $L^1(\mu)$: If $F_0 \in L^p(\mu)$ for $p > 1$ then the action of F_0 on an element $\psi \in (\mathcal{S})^1$ is given by

(1.27) $$\langle F_0, \psi \rangle = E[F_0 \psi] = \int_{\mathcal{S}'} F_0(\omega) \psi(\omega) d\mu(\omega),$$

so if $\psi \equiv 1$ then $\langle F_0, \psi \rangle = \langle F_0, 1 \rangle$ gives us the expectation of F_0. On the other hand, if a general $F \in (\mathcal{S})^{-1}$ has the chaos expansion

$$F = \sum_\alpha b_\alpha H_\alpha$$

then by (1.21) we have $\langle F, 1 \rangle = b_0 = \tilde{F}(0)$. Hence we define

(1.28) $$\tilde{F}(0) = b_0 = \langle F, 1 \rangle$$

to be the *generalized expectation* of $F \in (\mathcal{S})^{-1}$.

From now on we will use the notation

(1.29) $$E[F] = E_\mu[F] := \tilde{F}(0)$$

for the generalized expectation of $F \in (\mathcal{S})^{-1}$. Note that with this definition we have

(1.30) $$E[F \diamond G] = E[F] \cdot E[G] \quad \text{for all} \quad F, G \in (\mathcal{S})^{-1}.$$

and

(1.31) $$E[\mathrm{Exp}(X)] = \exp E[X] \quad \text{for all} \quad X \in (\mathcal{S})^{-1}.$$

More generally, from the chaos expansion of F we see that the Hermite transform gives us all the *actions*

$$\langle F, \langle \cdot, \phi \rangle^{\diamond n} \rangle$$

of $F \in (\mathcal{S})^{-1}$ on $\langle \cdot, \phi \rangle^{\diamond n} \in (\mathcal{S})^{1}$. Therefore, although F need not exist as a random variable, it exists as a stochastic distribution: Given a stochastic test function we can compute its associated average.

$(\mathcal{S})^{-1}$ processes

The Kondratiev space $(\mathcal{S})^{-1}$ of stochastic distributions turns out to be the right space to work in, not just for the Burgers equation discussed in this paper, but for several stochastic (partial or ordinary) differential equations. See e.g. [B 3], [HLØUZ 4]. Therefore, the solution we seek will be a function

$$u = u(t, x) : \mathbf{R}^{1+n} \to (\mathcal{S})^{-1},$$

which may be regarded as a (stochastic) distribution valued stochastic process. We call such functions $(\mathcal{S})^{-1}$ *processes*.

The *derivative* of an $(\mathcal{S})^{-1}$ process $f(t)$ w.r.t. t at $t = t_0 \in \mathbf{R}$ will then (if it exists) be the element $\eta = \eta(t_0) \in (\mathcal{S})^{-1}$ with the property that

$$(1.31) \qquad \frac{f(t_0 + h) - f(t_0)}{h} \to \eta(t_0) \quad \text{in} \quad (\mathcal{S})^{-1} \quad \text{as} \quad h \to 0.$$

If this holds we write $\eta(t_0) = \frac{df}{dt}(t_0)$ (or $\frac{\partial f}{\partial t}(t_0)$). By the characterization of the topology of $(\mathcal{S})^{-1}$ in terms of the Hermite transform (Lemma 1.7) this is equivalent to

$$(1.32) \qquad \frac{\tilde{f}(t_0 + h; z) - \tilde{f}(t_0; z)}{h} \to \tilde{\eta}(t_0; z)$$

pointwise boundedly for $z \in \mathbf{B}_q(\delta)$; for some $q < \infty, \delta > 0$.

For this it suffices that

$$(1.33) \qquad \frac{d}{dt}\tilde{f}(t; z) = \tilde{\eta}(t; z) \quad \text{for } t = t_0$$

pointwise for each $z \in \mathbf{B}_q(\delta)$, if we also have that

$$t \to \frac{d}{dt}\tilde{f}(t; z)$$

is continuous and uniformly bounded, for $z \in \mathbf{B}_q(\delta)$ and t in a neighbourhood of t_0. For if this holds, we can write

$$\frac{\tilde{f}(t_0 + h; z) - \tilde{f}(t_0; z)}{h} = \frac{1}{h} \int\limits_{t_0}^{t_0 + h} \frac{d}{ds}\tilde{f}(s; z)ds \quad \text{for small } h$$

and therefore this expression is uniformly bounded for $z \in \mathbf{B}_q(\delta)$ as $h \to 0$. If $\frac{df}{dt}$ exists and is t-continuous, we say that the $(\mathcal{S})^{-1}$ process $f(t)$ is C^1. A similar notation, C^k, is used for higher order derivatives, $k = 2, 3, \cdots$ (and for continuity if $k = 0$) and for several variables, $C^{k_1, k_2, \cdots, k_m}$.

In [HLØUZ 1-2] a different kind of solution concept is studied: An L^p *functional process* $(p \geq 1)$ is a map

$$X : S \times \mathbf{R}^d \times \mathcal{S}' \to \mathbf{R}$$

such that

(i) $x \to X(\phi, x, \omega)$ is (Borel) measurable for all $\phi \in \mathcal{S}, \omega \in \mathcal{S}'$
  and
(ii) $\omega \to X(\phi, x, \omega)$ belongs to $L^p(\mu)$ for all $\phi \in \mathcal{S}, x \in \mathbf{R}^d$.

Intuitively, $X(\phi, x, \omega)$ is the result of measuring the quantity X using the test function ("window") ϕ shifted to the point x and in the "experiment" ω.

EXAMPLE. White noise W may be regarded as an L^p functional process (for any $p < \infty$) by putting

$$W(p, x, \omega) = W_{\phi_x}(\omega),$$

where

$$\phi_x(y) = \phi(y - x).$$

REMARK For L^1 functional processes the definition of Wick product must be extended, since $L^1(\mu)$ is not contained in $(\mathcal{S})^{-1}$. See [HLØUZ 1].

§2. FROM THE STOCHASTIC BURGERS EQUATION TO THE STOCHASTIC HEAT EQUATION.

This transformation was performed in [HLØUZ 2], but only in the context of functional processes and hence with more assumptions than will be needed within $(\mathcal{S})^{-1}$:

THEOREM 2.1 (The Wick-Cole-Hopf transformation (I))

Let $N = N(t, x)$ be an $(\mathcal{S})^{-1}$-valued $C^{0,1}$-process and define

$$(2.1) \qquad\qquad w = -\nabla N.$$

Assume that there exists an $(\mathcal{S})^{-1}$-valued $C^{1,3}$-process $X(t, x)$ such that

$$(2.2) \qquad\qquad u = -\nabla X$$

solves the multidimensional Burgers equation

$$(2.3) \qquad \begin{cases} \frac{\partial u_k}{\partial t} + \lambda \sum_{j=1}^{n} u_j \diamond \frac{\partial u_k}{\partial x_j} = \nu \Delta u_k + w_k(t, x); \ t > 0; \ x \in \mathbf{R}^n \\ u_k(0, x) = g_k(x); \ 1 \le k \le n \end{cases}$$

Then the *Wick-Cole-Hopf transform* Y of u defined by

$$(2.4) \qquad\qquad Y := \mathrm{Exp}(\frac{\lambda}{2\nu} X)$$

solves the stochastic heat equation

(2.4)
$$\begin{cases} \frac{\partial Y}{\partial t} = \nu \Delta Y + \frac{\lambda}{2\nu} Y \diamond [N + C]; \ t > 0; \ x \in \mathbf{R}^n \\ Y(0, x) = \text{Exp}(\frac{\lambda}{2\nu} X(0, x)) \end{cases}$$

for some t-continuous $(\mathcal{S})^{-1}$-valued process $C(t)$ (independent of x).

Proof. The proof in the present $(\mathcal{S})^{-1}$-case follows the proof in [HLØUZ 2] with only minor modifications. For completeness we give the argument:

Substituting (2.1) and (2.3) in (2.2) we get

(2.5) $-\dfrac{\partial}{\partial x_k}\Big(\dfrac{\partial X}{\partial t}\Big) + \lambda \sum_j \dfrac{\partial X}{\partial x_j} \diamond \dfrac{\partial}{\partial x_j}\Big(\dfrac{\partial X}{\partial x_k}\Big) = -\nu \sum_j \dfrac{\partial^2}{\partial x_j^2}\Big(\dfrac{\partial X}{\partial x_k}\Big) - \dfrac{\partial N}{\partial x_k}$

or

(2.6) $\dfrac{\partial X}{\partial t} = \dfrac{\lambda}{2} \sum_j \Big(\dfrac{\partial X}{\partial x_j}\Big)^{\diamond 2} + \nu \Delta X + N + C,$

where $C = C(t)$ is a t-continuous, x-independent $(\mathcal{S})^{-1}$-process.

Basic Wick calculus rules give that

(2.7) $\dfrac{\partial Y}{\partial t} = \dfrac{\lambda}{2\nu} Y \diamond \dfrac{\partial X}{\partial t},$

(2.8) $\dfrac{\partial Y}{\partial x_j} = \dfrac{\lambda}{2\nu} Y \diamond \dfrac{\partial X}{\partial x_j}$

Hence

$$\Delta Y = \sum_j \dfrac{\partial}{\partial x_j}\Big(\dfrac{\partial Y}{\partial x_j}\Big) = \sum_j \dfrac{\partial}{\partial x_j}\Big(\dfrac{\lambda}{2\nu} Y \diamond \dfrac{\partial X}{\partial x_j}\Big)$$

(2.9) $$= \sum_j \Big(\dfrac{\lambda}{2\nu}\Big)^2 Y \diamond \Big(\dfrac{\partial X}{\partial x_j}\Big)^{\diamond 2} + \sum_j \dfrac{\lambda}{2\nu} Y \diamond \dfrac{\partial^2 X}{\partial x_j^2}$$

$$= \dfrac{\lambda}{2\nu} Y \diamond \Big[\dfrac{\lambda}{2\nu} \sum_j \Big(\dfrac{\partial X}{\partial x_j}\Big)^{\diamond 2} + \Delta X\Big]$$

Now apply (2.7), (2.6) and (2.9) to get

$$\dfrac{\partial Y}{\partial t} = \dfrac{\lambda}{2\nu} Y \diamond \Big[\dfrac{\lambda}{2} \sum_j \Big(\dfrac{\partial X}{\partial x_j}\Big)^{\diamond 2} + \nu \Delta X + N + C\Big]$$

$$= \dfrac{\lambda}{2\nu} Y \diamond \Big[\dfrac{\lambda}{2\nu} \sum_j \Big(\dfrac{\partial X}{\partial x_j}\Big)^{\diamond 2} + \Delta X\Big]\nu + \dfrac{\lambda}{2\nu} Y \diamond [N + C]$$

$$= \nu \Delta Y + \dfrac{\lambda}{2\nu} Y \diamond [N + C], \quad \text{as claimed.}$$

§3. $(\mathcal{S})^{-1}$ - SOLUTION OF THE STOCHASTIC HEAT EQUATION

We now consider the stochastic heat equation (2.4) obtained by performing the Wick-Cole-Hopf transformation:

THEOREM 3.1
Suppose that $H(t, x)$ and $f(x)$ are continuous $(\mathcal{S})^{-1}$-processes. Then the stochastic heat equation

$$(3.1) \qquad \begin{cases} \frac{\partial Y}{\partial t} = \nu \Delta Y + H \diamond Y \; ; \; t > 0, \; x \in \mathbf{R}^n \\ Y(0, x) = f(x) \; ; \; x \in \mathbf{R}^n \end{cases}$$

has the unique $(\mathcal{S})^{-1}$-solution

$$(3.2) \qquad Y(t, x) = \hat{E}^x[f(b_{\alpha t}) \diamond \mathrm{Exp}(\int_0^t H(s, b_{\alpha s})ds)]$$

where $\alpha = \sqrt{2\nu}$, $(b_t, \hat{P}^x)$ is standard Brownian motion in $\mathbf{R}^n$ and $\hat{E}^x$ denotes expectation with respect to $\hat{P}^x$.

Proof. Taking Hermite transforms of (3.1) we get the equation

$$(3.3) \qquad \begin{cases} \frac{\partial \tilde{Y}}{\partial t} = \nu \Delta \tilde{Y} + \tilde{H} \cdot \tilde{Y} \;\; ; \; t > 0, \; x \in \mathbf{R}^n, \; z \in \mathbf{B}_q(\delta) \\ \tilde{Y}(0, x) = \tilde{f}(x) \;\; ; \; x \in \mathbf{R}^n, \; z \in \mathbf{B}_q(\delta) \end{cases}$$

where

$$\tilde{Y} = \tilde{Y}(t, x) = \tilde{Y}(t, x; z_1, z_2, \cdots) \;\; ; \quad z \in \mathbf{B}_q(\delta)$$

denotes the Hermite transform of Y etc. and $\mathbf{B}_q(\delta)$ is some neighbourhood of 0 in $\mathbf{C}^{\mathbf{N}}$, as defined in §1. Fix $z \in \mathbf{B}_q(\delta)$. Then by the complex version of the Feynman-Kac formula the solution $\tilde{Y}(t, x, z)$ of (3.3) can be written

$$(3.4) \qquad \tilde{Y}(t, x; z) = \hat{E}^x[\tilde{f}(b_{\alpha t}; z) \exp[\int_0^t \tilde{H}(s, b_{\alpha s}; z)ds]]$$

with $(b_t, \hat{P}^x)$ as described above.

This is the Hermite transform of the $(\mathcal{S})^{-1}$ process

$$Y(t, x) = \hat{E}^x[f(b_{\alpha t}) \diamond \mathrm{Exp}(\int_0^t H(s, b_{\alpha s})ds]],$$

and the proof is complete.

REMARK. The equation

$$\frac{\partial u}{\partial t} = \Delta u + W \diamond u, \text{ where } W \text{ is white noise,}$$

has been studied in [NZ]. They prove the existence of a solution of a type they call *generalized Wiener functionals.*

§4. FROM THE STOCHASTIC HEAT EQUATION TO THE STOCHASTIC BURGERS EQUATION

Using the fact that an analytic function composed with a Hermite transform (of an $(\mathcal{S})^{-1}$-element) is again the Hermite transform (Corollary 1.9), we can now construct the inverse of the transformation in §2:

THEOREM 4.1 (The Wick-Cole-Hopf transformation (II))

Suppose that $Y(t,x)$ is an $(\mathcal{S})^{-1}$-process which solves the stochastic heat equation

$$(4.1) \qquad \begin{cases} \frac{\partial Y}{\partial t} = \nu \Delta Y + H \diamond Y \; ; \quad t > 0, x \in \mathbf{R}^n \\ Y(0,x) = f(x) \; ; \quad x \in \mathbf{R}^n \end{cases}$$

where $f(x), H(t,x)$ are given $(\mathcal{S})^{-1}$-processes, continuously differentiable w.r.t x, $H(t,x)$ continuous w.r.t. t and (see (1.29))

$$(4.2) \qquad E_\mu[f(x)] > 0 \quad \text{for all} \quad x \in \mathbf{R}^n$$

Then

$$(4.3) \qquad u(t,x) := -\frac{2\nu}{\lambda} \nabla(\text{Log } Y(t,x)) \in (\mathcal{S})^{-1} \quad \text{for all} \quad t \geq 0, x \in \mathbf{R}^n,$$

where Log denotes "Wick-log" (Example 1.10d)) and the gradient is taken with respect to x (in $(\mathcal{S})^{-1}$).

Moreover, $u = (u_1, \cdots, u_n)$ solves the stochastic Burgers equation

$$(4.4) \qquad \begin{cases} \frac{\partial u_k}{\partial t} + \lambda \sum_{j=1}^{n} u_j \diamond \frac{\partial u_k}{\partial x_j} = \nu \Delta u_k + w_k \; ; \; t > 0, \; x \in \mathbf{R}^n \\ u_k(0,x) = g_k(x) \; ; \; x \in \mathbf{R}^n, \end{cases}$$

where

$$(4.5) \qquad w_k(t,x) = -\frac{2\nu}{\lambda} \cdot \frac{\partial H}{\partial x_k}(t,x),$$

and

$$(4.6) \qquad g_k(x) = -\frac{2\nu}{\lambda} f(x)^{\diamond(-1)} \diamond \frac{\partial f}{\partial x_k}(x); \quad 1 \le k \le n$$

Proof. By Theorem 3.1 the solution Y of (4.1) is given by

$$(4.7) \qquad Y(t,x) = \hat{E}^x[f(b_{\alpha t}) \diamond \mathrm{Exp}(\int_0^t H(s, b_{\alpha s})ds)] \quad ; \quad \alpha = \sqrt{2\nu}.$$

Therefore, by (1.30)

$$E_\mu[Y(t,x)] = \hat{E}^x[E_\mu[f(b_{\alpha t})] \cdot \exp(\int_0^t E_\mu[H(s, b_{\alpha s})]ds)] > 0$$

for all t, x.

We conclude that the Wick-log of Y, $X = \mathrm{Log}\, Y$, exists in $(\mathcal{S})^{-1}$, by Example 1.10d. We can therefore reverse the argument in the proof of Theorem 2.1:

Put

$$(4.8) \qquad X(t,x) = \frac{2\nu}{\lambda} \mathrm{Log}\, Y(t,x)$$

and

$$(4.9) \qquad u(t,x) = -\nabla X(t,x) \quad \text{(gradient w.r.t. } x\text{)}$$

Then

$$(4.10) \qquad Y(t,x) = \mathrm{Exp}(\frac{\lambda}{2\nu} X(t,x)),$$

so by (2.7), (2.9) we get

$$\frac{\partial u_k}{\partial t} + \lambda \sum_{j=1}^n u_j \diamond \frac{\partial u_k}{\partial x_j} - \nu \Delta u_k - w_k$$

$$= -\frac{\partial}{\partial t}(\frac{\partial X}{\partial x_k}) + \lambda \sum_{j=1}^n \frac{\partial X}{\partial x_j} \diamond \frac{\partial}{\partial x_j}(\frac{\partial X}{\partial x_k}) + \nu \sum_{j=1}^n \frac{\partial^2}{\partial x_j^2}(\frac{\partial X}{\partial x_k}) - \frac{2\nu}{\lambda}\frac{\partial H}{\partial x_k}$$

$$= \frac{\partial}{\partial x_k}[-\frac{\partial X}{\partial t} + \frac{\lambda}{2}\sum_{j=1}^n(\frac{\partial X}{\partial x_j})^{\diamond 2} + \nu\Delta X - \frac{2\nu}{\lambda}H]$$

$$= \frac{\partial}{\partial x_k}[Y^{\diamond(-1)} \diamond (-\frac{2\nu}{\lambda}\frac{\partial Y}{\partial t} + \frac{2\nu^2}{\lambda}\Delta Y - \frac{2\nu}{\lambda}H \diamond Y)]$$

$$= \frac{2\nu}{\lambda} \cdot \frac{\partial}{\partial x_k}[Y^{\diamond(-1)} \diamond (-\frac{\partial Y}{\partial t} + \nu\Delta Y - H \diamond Y)] = 0,$$

which shows the first part of (4.4).

Finally, to prove the second part observe that

$$u_k(0,x) = -\frac{\partial X}{\partial x_k}(0,x) = -\frac{2\nu}{\lambda}Y(0,x)^{\diamond(-1)} \diamond \frac{\partial Y}{\partial x_k}(0,x)$$

$$= -\frac{2\nu}{\lambda}f(x)^{\diamond(-1)} \diamond \frac{\partial f}{\partial x_k}(x), \quad \text{as claimed.}$$

§5. AN EXISTENCE AND UNIQUENESS RESULT

Combining the results of §2-4 we obtain the following existence and uniqueness result for the stochastic Burgers equation:

THEOREM 5.1

a) (Existence): Let $h(x), N(t,x)$ be $(\mathcal{S})^{-1}$ processes which are continuously differentiable w.r.t. x, $N(t,x)$ continuous w.r.t. t and

$$(5.1) \qquad\qquad E_\mu[h(x)] > 0 \quad \text{for all} \quad x \in \mathbf{R}^n$$

Then there exists an $((\mathcal{S})^{-1})^n$-process $u = (u_1, \cdots, u_n)$ which solves the stochastic Burgers equation

$$(5.2) \qquad \begin{cases} \frac{\partial u_k}{\partial t} + \lambda \sum_{j=1}^n u_j \diamond \frac{\partial u_k}{\partial x_j} = \nu\Delta u_k - \frac{\partial N}{\partial x_k} \ ; \ t > 0, \ x \in \mathbf{R}^n \\ u_k(0,x) = -\frac{\partial}{\partial x_k}\mathrm{Log}(h(x)) \ ; \ 1 \le k \le n, \end{cases}$$

namely

$$(5.3) \qquad\qquad u(t,x) = -\frac{2\nu}{\lambda}\nabla\mathrm{Log}\,Y(t,x),$$

where Y is given by (4.7), i.e.

$$(5.4) \qquad Y(t,x) = \hat{E}^x[h(b_{\alpha t})^{\diamond \frac{\lambda}{2\nu}} \diamond \mathrm{Exp}(\frac{\lambda}{2\nu}\int_0^t N(s,b_{\alpha s})ds)], \quad \alpha = \sqrt{2\nu}.$$

b) (Uniqueness): Moreover, this process u in (5.3)-(5.4) is the only solution of (5.2) of *gradient form*, i.e. which is the gradient w.r.t. x of some continuously x-differentiable $(\mathcal{S})^{-1}$ process.

Proof.

a) Apply Theorem 4.1 to the case when

$$(5.5) \qquad H(t,x) = \frac{\lambda}{2\nu} N(t,x) \quad , \quad f(x) = h(x)^{\diamond \frac{\lambda}{2\nu}}$$

b) If u solves (5.2) and

$$u(t,x) = -\nabla X(t,x)$$

then by Theorem 2.1 the process $Y = \text{Exp}(\frac{\lambda}{2\nu}X)$ solves (3.1) with $H = \frac{\lambda}{2\nu}[N + C]$ for some x-independent $C(t) \in (\mathcal{S})^{-1}$, and with

$$f(x) = \text{Exp}(\frac{\lambda}{2\nu}X(0,x)).$$

By Theorem 3.1 Y is unique. Then $X = \frac{2\nu}{\lambda}\text{Log } Y$ is unique up to a constant and therefore u is unique. That completes the proof.

Acknowledgements

We are grateful to Y. Kondratiev and G. Våge for valuable comments.

This work is partially supported by VISTA, a research cooperation between The Norwegian Academy of Science and Letters and Den Norske Stats Olje-selskap A.S. (Statoil).

REFERENCES

[AKS] S. Albeverio, J. Kondratiev and L. Streit: Spaces of white noise distributions: Constructions, Descriptions, Applications II. Manuscript 1993.

[B 1] F.E. Benth: Integrals in the Hida distribution space $(\mathcal{S})^*$. In T. Lindstrøm, B. Øksendal and A. S. Ustunel (editors): Stochastic Analysis and Related Topics. Gordon & Breach 1993, 89-99.

[B 2] F.E. Benth: A note on population growth in a crowded stochastic environment. Manuscript, University of Oslo 1993.

[BCJ-L] L. Bertini, N. Cancrini and G. Jona-Lasinio: The stochastic Burgers equation. Manuscript 1993.

[DDT] G. DaPrato, A. Debussche and R. Teman: Stochastic Burgers equation.
Preprint, Scuola Normale Superiore Pisa 1993.

[DG] G. DaPrato and D. Gatarek: Stochastic Burgers equation with correlated noise. Preprint, Scuola Normale Superiore Pisa 1994.

[GHLØUZ] H. Gjessing, H. Holden, T. Lindstrøm, J. Ubøe and T.-S. Zhang: The Wick product. In H. Niemi, G. Högnäs, A.N. Shiryaev and A. Melnikov (editors): "Frontiers in Pure and Applied Probability", Vol. 1. TVP Publishers, Moscow, 1993, pp. 29-67.

[GjHØUZ] J. Gjerde, H. Holden, B. Øksendal, J. Ubøe and T.-S. Zhang: An equation modelling transport of a substance in a stochastic medium. Manuscript 1993.

[GV] I.M. Gelfand and N.Y. Vilenkin: Generalized Functions, Vol. 4: Applications of Harmonic Analysis. Academic Press 1964 (English translation).

[H] T. Hida: Brownian Motion. Springer-Verlag 1980.

[HI] T. Hida and N. Ikeda: Analysis on Hilbert space with reproducing kernel arising from multiple Wiener integral. Proc. Fifth Berkeley Symp. Math.Stat.Probab. II, part 1 (1965), 117-143.

[HKPS] T. Hida, H.-H. Kuo, J. Potthoff and L. Streit: White Noise Analysis. Kluwer 1993.

[HLØU 1] H. Holden, T. Lindstrøm, B. Øksendal and J. Ubøe: Discrete Wick calculus and stochastic functional equations. Potential Analysis 1 (1992), 291-306.

[HLØU 2] H. Holden, T. Lindstrøm, B. Øksendal and J. Ubøe: Discrete Wick products. In T. Lindstrøm, B. Øksendal and A.S. Ustunel (editors): Stochastic Analysis and Related Topics. Gordon & Breach 1993, pp. 123-148.

[HLØUZ 1] H. Holden, T. Lindstrøm, B. Øksendal, J. Ubøe and T.-S. Zhang: Stochastic boundary value problems: A white noise functional approach. Probab. Th. Rel. Fields 95 (1993), 391-419.

[HLØUZ 2] H. Holden, T. Lindstrøm, B. Øksendal, J. Ubøe and T.-S. Zhang: The Burgers equation with a noisy force. Comm. PDE 19 (1994), 119-141.

[HLØUZ 3] H. Holden, T. Lindstrøm, B. Øksendal, J. Ubøe and T.-S. Zhang: A comparison experiment for Wick multiplication and ordinary multiplication. In T. Lindstrøm, B. Øksendal and A.S. Ustunel (editors): Stochastic Analysis and Related Topics. Gordon and Breach 1993, pp. 149-159.

[HLØUZ 4] H. Holden, T. Lindstrøm, B. Øksendal, J. Ubøe and T.-S. Zhang: The pressure equation for fluid flow in a stochastic medium. Potential Analysis (to appear).

[HiP] E. Hille and R.S. Phillips: Functional Analysis and Semigroups. Amer. Math. Soc. Colloq. Publ. 31 (1957).

[K] Y. Kondratiev: Generalized functions in problems of infinite dimensional analysis. Ph.D. thesis, Kiev University 1978.

[KLS] Y. Kondratiev, P. Leukert and L. Streit: Wick calculus in Gaussian analysis. Manuscript 1994.

[LØU 1] T. Lindstrøm, B. Øksendal and J. Ubøe: Stochastic differential equations involving positive noise. In M. Barlow and N. Bingham (editors): Stochastic Analysis. Cambridge Univ. Press 1991, pp. 261-303.

[LØU 2] T. Lindstrøm, B. Øksendal and J. Ubøe: Wick multiplication and Ito-Skorohod stochastic differential equations. In S. Albeverio et al (editors): Ideas and Methods in Mathematical Analysis, Stochastics, and Applications. Cambridge Univ. Press 1992, pp. 183-206.

[LØU 3] T. Lindstrøm, B. Øksendal and J. Ubøe: Stochastic modelling of fluid flow in porous media. In S. Chen and J. Yong (editors): Control Theory, Stochastic Analysis and Applications. World Scientific 1991, pp. 156-172.

[NZ] D. Nualart and M. Zakai: Generalized Brownian functionals and the solution to a stochastic partial differential equation. J. Functional Anal. 84 (1989), 279-296.

[Ø] B. Øksendal: Stochastic partial differential equations and applications to hydrodynamics. Submitted to L. Streit (editor): Stochastic Analysis and Applications in Physics (to appear).

[Z] T.-S. Zhang: Characterizations of white noise test functions and Hida distributions. Stochastics 41 (1992), 71-87.

A Weak Interaction Epidemic among Diffusing Particles

by

Ingemar KAJ (*)

Department of Mathematics, Uppsala University
Box 480, S-751 06 Uppsala, Sweden
ingemar.kaj@math.uu.se

ABSTRACT. - A multi-type system of n particles performing spatial motions given by a diffusion process on R^d and changing types according to a general jump process structure is considered. In terms of their empirical measure the particles are allowed to interact, both in the drift of the diffusions as well as in the jump intensity measure for the type motions. In the limit $n \to \infty$ we derive a principle of large deviations from the McKean-Vlasov equation satisfied by the empirical process of the system. The resulting rate function is shown to admit convenient representations.

In particular, the set-up covers a measure-valued model for an epidemic of SIR-type among spatially diffusing individuals. The infection rate is then proportional to the number of infective individuals and their distances to the susceptible one.

(*) Research supported by a Swedish Natural Sciences Research Council grant, contract F-FU 09481.

1. INTRODUCTION

1.1 Purpose. The purpose of this report is to provide a multi-type extension, allowing weak interaction in both space and type, of the well-known results of Dawson and Gärtner (1987) [DG] regarding large deviations from the McKean-Vlasov limit for weakly interacting diffusions. This is achieved by integrating more systematically the previous work Djehiche and Kaj (1994) [DK], in which a large deviation result is derived for a class of measure-valued jump processes, with the setting of the Dawson-Gärtner large deviation principle. Necessarilly, some aspects of such an extension will be mere notational rather than substantial. We will try to focus on those parts that are less evident and to point out some techniques from [DK] which can be used as an alternative to those of [DG].

We do not strive for maximal generality, the intention is rather to demonstrate the broad scope of the path-valued approach to large deviations as developed in [DG], in particular to show that it encompasses some natural models for spatial epidemics. As a motivation for this work the further question then arises whether one can apply large deviation techniques to study threshold limit theorems for epidemics, see e.g. Martin-Löf (1987) and Andersson and Djechiche (1994), in a similar manner as Dawson and Gärtner (1989) apply their results to study the phase-transition behaviour of a mean field interaction Curie-Weiss model.

1.2 Model. Fix a set $E \subset R^d$ and let $\mathcal{B}_E$ denote the σ-algebra on E generated by the Borel subset topology. The product space $R^d \times E$ is the one-particle state space and we consider paths of the form $t \mapsto x_t = (u_t, z_t)$, $t \in I := [0, T]$, where u_t is a diffusion process on R^d and z_t a Markov jump process with piecewise constant and rightcontinuous trajectories taking values in E. We interpret a system $x = (x^1, \ldots, x^n)$ of n such particles by saying that at time $t \in I$ the ith particle occupies the position u_t^i and is of type z_t^i. This information is captured by the empirical distribution. Hence, let $\mathcal{M}$ denote the set of probability measures on $R^d \times E$ equipped with the topology of weak convergence and total variation norm $\|\mu\|$. The point measure

$$\mu_t^n = \sum_{i=1}^n \delta_{x_t^i},$$

represents a realization of the state at time t.

To introduce the space and type motions we quote [DG] and [DK].respectively. However, we restrict to the time-homogeneous case. Furthermore, to simplify the presentation we restrict the class of diffusions significantly to the case of bounded drift and covariance coefficients. Let $\mathcal{C}^{2,0}(R^d \times E)$ denote the set of bounded continuous functions $f(x) = f(u, z)$ twice continuously differentiable in u. For a given n-particle state x, let

$$\mathcal{L}f^i(x) = \mathcal{L}f(x^i; \mu) = \frac{1}{2} \sum_{k,l=1}^d a^{k,l}(u^i) \frac{\partial^2 f(u^i, z^i)}{\partial u^k \partial u^l} + \sum_{k=1}^d b^k(u^i; \mu) \frac{\partial f(u^i, z^i)}{\partial u^k}$$

denote the diffusion generator for $t \mapsto u_t^i$. Assume that the matrix $\{a^{k,l}(u)\}$ is strictly positive definite for each $u \in R^d$ and that for each k, l the map $a^{k,l} : R^d \to R$ is bounded and locally Hölder continuous. Moreover, assume that the functions $(x, \mu) \mapsto b^k(x, \mu) : R^d \times \mathcal{M} \to R$ are bounded and continuous in x and uniformly continuous in μ. Following [DG], introduce the notations

$$(\nabla_a f)^k = \sum_{l=1}^d a^{k,l} \frac{\partial f}{\partial u^k}, \quad (\nabla_a f, \nabla_a g) = \sum_{k,l=1}^d a^{k,l} \frac{\partial f}{\partial u^k} \frac{\partial g}{\partial u^l}$$

and
$$|\nabla_a f|^2 = (\nabla_a f, \nabla_a f).$$

Regarding the type motion, let the jump measure on E have the form

$$\nu^i(x, dy) = \nu(x^i, dy\,; \mu) = \gamma(x^i; \mu)\,\pi(x^i, dy), \quad x \in (R^d \times E)^n, \quad y \in E,$$

where the jump intensity measure $\gamma(x^i; \mu)\,dt$ gives the rate of a change from type z^i for a particle at u^i when the system is in state x. Moreover, if a change of type occurs then it is of the form $z^i \to y \in B$ with probability $\int_B \pi(x^i, dy)$, all $B \in \mathcal{B}_E$. Assume $\mu \mapsto \gamma(\cdot, \mu)$ is uniformly continuous and assume there is a constant C such that for each n and $1 \le i \le n$

$$\sup_{x \in E^n} \int_E \nu^i(x, dy) \le C.$$

Write

$$\Delta f(x, y) := f(u, y) - f(u, z), \quad x = (u, z) \in R^d \times E, \ y \in E.$$

and define

$$\mathcal{A}f(x^i; \mu) = \int_E \Delta f(x^i, y)\,\nu^i(x, dy).$$

Let $\mathcal{C}^{1,2,0}(I \times R^d \times E)$ denote the set of bounded continuous functions $f_t(x) = f_t(u, z)$ continuously differentiable in t and twice continuously differentiable in u. The total motion generator acting on $\mathcal{C}^{1,2,0}(I \times R^d \times E)$ can then be written

$$\left(\tfrac{\partial}{\partial t} + \mathcal{L} + \mathcal{A}\right) f_t = \frac{\partial f_t}{\partial t} + \frac{1}{2} \sum_{k,l=1}^d a^{k,l} \frac{\partial^2 f_t}{\partial u^k \partial u^l} + \sum_{k=1}^d b^k \frac{\partial f_t}{\partial u^k} + \int_E \Delta f_t\,d\nu.$$

1.3 Martingale problem. We introduce the empirical distribution process

$$t \mapsto X_t^n = \frac{1}{n} \sum_{i=1}^n \delta_{x_t^i} \in \mathcal{M}, \quad t \in I.$$

Denote the natural duality between $\mathcal{M}$ and the set $\mathcal{C}$ of bounded continuous functions on $R^d \times E$ by

$$\langle \mu, f \rangle = \int f(u, y)\,\mu(du \otimes dy), \quad f \in \mathcal{C}, \quad \mu \in \mathcal{M}.$$

Let $\mathcal{D}(I, \mathcal{M})$ denote the path space of càdlàg functions from I into $\mathcal{M}$ furnished with the usual Skorokhod topology and $\mathcal{C}(I, \mathcal{M})$ its subset of uniformly continuous paths. The trajectories of X are realized in the canonical probability space $[\mathcal{D}(I, \mathcal{M}), (\mathcal{F}_t)_{t \in I}]$, where $\mathcal{F}_t$, $t \in I$, denotes the filtration of σ-algebras generated by the process $t \mapsto x_t$.

Proposition 1.1. *Suppose that weakly interacting generators $\mathcal{L}f(x;\mu)$ and $\mathcal{A}f(x;\mu)$, $x \in R^d \times E$, are given such that the diffusion coefficients a and b and the jump measure ν satisfies the conditions stated above. For each n and each $\mu_0 \in \mathcal{M}$, there exists a measure $\mathcal{P}^n$ with $\mathcal{P}^n(X_0^n = \mu_0) = 1$ such that for each f in $\mathcal{C}^{1,2,0}(I \times R^d \times E)$,*

$$\langle X_t^n, f_t \rangle = \langle X_0^n, f_0 \rangle + \int_0^t \left\langle X_r^n, \tfrac{\partial}{\partial r} f_r + \mathcal{L}f_r(\cdot\,; X_r^n) + \mathcal{A}f_r(\cdot\,; X_r^n) \right\rangle dr + M_t^f \,,$$

where $M^f = (M_t^f)_{t \in I}$ is a $(\mathcal{P}^n, \mathcal{F}_t)$-martingale. For f, g in $\mathcal{C}^{1,2,0}(I \times R^d \times E)$ the predictable quadratic variation of M^f and M^g is given by

$$\langle\!\langle M_\cdot^f, M_\cdot^g \rangle\!\rangle_t = \frac{1}{n} \int_0^t \Big\langle X_r^n, (\nabla_a f_r(\cdot), \nabla_a g_r(\cdot))$$

$$+ \int \mathcal{A}f_r(\cdot, y) \mathcal{A}g_r(\cdot, y)\, \nu(\cdot, dy\,; X_r^n) \Big\rangle dr \,.$$

Suppose X_0^n converges weakly to some $\eta_0 \in \mathcal{M}$. Then X^n converges weakly in $\mathcal{D}(I, \mathcal{M})$ to a deterministic path $\eta \in \mathcal{C}(I, \mathcal{M})$ which is the unique solution of the McKean-Vlasov equation

$$(1.2) \qquad \langle \eta_t, f_t \rangle = \langle \eta_0, f_0 \rangle + \int_0^t \left\langle \eta_r, \tfrac{\partial}{\partial r} f_r + \mathcal{L}f_r(\cdot\,; \eta_r) + \mathcal{A}f_r(\cdot\,; \eta_r) \right\rangle dr \,.$$

PROOF; REFERENCES: For the diffusion part see [DG], section 5.1, noting that we avoid the main difficulties regarding the inductive topology setting by imposing bounded coefficients. Similarly, the assumption of bounded jump rates considerably simplifies the task of showing that the martingale problem is well-posed. In the pure jump case, a more general situation was studied by Feng (1994) using the inductive topology approach. For complete proofs under general assumptions in closely related models including both continuous and jump processes, see e.g. Oelschläger (1984) and Graham (1992). □

We next consider the general martingale problem

$$F(X_t^n) = F(X_0^n) + \int_0^t \mathcal{G}^n F(X_s^n)\, ds + M_t^F \,,$$

for functionals $F(\mu) = F(\langle \mu, f \rangle)$, $F \in \mathcal{C}^2(R)$, $\mu \in \mathcal{M}$ and $f \in \mathcal{C}^{2,0}(R^d \times E)$, where M_t^F is a $(\mathcal{P}^n, \mathcal{F}_t)$-local martingale and the infinitesimal measure generator $\mathcal{G}^n$ acts on probability measures $\mu \in \mathcal{M}$ by

$$\mathcal{G}^n F(\mu) = \langle \mu, \mathcal{L}f(\mu) \rangle F'(\langle \mu, f \rangle) + \frac{1}{2n} \langle \mu, |\nabla_a f|^2 \rangle F''(\langle \mu, f \rangle)$$

$$+ n \left\langle \mu, \int \Big(F(\mu + \tfrac{1}{n}\delta_y - \tfrac{1}{n}\delta_\cdot) - F(\mu) \Big) \nu(\cdot, dy\,; \mu) \right\rangle .$$

The corrresponding Hamiltonian operator is defined as

$$\mathcal{H}^n(\mu, f) = e^{-\langle \mu, f \rangle} \, \mathcal{G}^n \, e^{\langle \mu, f \rangle}.$$

Hence

$$\mathcal{H}^n(\mu, f) = \langle \mu, \mathcal{L}f(\mu) \rangle + \frac{1}{2n}\langle \mu, |\nabla_a f|^2 \rangle + \left\langle \mu, \int n(e^{\Delta f/n} - 1)\, d\nu(\mu) \right\rangle.$$

It turns out that the scaled Hamiltonian

(1.3)
$$\mathcal{H}(\mu, f) := \lim_{n \to \infty} \frac{1}{n}\mathcal{H}^n(\mu, nf)$$

$$= \langle \mu, \mathcal{L}f(\mu) \rangle + \frac{1}{2}\langle \mu, |\nabla_a f|^2 \rangle + \left\langle \mu, \int (e^{\Delta f} - 1)\, d\nu(\mu) \right\rangle,$$

is a central quantity for the large deviation result (in this case $\frac{1}{n}\mathcal{H}^n(\mu, nf)$ is independent of n).

2. Large deviations result

2.1 Weak form McKean-Vlasov equation.

By considering $\mathcal{M}$ as a subset of the Schwartz space $\mathcal{D}'$ of real distributions on $R^d \times E$, the notion of absolutely continuous maps $t \mapsto \mu_t \in \mathcal{D}'$ can be made precise, see [DG], section 4.1. For a discussion relevant to jump processes, see [DK], section 2.1. We will not recall these matters here, but use freely the time derivatives

$$\dot{\mu}_t = \lim_{h \to 0} h^{-1}\big(\mu_{t+h} - \mu_t\big) \qquad \text{for almost all } t \in I,$$

existing in the distribution sense and satisfying the integration by parts formula

$$\int_0^t \langle \dot{\mu}_r, f_r \rangle \, dr = \langle \mu_t, f_t \rangle - \langle \mu_0, f_0 \rangle - \int_0^t \langle \mu_r, \tfrac{\partial}{\partial r} f_r \rangle \, dr.$$

Here $f \in \mathcal{C}_0^\infty(I \times R^d \times E)$, the set of smooth test functions compactly supported on $R^d \times E$. Introduce a formal adjoint $\mathcal{L}^*(a, b)$ of the diffusion generator $\mathcal{L}$ with coefficients a and b and a formal adjoint $\mathcal{A}^*(\nu)$ of the type generator $\mathcal{A}$ corresponding to the jump measure ν. The McKean-Vlasov equation (1.2) takes the weak form

$$\dot{\eta}_t = \mathcal{L}^*(a, b)\eta_t + \mathcal{A}^*(\nu)\eta_t, \quad \text{almost all } \quad t \in I,$$

where again b and ν may depend interactively on η_t.

Next we introduce for any given path $t \to \mu_t \in \mathcal{M}$ two function spaces. First the set $L^2(I, \mu) = L^2(I \times R^d \times E; \, dr\, \mu_r(dx))$ of square-integrable functions $f : I \times R^d \times E \to R^d$ with

$$\|f\|_2^2 = \int_0^T \langle \mu_r, |f_r|^2 \rangle \, dr < \infty,$$

where $|\cdot|$ refers to the 'Riemannian structure' generated by the matrix a as in section 1.2. Second the Orlicz space

$$\mathcal{O}(I,\mu) = \mathcal{O}\big(I \times R^d \times E^2 \, ; \, ds\, \mu_s(dx)\, n(x,dy)\big)$$

of strictly positive functions g on $I \times R^d \times E$ for which the norm expression

$$\|g\| = \inf\Big\{\theta > 0 : \int_I \Big\langle \mu_r, \int \big(\tfrac{g_r}{\theta}\log\tfrac{g_r}{\theta} - \tfrac{g_r}{\theta} + 1\big)\, d\nu\Big\rangle\, dr \le 1\Big\}$$

is finite.

Given a, b, ν and the unique solution $t \mapsto \eta_t$, according to Proposition 1.1, we now define a set H consisiting of all deterministic paths $t \mapsto \mu_t$ with the following properties:

(i) $\mu_0 = \eta_0$,

(ii) the map $t \mapsto \mu_t$ defined on I is absolutely continuous,

(iii) there exists $h \in L^2(I,\mu)$ and $g \in \mathcal{O}(I,\mu)$ such that the path μ is the unique solution of the weak McKean-Vlasov equation

$$\dot{\mu}_t = \mathcal{L}^*(a, b + ah_t)\mu_t + \mathcal{A}^*(g_t\,\nu)\mu_t, \quad \text{almost all} \quad t \in I,$$

that is,

$$\langle \dot{\mu}_t, f\rangle = \Big\langle \mu_t, \, \mathcal{L}f(\mu_t) + h_t \cdot \nabla_a f + \int \Delta f\, g_t\, d\nu(\mu_t)\Big\rangle,$$

for $f \in \mathcal{C}_0^\infty(R^d \times E)$ and almost all t. We call H the set of paths admissible for η.

2.2 A large deviation theorem. The following is a version of Theorems I and II in [DK] for the extended interacting multitype diffusions model.

Theorem 2.1. *Suppose a, b and ν are given as above. Fix $\eta_0 \ne 0$ and let η be the McKean-Vlasov limit solution of equation (1.2). For $\mu \in \mathcal{D}(I,\mathcal{M})$ with $\mu_0 = \eta_0$ and such that $t \mapsto \mu_t$ is absolutely continuous define*

$$S(\mu) = \int_0^T \sup_{f\in\mathcal{C}_0^\infty(R^d\times E)}\Big\{\langle\dot{\mu}_t, f\rangle - \mathcal{H}(\mu_t, f)\Big\}\, dt,$$

where $\mathcal{H}(\mu, f)$ is the scaled Hamiltonian defined in (1.3). Put $S(\mu) = \infty$ otherwise. For any measurable set A in $\mathcal{D}(I,\mathcal{M})$, put

$$\mathcal{P}^n(A) = \mathcal{P}^n(X^n \in A | X_0^n = \eta_0).$$

Then

(i) *for each open subset G of $\mathcal{D}(I, \mathcal{M})$,*

$$\liminf_{n \to \infty} \frac{1}{n} \log \mathcal{P}^n(G) \geq - \inf_{\mu \in G} S(\mu),$$

(ii) *for each closed subset F of $\mathcal{D}(I, \mathcal{M})$,*

$$\limsup_{n \to \infty} \frac{1}{n} \log \mathcal{P}^n(F) \leq - \inf_{\mu \in F} S(\mu),$$

(iii) *the level sets $\{\mu \in \mathcal{D}(I, \mathcal{M}) : S(\mu) \leq N\}$, $N > 0$, are compact.*

Moreover,
(2.2)

$$S(\mu) = \sup_f \left\{ \langle \mu_T, f_T \rangle - \langle \mu_0, f_0 \rangle - \int_0^T \langle \mu_r, \tfrac{\partial}{\partial r} f_r \rangle \, dr - \int_0^T \mathcal{H}(\mu_r, f_r) \, dr \right\},$$

where the supremum is over all $f \in \mathcal{C}^{1,2,0}(I \times R^d \times E)$. Finally, $S(\mu) < \infty$ if and only if $\mu \in H$, the set of paths which are admissible for η, and if $\mu \in H$ then

$$(2.3) \qquad S(\mu) = \int_0^T \left\langle \mu_r, \frac{1}{2}|h_r|^2 + \int (g_r \log g_r - g_r + 1) \, d\nu \right\rangle dr.$$

Remark. The basic representation of S in [DG] (when $g = 1$) is in terms of the dual form of the squared $L^2(\mu_r)$-norm of h:

$$S(\mu) = \frac{1}{2} \int_0^T \sup_{f \in C_0^\infty(R^d)} \frac{\langle \mu_r, h \cdot \nabla_a f \rangle^2}{\langle \mu_r, |\nabla_a f|^2 \rangle}$$

$$= \frac{1}{2} \int_0^T \sup_{f \in C_0^\infty(R^d)} \frac{\langle \dot{\mu}_r - \mathcal{L}^*(a, b) \mu_r, f \rangle^2}{\langle \mu_r, |\nabla_a f|^2 \rangle}.$$

We do not know of a similar representation in the multitype case.

2.3 SIR-epidemics. Consider the case when E is discrete and only consists of three states $E = [\text{susc, inf, rem}]$ for the possible types S=*susceptible*, I=*infective* or R=*removed* and the space motion extends with a cemetry position † to $\dot{R}^d = R^d \cup \dagger$. The motion $t \mapsto x_t^i$ signifies that among n individuals the ith follows the type cycle *susc→inf→rem* while diffusing according to u_t^i, until z_t^i hits *rem* when it is brought to position †. Using self-contained notations the empirical process may be written

$$\frac{1}{n} \sum_{i=1}^n \delta_{x^i} = \frac{1}{n} \left(\sum_{i=1}^{|S|} \delta_{u^i}^{(S)} + \sum_{j=1}^{|I|} \delta_{u^j}^{(I)} + \sum_{k=1}^{|R|} \delta_\dagger \right).$$

Furthermore, if f is a bounded continuous function on $\dot{R}^d \times E$ put $\varphi_1(u) = f(u, susc)$, $\varphi_2(u) = f(u, inf)$ and extend by letting $f(u, rem) = f(\dagger, rem) = 0$. Then write $\varphi = (\varphi_1, \varphi_2)$ and

$$\langle X_t^n, \varphi \rangle = \langle S_t^n, \varphi_1 \rangle + \langle I_t^n, \varphi_2 \rangle .$$

In general, $\langle \mu, \varphi \rangle = \langle \mu^1, \varphi_1 \rangle + \langle \mu^2, \varphi_2 \rangle$, where μ^1 and μ^2 are measures on R^d.

To define the type interaction, let

$$\lambda(u, v) : R^d \times R^d \mapsto R,$$

be a bounded, continuous and nonnegative function. Then introduce a function $\Phi(u, v) = (\Phi_1(u, v), \Phi_2(u, v))$, $u, v \in R^d$, by setting

$$\Phi_1(u, v) = 0, \quad \Phi_2(u, v) = \lambda(u, v).$$

Because the type only changes from susceptible to infective and from infective to removed, the jump measure ν is determined by the rate function $\gamma(x^i; \mu)$. We choose the following weakly interacting rates

$$\gamma((susc, u^i); \mu) = \langle \mu, \Phi(u^i, \cdot) \rangle = \langle \mu^2, \lambda(u^i, \cdot) \rangle, \quad \gamma((inf, u^i); \mu) = \rho > 0.$$

Hence, if X^n is the state at some fixed time when there are $|I|$ infective particles at positions u^j, then

$$\nu((susc, u^i), inf; X^n) = \langle I^n, \lambda(u^i, \cdot) \rangle = \frac{1}{n} \sum_{j=1}^{|I|} \lambda(u^i, u^j)$$

is the rate at which a susceptible particle at position u^i contracts the disease. By the second assumption we have adopted the standard model of constant removal rate

$$\nu((inf, u^i), rem; X^n) = \rho.$$

To concretize the martingale problem for this model one should note that now

$$\int_0^t \left\langle X_r^n, \mathcal{A}f_r(\cdot; X_r^n) \right\rangle dr = \int_0^t \left\langle S_s^n \otimes I_s^n, \lambda(\varphi_2 - \varphi_1) \right\rangle ds - \varrho \int_0^t \langle I_s^n, \varphi_2) \rangle ds,$$

and

$$\int_0^t \left\langle X_r^n, \int \Delta f_r(\cdot, y) \Delta g_r(\cdot, y) \nu(\cdot, dy; X_r^n) \right\rangle dr$$

$$= \int_0^t \left\langle S_s^n \otimes I_s^n, \lambda(\varphi_2 - \varphi_1)^2 \right\rangle ds + \varrho \int_0^t \left\langle I_s^n, \varphi_2^2 \right\rangle ds.$$

Moreover,

$$\mathcal{H}(\mu, \varphi) = \langle \mu, \mathcal{L}\varphi(\mu) \rangle + \frac{1}{2} \langle \mu, |\nabla_a \varphi|^2 \rangle$$
$$+ \langle \mu^1 \otimes \mu^2, \lambda(e^{\varphi_2 - \varphi_1} - 1) \rangle + \varrho \langle \mu^2, (e^{-\varphi_2} - 1) \rangle.$$

Therefore, as a corollary of Theorem 2.1 we obtain a large deviation principle for the SIR-epidemics with rate function

$$S(\mu) = \int_0^T \langle \mu_r, |h_r|^2 \rangle$$

$$+ \int_0^T \left(\langle \mu_r^1 \otimes \mu_r^2, \lambda(g_r^1 \log g_r^1 - g_r^1 + 1) \rangle + \varrho \langle \mu_r^2, (g_r^2 \log g_r^2 - g_r^2 + 1) \rangle \right) dr.$$

Here $h \in L^2(\mu)$ and $g = (g^1, g^2) \in \mathcal{O}(\mu, \nu)$ are such that for smooth φ, μ is a solution of

$$\langle \dot{\mu}_t, \varphi \rangle = \langle \mu_t, \mathcal{L}\varphi(\mu_t) + h_t \cdot \nabla_a \varphi \rangle + \langle \mu_t^1 \otimes \mu^2, \lambda(\varphi_2 - \varphi_1) g_t^1 \rangle - \langle \mu_t^2, \varphi_2 g_t^2 \rangle.$$

3. Proofs

Section 3.1 is devoted to properties of the integral S and the space H which can be derived without reference to large deviation estimates. Then we start to prove that the function $S(\mu)$ actually is a large deviation rate function. According to the general Theorems 5.2 and 5.3 in [DG] it suffices to derive local lower and upper bounds and check the exponential tightness property. In sections 3.2-3.5 we give proofs for the independent case with locally frozen interaction. The change-of-measure applied to "switch on" true interaction is the topic of section 3.6.

3.1 Representation of S. We show first that $S(\mu)$ has the representation (2.2).

For fixed $\mu \in \mathcal{D}(I, \mathcal{M})$, introduce for $f \in \mathcal{C}^{1,2,0}(I \times R^d \times E)$ and $0 \leq s \leq t \leq T$ the functionals

$$J_{s,t}(f) = \langle \mu_t, f_t \rangle - \langle \mu_s, f_s \rangle - \int_s^t \langle \mu_r, \tfrac{\partial}{\partial r} f_r \rangle \, dr - \int_0^t \mathcal{H}(\mu_r, f_r) \, dr,$$

with the Hamiltonian $\mathcal{H}$ defined in (1.3). Similarly, put

$$\ell_{s,t}(f) := \langle \mu_t, f_t \rangle - \langle \mu_s, f_s \rangle - \int_s^t \langle \mu_r, \tfrac{\partial}{\partial r} f_r + \mathcal{L} f_r + \mathcal{A} f_r \rangle \, dr.$$

Hence

$$(3.1) \qquad J_{0,T}(f) = \ell_{0,T}(f) - \int_0^T \left\langle \mu_r, \tfrac{1}{2} |\nabla_a f_r|^2 + \int \tau(\Delta f_r) \, d\nu \right\rangle dr.$$

Observe that for all $f \in \mathcal{C}_0^\infty(I \times R^d \times E)$,

$$\ell_{s,t}(f) = \int_s^t \left(\langle \dot{\mu}_r, f_r \rangle - \langle \mu_r, \mathcal{L}f_r + \mathcal{A}f_r \rangle \right) dr \,.$$

By approximation this implies that $\ell_{s,t}(f)$ can be obtained from $\ell_{0,T}(f)$ by restricting to f non-zero on $[s,t]$ only. Now, for any smooth f,

$$S(\mu) = \int_0^T \sup_{f \in \mathcal{C}_0^\infty(R^d \times E)} \left\{ \langle \dot{\mu}_t, f \rangle - \mathcal{H}(\mu_t, f) \right\} dt$$

$$\geq \int_I \left(\langle \dot{\mu}_r, f_r \rangle - \mathcal{H}(\mu_r, f_r) \right) dr = J_{0,T}(f) \,.$$

Hence the right hand side of (2.2) is bounded by $S(\mu)$:

$$\sup_{f \in \mathcal{C}^{1,2,0}(I \times R^d \times E)} J_{0,T}(f) \leq S(\mu) \,.$$

Before proving the opposite inequality we introduce some more notation. The Orlicz space $\mathcal{O}(I, \mu)$ is obtained from the pair of Young functions

$$\tau(t) = e^t - t - 1 \,, \qquad \tau^*(s) = (s+1)\log(s+1) - s \,, \qquad s > -1 \,.$$

In fact, τ and τ^* define a pair of topologically dual Orlicz spaces $(L^\tau(I, \mu), \|\cdot\|_\tau)$ and $(L^{\tau^*}(I, \mu), \|\cdot\|_{\tau^*})$ of functions on $I \times R^d \times E^2$. Then $\mathcal{O}(\mu, \nu)$ consists of all functions $g = 1 + \tilde{h}$ with $\tilde{h}$ in L^{τ^*}.

Consider the product space $L^2(I, \mu) \times L^\tau(I, \mu)$ equipped with the norm

$$\|(f, \tilde{f})\| = \max \left\{ \|f\|_2, \|\tilde{f}\|_\tau \right\}$$

and its dual product $L^2(I, \mu) \times L^{\tau^*}(I, \mu)$ with norm

$$\|(h, \tilde{h})\|_* = \|h\|_2 + \|\tilde{h}\|_{\tau^*} \,.$$

For our application typical elements of $L^2(I, \mu) \times L^\tau(I, \mu)$ will be (equivalence classes) of the form $(\nabla_a f, \Delta f)$, in one-to-one correspondence with $f \in \mathcal{C}^{1,2,0}(I \times R^d \times E)$.

Lemma 3.2.

$$S(\mu) \leq \sup_{f \in \mathcal{C}^{1,2,0}(I \times R^d \times E)} J_{0,T}(f) \,.$$

PROOF: For $f \in \mathcal{C}^{1,2,0}(I \times R^d \times E)$ put $c = \|(\nabla_a f, \Delta f)\|$. By (3.1),

$$\frac{1}{c}\ell_{0,T}(f) \leq J_{0,T}(f/c) + \int_0^T \left\langle \mu_r, \frac{1}{2}\Big|\frac{\nabla_a f_r}{\|\nabla_a f\|_2}\Big|^2 + \int \tau\Big(\frac{\Delta f_r}{\|\Delta f\|_\tau}\Big) d\nu \right\rangle dr$$

$$\leq J_{0,T}(f/c) + \frac{1}{2} + 1 \,.$$

We can assume that the supremum on the right side in the lemma is finite. Hence
$$\ell_{0,T}(f) \leq \text{const } \|(\nabla_a f, \Delta f)\|,$$

and therefore $\ell_{0,T}$ can be viewed as a bounded linear functional on the linear space L_{diff} of all pairs $(\nabla_a f, \Delta f)$ in $L^2(I,\mu) \times L^\tau(I,\mu)$.

By the Riesz representation theorem there exists a unique element $(h, \tilde{h})$ in the closure in $L^2(I,\mu) \times L^{\tau^*}(I,\mu)$ of the dual linear space of L_{diff}, such that

$$\ell_{0,T}(f) = \int_0^T \Big\langle \mu_r, \, h_r \cdot \nabla_a f_r + \int \tilde{h}_r \Delta f_r \, d\nu \Big\rangle \, dr, \quad f \in \mathcal{C}^{1,2,0}(I \times R^d \times E).$$

In particular, for $f \in \mathcal{C}_0^\infty(R^d \times E)$ and $0 \leq s < t \leq T$ and putting $g = \tilde{h}+1$,

$$\langle \mu_t, f \rangle = \langle \mu_s, f \rangle + \int_s^t \Big\langle \mu_r, \, \mathcal{L}f + h_r \cdot \nabla_a f + \int g_r \Delta f \, d\nu \Big\rangle \, dr.$$

From this it follows that μ is absolutely continuous, c.f. [DG]. Moreover, for almost all t, we have the McKean-Vlasov equation

$$(3.3) \qquad \langle \dot{\mu}_t, f \rangle = \Big\langle \mu_t, \, \mathcal{L}f + h_t \cdot \nabla_a f + \int \Delta f \, g_t \, dn_r \Big\rangle.$$

Now, by (3.3) and (1.3), computing Legendre transforms

$$S(\mu) = \int_0^T \sup_{f \in \mathcal{C}_0^\infty(R^d \times E)} \Big\langle \mu_r, \, h_r \cdot \nabla_a f + \int \tilde{h}_r \Delta f \, d\nu$$
$$- \frac{1}{2}|\nabla_a f|^2 - \int \tau(\Delta f) \, d\nu \Big\rangle \, dr$$
$$\leq \int_0^T \Big\langle \mu_r, \, \frac{1}{2}|h_r|^2 + \int \tau^*(\tilde{h}_r) \, d\nu \Big\rangle \, dr.$$

However, for smooth f we have also

$$J_{0,T}(f) = \int_0^T \Big\langle \mu_r, \, \frac{1}{2}|h_r|^2 + \int \tau^*(\tilde{h}_r) \, d\nu \Big\rangle \, dr$$
$$- \int_0^T \Big\langle \mu_r, \, \frac{1}{2}|h_r - \nabla_a f|^2 + \int (\tau^*(\tilde{h}_r) + \tau(\Delta f_r) - h_r \Delta f_r) \, dn_r \Big\rangle \, dr.$$

By Young's inequality and approximation in $L^2(I,\mu) \times L^\tau(I,\mu)$, therefore

$$(3.4) \qquad \sup_{f \in \mathcal{C}^{1,2,0}(I \times R^d \times E)} J_{0,T}(f) = \int_0^T \Big\langle \mu_r, \, \frac{1}{2}|h_r|^2 + \int \tau^*(\tilde{h}_r) \, d\nu \Big\rangle \, dr,$$

which finishes the proof of the lemma. $\qquad \square$

Lemma 3.5. *For any* $\mu \in \mathcal{D}(I, \mathcal{M})$, *absolutely continuous and with* $\mu_0 = \eta_0$, $\mu \in H$ *if and only if* $S(\mu) < \infty$. *If* $\mu \in H$ *then* (2.3) *holds.*

PROOF: Suppose $S(\mu) < \infty$. Then the supremum of the functional $J_{0,T}$ in (2.2) is finite. It was then a byproduct of the proof of the previous lemma that μ solves equation (3.3) for h and g such that μ is admissible for η, that is, $\mu \in H$.

Conversely, suppose $\mu \in H$. Then there exists $h \in L^2(I, \mu)$ and $g \in \mathcal{O}(I, \mu)$, with $\tilde{h} = g - 1 \in L^{\tau^*}(I, \mu)$, such that (3.3) and hence (3.4) hold. But since τ^* has the growth property

$$\sup_t \tau^*(at)/\tau^*(t) < \infty \quad \text{for some } a > 1,$$

the integral

$$\int_0^T \left\langle \mu_r, \int \tau^*(\tilde{h}_r)\, d\nu \right\rangle dr = \int_0^T \left\langle \mu_r, \left(\int g_r \log g_r - g_r + 1 \right) d\nu \right\rangle dr$$

is finite, see e.g. Neveu (1975), Appendix. Therefore, by (3.4), $S(\mu) < \infty$ and (2.3) holds.

$\square$

3.2 Independent case, Cramér transformation.

We start to prove Theorem 1.1 in the case when the processes $t \mapsto x_t^i$ are independent. Hence fix a path $\hat{\mu}$ in $\mathcal{D}(I, \mathcal{M})$ and consider the system with locally frozen interaction $\hat{\mu}$. This means that the diffusion generator $\mathcal{L}$ has variance matrix a and drift vector $b(\hat{\mu})$ and the jump measure is of the form $\hat{\nu} = \nu(x^i, y; \hat{\mu})$.

For $\mu \in \mathcal{D}(I, \mathcal{M})$ and $f \in \mathcal{C}^{1,2,0}(I \times R^d \times E)$, define

$$K_t(\mu, f) = \langle \mu_t, f_t \rangle - \langle \mu_0, f_0 \rangle - \int_0^t \left\langle \mu_r, \tfrac{\partial}{\partial r} f_r \right\rangle dr,$$

and consider the signed measure paths

$$f \mapsto K_t^{n,f} := K_t(X^n, f), \quad t \in I.$$

Then for each $\mathcal{F}_t$-predictable and $\mathcal{P}^n$- *a.s.* bounded function $(c_t)_{t \in I}$,

$$\xi_t^{n,f}(c) := \exp\left\{ n \int_0^t c_r\, dK_r^{n,f} - n \int_0^t \mathcal{H}_r(X_r^n, c_r f_r)\, dr \right\}, \quad t \in I,$$

is a $(\mathcal{P}^n, \mathcal{F}_t)$-martingale. The martingale (or for more general c, the local martingale) $\xi_t^{n,f}(c)$ is called the Esscher-Cramér transform or generalized Cramér transform. The relation

$$\xi_t^{n,f}(c) = \frac{d\mathcal{P}_\lambda^n}{d\mathcal{P}^n}\bigg|_{\mathcal{F}_t},$$

defines a new probability measure $\mathcal{P}_c^n$ which is equivalent to $\mathcal{P}^n$ and under $\mathcal{P}_c^n$ subsequent martingales (local martingales) can be derived. In fact, introduce the derivatives

$$\mathcal{H}_t'(\mu, cf) = \langle \mu, \mathcal{L}f \rangle + c \langle \mu, |\nabla_a f|^2 \rangle + \Big\langle \mu, \int \Delta f e^{c\,\Delta f}\, d\widehat{v} \Big\rangle,$$

$$\mathcal{H}_t''(\mu, cf) = \langle \mu, |\nabla_a f|^2 \rangle + \Big\langle \mu, \int (\Delta f)^2 e^{c\,\Delta f}\, d\widehat{v} \Big\rangle.$$

Then, for each $(\beta_t)_{t \in I}$, $\mathcal{F}_t$-predictable and $\mathcal{P}^n$- *a.s.* bounded, the process

$$M_t^n = \int_0^t \beta_r\, dK_r^{n,f} - \int_0^t \beta_r\, \mathcal{H}_r'(X_r, c_r f_r)\, dr$$

is a $(\mathcal{P}_c^n, \mathcal{F}_t)$-martingale on I with predictable quadratic variation

$$\langle\!\langle M^n \rangle\!\rangle_t = \frac{1}{n} \int_0^t \beta_r^2\, \mathcal{H}_r''(X_r, c_r f_r)\, dr\,,$$

compare [DK], Lemmas 4.1 and 4.2.

Next we identify a specific predictable function $c = \lambda^n$ for which $\mathcal{P}_{\lambda^n}^n$ will be used to derive the lower large deviation bound. As usual $f^+ = f \vee 0$ and $f^- = (-f) \vee 0$ denote positive and negative parts of a realvalued function f. The notation $f^{\pm}$ refers to either of the two functions.

Lemma 3.6. *Let $t \mapsto \mu_t$ be absolutely continuous. Fix a function $f \in \mathcal{C}_0^{1,2,0}(I \times R^d \times E)$ which satisfies*

$$\gamma^{\pm}(f) := \inf_{t \in I} \Big\{ |\nabla_a f_t|^2 + \int (\Delta f_t^{\pm})^2\, d\widehat{v}_t \Big\} \geq c_1 > 0\,,$$

$$|\langle \dot{\mu}_t, f_t \rangle| \leq c_2\,, \quad \text{for almost all} \quad t \in I.$$

Then there exists a unique function $\lambda_t = \lambda(\mu_t, \dot{\mu}_t, f_t, t)$, such that for almost all $t \in I$

$$\langle \dot{\mu}_t, f_t \rangle = \mathcal{H}'(\mu_t, \lambda_t f_t)\,,$$

and a unique progressively measurable process $t \mapsto \lambda_t^n = \lambda(X_t^n, \dot{\mu}_t, f_t, t)$, almost all $t \in I$, with

$$\langle \dot{\mu}_t, f_t \rangle = \mathcal{H}'(X_t^n, \lambda_t^n f_t)\,.$$

Moreover, there is a constant K such that for all n

$$|\lambda_t| \vee |\lambda_t^n| \leq \Big(|\langle \dot{\mu}_t, f_t \rangle| + |\langle \mu_t, Af_t + \mathcal{L}f_t \rangle| \Big) / (\gamma^+ \wedge \gamma^-) \leq K\,,$$

and λ^n is continuous in X^n.

PROOF: This is a straightforward extension of [DK], section 4.2. We have indicated the inequality which we use together with the uniform bounds on a, b and ν to obtain the constant K. □

Two of the martingales M^n with $c = \lambda^n$ are used in the proofs below. First, taking $\beta_t = \lambda^n_{t-}$,

$$M_t^{1,n} := \int_0^t \lambda^n_{r-} \left(dK_r^{n,f} - \mathcal{H}'_r(X_r^n, \lambda_r^n f_r) \, dr \right)$$

$$(3.7) \qquad = \int_0^t \lambda^n_{r-} \left(dK_r^{n,f} - \langle \dot{\mu}_r, f_r \rangle \, dr \right)$$

$$\langle\!\langle M_\cdot^{1,n} \rangle\!\rangle_t = \frac{1}{n} \int_0^t (\lambda_r^n)^2 \, \mathcal{H}''_r(X_r^n, \lambda_r^n f_r) \, dr \, .$$

Second, take $\beta_t = 1$ and $f_t = \varphi$. Since then

$$M_t^{2,n} := K_t^{n,\varphi} - \int_0^t \mathcal{H}'_r(X_r^n, \lambda_r^n \varphi) \, dr = \langle X_t^n, \varphi \rangle - \langle \mu_t, \varphi \rangle$$

is a martingale under $\mathcal{P}^n_{\lambda^n}$, we have $E^n_{\lambda^n} \langle X_t, \varphi \rangle = \langle \mu_t, \varphi \rangle$. Also

$$\langle\!\langle M_\cdot^{2,n} \rangle\!\rangle_t = \frac{1}{n} \int_0^t \mathcal{H}''_r(X_r^n, \lambda_r^n \varphi) \, dr \, .$$

Here

$$(3.8) \qquad \mathcal{H}''_r(X_r^n, \lambda_r^n \varphi) = \left\langle X_r^n, \, |\nabla_a \varphi|^2 + \int (\Delta f)^2 e^{\lambda^n \Delta f} \, d\hat{\nu} \right\rangle \leq K_1 \, .$$

3.3 Lower bound in the independent case. The following proposition is the analog of [DK], Proposition 4.6. To demonstrate the method we repeat the proof partially.

Proposition 3.9. *Fix $\mu \in \mathcal{D}(I, \mathcal{M})$ and let V be an open neighbourhood of μ. Then*

$$\liminf_{n \to \infty} \frac{1}{n} \log \mathcal{P}^n(V) \geq -S(\mu) \, .$$

We can suppose $S(\mu) < \infty$ and $\mu_0 = \eta_0$. In the course of the proof of Lemma 3.2 is was seen that μ is then absolutely continuous. Take $f \in \mathcal{C}_0^{1,2,0}(I \times R^d \times E)$ as in Lemma 3.6. The proof of the following lemma is identical to [DK], Lemma 4.7.

Lemma 3.10. *Put*

$$\Sigma_T^n = \int_0^T \left\{ \left| \mathcal{H}_r(\mu_r, \lambda_r f_r) - \mathcal{H}_r(X_r^n, \lambda_r^n f_r) \right| + \left| \langle \dot{\mu}_r, f_r \rangle \right| \left| \lambda_r^n - \lambda_r \right| \right\} dr .$$

Then

$$\left| \int_0^T \lambda_r^n \, dK_r^{n,f} - \int_0^T \mathcal{H}_r(X_r^n, \lambda_r^n f_r) \, dr \right| \leq S(\mu) + \left| M_T^{1,n} \right| + \Sigma_T^n .$$

For $\delta > 0$, set

$$V_\delta = \left\{ \nu \in V : \sup_{t \in I} \|\nu_t - \mu_t\| < \delta \right\} .$$

The next two lemmas are also direct adaptions from [DK], Lemma 4.7 and 4.8. The proof of the first lemma uses the continuity of λ^n in X^n and the second lemma is an application of Lenglart-Rebolledo's inequality for $M^{2,n}$, using the bound (3.8).

Lemma 3.11. *For any $\varepsilon > 0$ there is a $\delta_0 > 0$ such that if $\delta \leq \delta_0$ and $X^n \in V_\delta$, then $\Sigma_T^n \leq \varepsilon$.*

Lemma 3.12. *For every $\delta > 0$,*

$$\lim_{n \to \infty} \mathcal{P}_{\lambda^n}^n \left(X^n \notin V_\delta \right) = 0 .$$

PROOF OF PROPOSITION 3.9: By Lemma 3.10,

$$\xi_T^{n,f}(\lambda^n) \leq e^{n \, S(\mu)} \exp n \left\{ \left| M_T^{1,n} \right| + \Sigma_T^n \right\} .$$

Therefore

$$\mathcal{P}^n(V) = \int_{X^n \in V} \xi_T^n(\lambda^n)^{-1} \, d\mathcal{P}_{\lambda^n}^n \geq \exp\{-n \, S(\mu)\} \, \Delta^n ,$$

where

$$\Delta^n = \int_{X^n \in V} \exp \left\{ -n \left| M_T^{1,n} \right| - n \, \Sigma_T^n \right\} d\mathcal{P}_{\lambda^n}^n .$$

Restrict to V_δ and further to the set of paths for which $\left| M_T^{1,n} \right| \leq \delta$. Then

$$\Delta^n \geq e^{-n \, \delta} \int_{X^n \in V_\delta, \, |M_T^{1,n}| \leq \delta} e^{-n \, \Sigma_T^n} \, d\mathcal{P}_{\lambda^n}^n .$$

For $\varepsilon > 0$ and $\delta \leq \delta_0$, by Lemma 3.11

$$\Delta^n \geq e^{-n(\delta+\varepsilon)} \mathcal{P}_{\lambda^n}^n \left(X^n \in V_\delta \,,\, |M_T^{1,n}| \leq \delta \right).$$

Now

$$\mathcal{P}_{\lambda^n}^n \left(X^n \in V_\delta \,,\, |M_T^{1,n}| \leq \delta \right) \geq \mathcal{P}_{\lambda^n}^n \left(X^n \in V_\delta \right) - \mathcal{P}_{\lambda^n}^n \left(|M_T^{1,n}| > \delta \right).$$

By Doob's inequality for the martingale in (3.7) using (3.8),

$$\mathcal{P}_{\lambda^n}^n \left(|M_T^{1,n}| > \delta \right) \leq \frac{\delta^{-2}}{n} K^2 K_1 T.$$

By this estimate and Lemma 3.12 we can now choose n_0 and $\delta_n \to 0$, such that

$$\mathcal{P}_{\lambda^n}^n \left(X^n \in V_\delta \,,\, |M_T^{1,n}| \leq \delta \right) \geq \frac{1}{2}, \quad n \geq n_0.$$

Then

$$\frac{1}{n} \log \Delta^n \geq -(\delta_n + \varepsilon) - \frac{1}{n} \log 2 \to -\varepsilon, \quad n \to \infty,$$

and finally,

$$\liminf_{n \to \infty} \frac{1}{n} \log \mathcal{P}^n(V) \geq -S(\mu) + \limsup_{n \to \infty} \frac{1}{n} \log \Delta^n = -S(\mu) - \varepsilon.$$

Let ε go to zero to finish the proof. $\qquad\square$

3.4 Upper bound in the independent case.

Proposition 3.13. *Fix $\mu \in \mathcal{D}(I, \mathcal{M})$. For every $\delta > 0$, there exists an open neighbourhood V of μ such that*

$$\limsup_{n \to \infty} \frac{1}{n} \log \mathcal{P}^n(V) \leq -S(\mu) + \delta.$$

provided $S(\mu) < \infty$. If $S(\mu) = \infty$ the assertion holds with $-\delta$ on the right side of the inequality.

PROOF: We follow [DK], Proposition 4.7. Fix $\delta > 0$. For $\delta' > 0$, put

$$V = V(\delta') := \left\{ \nu : \sup_{t \in I} \sup_{f \in \mathcal{C}^{1,2,0}(I \times R^d \times E)} |\langle \nu_t - \mu_t, f_t \rangle| < \delta' \right\}.$$

In this proof we use the notation $J_{s,t}(\mu, f) = J_{s,t}(f)$. From section 3.2 we know that $\xi_t^{n,f}(1) = \exp n\, J_{0,t}(X^n, f)$ is a $\mathcal{P}^n$-martingale. Use $1 \geq E^n \left[\xi_T^{n,f}(1) \,;\, X^n \in V \right]$ to see

$$\exp -n\, J_{0,T}(\mu, f) \geq E^n \left[\exp -n \big| J_{0,T}(X^n, f) - J_{0,T}(\mu, f) \big| \,;\, X^n \in V \right].$$

But simple estimates show that for $X^n \in V$ there is a constant C_1 such that

$$\left| J_{0,T}(X^n, f) - J_{0,T}(\mu, f) \right| \leq C_1 \, \delta' \, .$$

Choose $\delta' = \delta/C_1$ to get a set V for which

$$\exp -n \, J_{0,T}(\mu, f) \geq e^{-n \, \delta} \, \mathcal{P}^n(V) \, .$$

Hence

$$\frac{1}{n} \log \mathcal{P}^n(V) \leq -J_{0,T}(\mu, f) + \delta$$

and therefore, by (2.2) if $S(\mu) < \infty$,

$$\limsup_{n \to \infty} \frac{1}{n} \log \mathcal{P}^n(V) \leq - \sup_{f \in \mathcal{C}^{1,2,0}(I \times R^d \times E)} J_{0,T}(\mu, f) + \delta = -S(\mu) + \delta \, .$$

If $S(\mu) = \infty$, by (2.2) we can find $f \in \mathcal{C}^{1,2,0}(I \times R^d \times E)$ such that $J_{0,T}(f) \geq 2\delta$. $\qquad\qquad \square$

3.5 Exponential tightness.
The basic estimate needed to prove compactness of the level sets for S is covered by the following

Proposition 3.14. *For each $m \geq 1$ there is a compact set E_m in $\mathcal{D}(I, \mathcal{M})$ such that*

$$\mathcal{P}(X^n \notin E_m) \leq \exp -n \, m \, .$$

In [DG], Lemma 5.6, a method was developed to obtain such bounds. For jump processes Feng (1994), Lemma 3.9, gave a similar result. Proposition 4.11 in [DK] follows these ideas very closely. It is tedious but straightforward to extend the proofs to the multitype situation. Since the ideas are clear from the given references we feel it is superfluous to repeat them here.

3.6 Large deviation estimates for the dependent case.
We have proved the theorem for jump measures $\widehat{n}(x, dy)$ with respect to some fixed $\widehat{\mu}$. Let $\widehat{\mathcal{P}}^n$ denote the law corresponding to the frozen interaction $\widehat{n}$ and let $\mathcal{P}^n$ now denote the probability law in the general case. The common method to lift this restriction and obtain the result for the interaction model is to follow Shiga and Tanaka (1985) and apply a Girsanov tranformation to go from $\widehat{\mathcal{P}}^n$ to $\mathcal{P}^n$. Again we are somewhat sketchy and refer to [DK], section 5, and the references there. However, we give the form of the derivative $d\mathcal{P}^n/d\widehat{\mathcal{P}}^n$ and a rather complete proof of the lower bound.

For an n-particle motion $x = (x_t^1, \ldots, x_t^n)_{t \in I}$, let

$$\xi_t^i = u_t^i - u_0^i - \int_0^t b(u_s^i, \widehat{\mu}_s^n) \, ds \, , \quad \langle\!\langle \xi_\cdot^i, \xi_\cdot^i \rangle\!\rangle_t = \int_0^t a(u_s^i) \, ds \, ,$$

denote the d-dimensional continuous $\widehat{\mathcal{P}}^n$-martingale corresponding to the diffusion part and $N_t(x^i)$ the counting process for the total number of jumps in E starting at x^i, i.e. $N_t(x^i)$ is compensated under $\widehat{\mathcal{P}}^n$ by $\int_0^t \int_E \widehat{\nu}(x_s^i, dy)\, ds$. Define

$$A_t^n(x) = \sum_{i=1}^n \left\{ \int_0^t \left(b(x_r^i \,;\, X_r^n) - b(x_r^i \,;\, \widehat{\mu}_r) \right) \cdot d\xi_r^i \right.$$
$$\left. + \int_0^t \log \frac{\gamma(x_{r-}^i \,;\, X_{r-}^n)}{\gamma(x_{r-}^i \,;\, \widehat{\mu}_{r-})}\, dN_r(x_r^i) \right\},$$

and

$$B_t^n(x) = \sum_{i=1}^n \int_0^t \left\{ \frac{1}{2} \left| b(x_r^i \,;\, X_r^n) - b(x_r^i \,;\, \widehat{\mu}_r) \right|^2 + \left(\gamma(x_r^i \,;\, X_r^n) - \gamma(x_r^i \,;\, \widehat{\mu}_r) \right) \right\} dr.$$

Moreover, for any real α, put

$$B_t^{n,(\alpha)}(x) = \sum_{i=1}^n \int_0^t \left\{ \frac{\alpha^2}{2} \left| b(x_r^i \,;\, X_r^n) - b(x_r^i \,;\, \widehat{\mu}_r) \right|^2 + \left(\frac{\gamma_r(x_r^i, X_r^n)}{\gamma_r(x_r^i, \widehat{\mu}_r)} \right)^\alpha - 1 \right\} dr.$$

Then by standard theory the change of measure from $\widehat{\mathcal{P}}^n$ to $\mathcal{P}^n$ is absolutely continuous with Radon-Nikodym derivative

$$\left. \frac{d\mathcal{P}^n}{d\widehat{\mathcal{P}}^n} \right|_{\mathcal{F}_t} = \exp\left\{ A_t^n - B_t^n \right\},$$

and, for any $\alpha \in R$,

$$Y_t^{(\alpha)} := \exp\{ \alpha A_t^n - B_t^{n,(\alpha)} \}, \quad t \in I,$$

is a $(\widehat{\mathcal{P}}^n, \mathcal{F}_t)$-local martingale and thus a supermartingale.

To prove the lower bound Proposition 3.9 in the general case use first the uniform continuity in μ of b and γ to find for all $\delta > 0$ a neighborhood W of $\widehat{\mu}$ with $W \subset V$ such that

$$|B_T^n| \leq n\delta \quad \text{uniformly on} \quad W$$

and

$$|B_T^{n,(-\alpha)}| \leq C(\alpha)\, \delta\, n \quad \text{uniformly on} \quad W,$$

where $C(\alpha)$ is a constant depending on α only. Then, for a pair of conjugate exponents $p, q > 1$,

$$\mathcal{P}^n(V) \geq \mathcal{P}^n(W) = \widehat{E}^n \left[\exp\left\{ A_T^n - B_T^n \right\} ; X^n \in W \right]$$
$$\geq e^{-n\delta}\, \widehat{E}^n \left[\exp A_T^n ; X^n \in W \right]$$
$$\geq e^{-n\delta}\, e^{-n\delta C(\alpha)/\alpha}\, \widehat{E}^n \left[\exp\left\{ A_T^n + \frac{p}{q} B_T^{n,(-q/p)} \right\} ; X^n \in W \right].$$

Thus, by Hölder's inequality and the supermartingale property of $Y_t^{(-q/p)}$,

$$\frac{1}{n}\log \mathcal{P}^n(V) \geq -\delta(1 + C(\alpha)/\alpha) + p\frac{1}{n}\log \widehat{\mathcal{P}}^n(W).$$

We know that Theorem 1.1 holds for $\widehat{\mathcal{P}}^n$. Hence

$$\limsup_{n\to\infty} \frac{1}{n}\log \mathcal{P}^n(V) \geq -\delta(1 + C(\alpha)/\alpha) - p\,S(\widehat{\mu}).$$

Take $\delta \to 0$ and then $p \to 1$ to finish the proof of Proposition 3.9 in the interacting case. The proof of Proposition 3.13 is carried out similarly, compare [DK], Lemma 5.3.

REFERENCES

[1] H. ANDERSSON, B. DJEHICHE, Limit theorems for spatial epidemics. (Preprint Trita 1994:0019, Royal Institute of Technology, Stockholm, 1994).

[2] D.A. DAWSON, J. GÄRTNER, Large deviations from the McKean-Vlasov limit for weakly interacting diffusions. *Stochastics* **20** (1987) 247-308.

[3] D.A. DAWSON, J. GÄRTNER, Large deviations, free energy functional and quasi-potential for a mean field model of interacting diffusions. *Mem. Am. Math. Soc.* **78** (1989).

[4] B. DJEHICHE, I. KAJ, The rate function for some measure-valued jump processes. Preprint, Uppsala University 1994. To appear: *Ann. Probab.*

[5] S. FENG, Large deviations for empirical process of mean field interaction particle system with unbounded jump. To appear: *Ann. Probab.* 1994.

[6] C. GRAHAM, McKean-Vlasov Ito-Skorohod equations, and nonlinear diffusions with discrete jump sets. *Stoch. Proc. Appl.* **40** (1992) 69-82.

[7] A. MARTIN-LÖF, Symmetric sampling procedures, general epidemic processes and their threshold limit theorems. *J. Appl. Probab.* **23** (1986) 265-282.

[8] J. NEVEU, Discrete-parameter martingales. (North-Holland, Amsterdam, 1975).

[9] K. OELSCHLÄGER, A martingale approach to the law of large numbers for weakly interacting stochastic processes. *Ann. Probab.* **12** (1984) 458-479.

[10] T. SHIGA, H. TANAKA, Central limit theorem for a system of Markovian particles with mean field interaction. *Z. Wahrscheinlichk. verw. Gebiete* **69** (1985) 439-459.

Noise and dynamic transitions

G.D. LYTHE
Department of Applied Mathematics and Theoretical Physics,
University of Cambridge, Cambridge CB3 9EW, UK

Abstract

A parabolic stochastic PDE is studied analytically and numerically, when a bifurcation parameter is slowly increased through its critical value. The aim is to understand the effect of noise on delayed bifurcations in systems with spatial degrees of freedom. Realisations of the nonautonomous stochastic PDE remain near the unstable configuration for a long time after the bifurcation parameter passes through its critical value, then jump to a new configuration. The effect of the nonlinearity is to freeze in the spatial structure formed from the noise near the critical value.

Introduction

Many physical systems undergo a transition from a spatially uniform state to one of lower symmetry. Such systems are commonly modelled by a simple differential equation which has a bifurcation parameter with a critical value at which there is an exchange of stability between steady states [1]. The model can be improved by adding noise [2], which can be understood as allowing for small rapidly varying effects neglected in the formulation of the model. Noise is also necessary to provide the initial symmetry-breaking which permits the system to choose one of the available lower-symmetry states.

If the bifurcation parameter is not a function of time, the effect of noise is to make a bifurcation ill-defined in an $\mathcal{O}(\epsilon)$ region near the critical value, where ϵ is the size of the noise [3]. A more dramatic effect is found if the bifurcation parameter is a function of time, corresponding to an experimental situation in which the critical value is not known in advance, so the parameter is slowly changed until a qualitative change is seen in the behaviour of the system [4].

In the case of the pitchfork bifurcation, described by the normal form

$$\dot{y} = gy - y^3, \tag{1}$$

181

it is normally said that, if $g > 0$, y will be found either at $y = \sqrt{g}$ or $y = -\sqrt{g}$. However, on solving the stochastic differential equation

$$dy = (gy - y^3)dt + \epsilon dw \tag{2}$$

with $g = \mu t$, starting with $y = 0$ at $g = g_0 < 0$, it is found that y remains near $y = 0$ until well after $g = 0$ (Figure 1). (The Wiener process – standard Brownian motion – is denoted by lower case w.) If μ is small, a typical trajectory consists of a long sojourn near $y = 0$, a sudden jump away from $y = 0$ and then relaxation towards $y^2 = g$. The value of g at which the jump occurs is a random variable whose probability distribution can be found by solving the linearised version of (2); the mean value is approximately $\sqrt{2\mu|\ln \epsilon|}$ and the width of the probability distribution is proportional to μ [5].

The phenomenon of delayed bifurcation just described and its sensitivity to noise has been studied theoretically and experimentally in the case of non-autonomous stochastic ordinary differential equations [6-9]; the corresponding phenomenon for partial differential equations, delayed transition, is the subject of this article. A dynamic transition is described by the following non-autonomous parabolic stochastic partial differential equation:

$$dY = (g(t)Y - Y^3 + \Delta Y)dt + \epsilon dW, \tag{3}$$

where $g = \mu t$ is slowly increased through 0, W is the Brownian sheet, $\Delta = \sum_{i=1}^{m} \frac{\partial^2}{\partial x_i^2}$ is the Laplacian in $\mathbf{R}^m$ (the derivatives exist only in the sense of generalised functions) and μ and ϵ are constants such that

$$\epsilon \ll \mu \ll 1. \tag{4}$$

In section 2 this initial value problem is solved using periodic boundary conditions in one space dimension, in which case Y is a stochastic process with values in the space of real-valued continuous functions on $[0, L]$, and has continuous mean $\langle Y_t(x) \rangle$ and correlation function $\langle Y_t(x)Y_t(x') \rangle$ [10-11]. In section 3 the corresponding equation in space dimension, m, greater than one is discussed.

The present work was motivated by the behaviour of some nonlinear systems of ordinary and partial differential equations whose dynamics are controlled by noise [12,13]. The controlling influence of noise arises because trajectories spend long times near a slow invariant manifold, and the dynamics near this manifold are similar to those of a dynamic bifurcation [5]. The most important parameter in the description of these systems is $\mu|\ln \epsilon|$, where $\frac{1}{\mu}$ is the timescale for the dynamics near the slow manifold and ϵ is the magnitude of the noise.

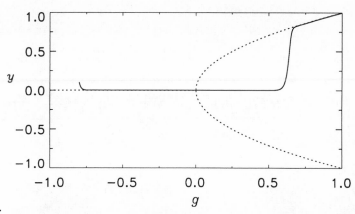

Fig. 1.

Dynamic pitchfork bifurcation with noise. The solid line is one trajectory of the ordinary SDE (2), with $\epsilon = 10^{-10}$, $\mu = 0.01$, and initial condition $y = 0.1$ at $g = -0.8$. Also shown, as dotted lines, are the loci of stable fixed points of the corresponding autonomous system (1). The trajectory lingers near $y = 0$ for a long time after g passes through 0; the value of g at which the jump towards $y = \pm\sqrt{g}$ occurs is a random variable with mean approximately equal to $\sqrt{2\mu|\ln\epsilon|}$.

2. Dynamic transition in one space dimension

Four configurations from a numerical realisation of (3) with $m = 1$ are shown in Figure 2. According to static analysis the zero configuration is unstable for $g > 0$, but Y typically remains much closer to 0 than to $Y^2 = g$ until well after $g = 0$. The first part of a dynamic transition, the loss of stability of the zero configuration, is thus controlled by the noise and the sweep rate, not by the nonlinearity.

If Y obeys (3), the linearised version of the stochastic differential equation for its kth Fourier mode, $u_t^{(k)} = \frac{1}{\sqrt{L}}\int_0^L Y_t(x')e^{ikx2\pi/L}dx'$, is

$$du^{(k)} = (g(t) - (\frac{2\pi}{L})^2 k^2)dt + \epsilon dw^{(k)}, \qquad k = 0, 1, 2, \ldots \qquad (5)$$

with each $w^{(k)}$ an independent Wiener process. Thus, for as long as nonlinear terms are unimportant, each Fourier mode evolves independently, satisfying an equation of the dynamic bifurcation type.

As the in ODE case, the new configuration (Figure 2(d)) appears abruptly at a value of g which is a random variable with mean approximately $\sqrt{2\mu|\ln\epsilon|}$. The effect of the cubic nonlinearity is to arrest the exponential growth of $\langle Y_t(x)Y_t(x)\rangle$ when it becomes $\mathcal{O}(1)$ and to freeze in the spatial structure formed during the slow sweep. Establishing when this happens makes it possible to answer another question which is not considered in the static version: what is the typical size of the spatial domains formed in the transition?

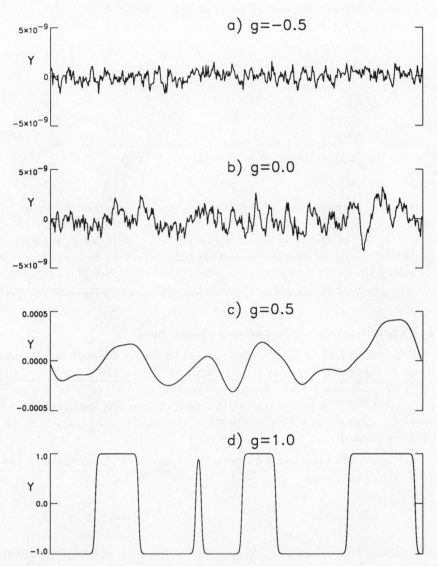

Fig. 2.

Dynamic transition in one space dimension. One numerically-generated realisation of the stochastic PDE (3) $Y_t(x)$, $x \in [0, L]$, is displayed at four times as the bifurcation parameter, $g = \mu t$, is slowly increased, passing through 0 at $t = 0$. (Note the different vertical scales.) It remains close to the zero configuration until well after $g = 0$. Nonlinear terms become important when $g \simeq \sqrt{2\mu |\ln \epsilon|}$; their effect is to freeze in the spatial structure formed, near $g = 0$, from the noise. The data shown here were generated using a finite difference method with 500 points. ($L = 300$, $\mu = 0.01$, $\epsilon = 10^{-10}$.)

To examine the dynamics of the loss of stability of the zero configuration, consider the solution of the linearised version of (3),

$$dY = (g(t)Y + \Delta Y)dt + \epsilon dW, \qquad (6)$$

with initial data $Y_t(x) = f(x)$ on $x \in [0, L]$ at $t = t_0 < 0$ when $g = g_0$:

$$Y(x,t) = \int_{)}^{L} G(t, t_0, x, x')f(x')dx' + \epsilon \int_{t_0}^{t} \int_{[0,L]} G(t, t_0, x, x')dW_t(x'). \qquad (7)$$

The first integral is an ordinary integral; the second a space-time Ito integral. The fundamental solution $G(t, t_0, x, x')$ is given by

$$G(t, t', x, x') = \frac{1}{\sqrt{4\pi(t - t')}} e^{\frac{1}{2}\mu(t^2 - t'^2)} \sum_{j=-\infty}^{\infty} e^{-\frac{(x-x'-jL)^2}{4(t-t')}}. \qquad (8)$$

The solution (7) prescribes the mean and correlation function of Y as a function of time. The mean value of Y is the first term, an ordinary integral which, for large $t - t_0$, is given by

$$\langle Y_t(x) \rangle = e^{\frac{1}{2}\mu(t^2 - t_0^2)}\bar{y} \qquad (9)$$

where $\bar{y}$ is a smoothed version of the initial data $f(x)$. In the rest of this article, it is assumed that $\bar{y} \leq \mathcal{O}(1)$ and $2\mu|\ln \epsilon| < g_0^2$. This ensures that $\langle Y_t \rangle^2 \ll \langle Y_t^2 \rangle$, ie Y is taken as a mean-zero Gaussian process. It is also assumed that $|t_0| > \frac{1}{\sqrt{\mu}}$, so that quasi-static equilibrium is reached before $g = 0$, and $L \gg \sqrt{t - t_0}$ so that, at any one x, only one term in (8) is important.

The correlation function is a mean square quantity which can be evaluated using the 'delta function' property of the Brownian sheet as an integrator:

$$\langle Y_t(x)Y_t(x') \rangle = \epsilon^2 \int_0^L \int_{t_0}^t G^2(t, s, x, x')ds dx'$$

$$= \epsilon^2 \int_{t_0}^t ds \frac{e^{\mu(t^2 - s^2)}}{\sqrt{8\pi(t - s)}} e^{-\frac{(x-x')^2}{8(t-s)}}. \qquad (10)$$

For times t such that $-t > 1/\sqrt{\mu}$ the nonautonomous correlation function (10) differs only by $\mathcal{O}(\mu)$ from the result obtained for fixed $g < 0$ if $t_0 \to \infty$ [14]:

$$\langle Y_t(x)Y_t(x') \rangle = \frac{\epsilon^2}{4\sqrt{|g|}} e^{-|x-x'|\sqrt{|g|}}. \qquad (11)$$

In the non-autonomous case, the correlation function remains finite as t passes through 0, and for large positive times $(t > 1/\sqrt{\mu})$, it is well approximated by:

$$\left\langle Y_t(x)Y_t(x')\right\rangle \simeq \epsilon^2 \frac{e^{\mu t^2}}{\sqrt{8\mu t}} e^{-(x-x')^2/8t}. \tag{12}$$

Note the Gaussian form of the spatial correlation and the existence of a characteristic length $\sqrt{8t}$ at time t. This formula is valid until $g \simeq \sqrt{2\mu|\ln\epsilon|}$, when the cubic nonlinearity becomes important. The correlation length at this time becomes the characteristic size of the spatial domains formed (Figure 2(d)).

3. Higher dimensions

Proceeding as in section 2, the expression obtained for the correlation function at time t is

$$\left\langle Y_t(x)Y_t(x')\right\rangle = \int_{t_0}^t ds \frac{e^{\mu(t^2-s^2)}}{(8\pi(t-s))^{\frac{m}{2}}} e^{-\frac{||x-x'||^2}{8(t-s)}}, \tag{13}$$

where $||x - x'||$ is the Euclidean norm on $\mathbf{R}^m$. The major difference from the case $m = 1$ is that the correlation function does not approach a finite limit as $||x - x'|| \to 0$. This is a general feature of second order stochastic PDEs in more than one space dimension [15,16]. On a finite grid, equation (6), with $g < 0$ fixed, is of interest because it generates a stochastic process with non-zero correlation length in both space and time [17]. The singularity in the correlation function at $x = x'$ manifests itself in the normalisation, which increases without a finite limit as the grid spacing is decreased. In the nonautonomous version the same divergence is found (logarithmic in $||x - x'||$ when $m = 2$), but for $g > 0$ the singularity is only noticeable on very small length scales. The nature of these divergences, in both discrete and continuous versions, is conveniently studied in Fourier space, where the linearised equation separates into uncoupled ordinary SDEs, each of which can be solved exactly.

The finite difference method for a parabolic SPDE consists of replacing the infinite dimensional system (3) by N^m ordinary SDEs on a grid of equally-spaced points in $[0, L]^m$. For the SPDE (3), the SDE at each point is

$$dY(x) = (gY(x) - Y^3(x))dt + \rho\tilde{\Delta}Y(x)dt + \rho^{\frac{m}{4}}\epsilon dw \tag{14}$$

where the discrete Laplacian $\tilde{\Delta}$ is defined by

$$\tilde{\Delta}Y(x) = \sum_{x'} Y(x') - 2mY(x) \tag{15}$$

and the sum is over the $2m$ nearest neighbours of x. Each of the N^m ordinary SDEs has an associated Wiener process, independent of all the others, and so can be numerically solved in the normal way [18]. Note the scaling of the magnitude of the noise added at each grid point with ρ. The smallest length scale resolved is roughly the distance between nearest neighbours, $\frac{1}{\sqrt{\rho}}$. As in numerical solution of deterministic parabolic PDEs, there is a maximum timestep, proportional to $\frac{1}{\rho}$, if the finite difference method is to be numerically stable [19].

Solved on a finite grid in this way, the behaviour of realisations of the non-autonomous version SPDE (3), with $m > 1$ and $g = \mu t$ as in section 2, is qualitatively similar to that in one space dimension [20]. That is, a Gaussian form of the spatial correlation function emerges from the slow sweep through $g = 0$,

$$\left\langle Y_t(x) Y_t(x') \right\rangle \simeq \epsilon^2 \sqrt{\frac{\pi}{\mu}} \frac{e^{\mu t^2}}{(8\pi t)^{\frac{m}{2}}} e^{-\|x-x'\|^2/8t}, \tag{16}$$

and metastable domains of positive and negative Y are formed when the non-linearity becomes important, at $g \simeq \sqrt{2\mu|\ln \epsilon|}$.

4. Conclusion

In systems with spatial degrees of freedom, the phenomenon of delayed bifurcation is accompanied by a noise-controlled formation of spatial structure. Here the stochastic PDE (3) was studied, whose realisations consist long sojourns near the zero configuration during which the shape of the spatial correlation function becomes Gaussian, followed by an abrupt change to the nonlinear régime, with the formation of domains of positive and negative Y. This spatial structure, formed near the critical value of the bifurcation parameter g when noise dominates, is frozen in at $g \simeq \sqrt{2\mu|\ln \epsilon|}$, when the characteristic length is $\sqrt{8t}$.

Acknowledgements

My thanks to the Alison Etheridge and all others responsible for the smooth running of the conference. I am grateful for an Overseas Research Students Award and financial support from Trinity College, Cambridge.

References

[1] Alan C. Newell, Thierry Passot and Joceline Lega *Order parameter equations for patterns* Ann. Rev. Fluid Mech. **25** 399–453 (1993)

[2] Frank Moss and P.V.E. McClintock *Noise and nonlinear dynamical systems* (Cambridge University Press, Cambridge, 1987)

[3] C. Meunier and A.D. Verga *Noise and Bifurcations* J. Stat. Phys. **50** 345–375 (1988)

[4] Christopher W. Meyer, Guenter Ahlers and David S. Cannell *Stochastic influences on pattern formation in Rayleigh-Bénard convection: Ramping experiments* Phys. Rev. A **44** 2514–2537 (1991)

[5] G.D. Lythe and M.R.E. Proctor, *Noise and slow-fast dynamics in a three-wave resonance problem* Phys. Rev. E. **47** 3122-3127 (1993).

[6] N.G. Stocks, R. Mannella and P.V.E. McClintock *Influence of random fluctuations on delayed bifurcations: The case of additive white noise* Phys. Rev. A **40** 5361–5369 (1989)

[7] J.W. Swift, P.C. Hohenberg and Guenter Ahlers *Stochastic Landau equation with time-dependent drift* Phys. Rev. A **43** 6572–6580 (1991)

[8] Claude Lobry *A propos du sens des textes mathématiques. Un example: la théorie des "bifurcations dynamiques"* Ann. Inst. Fourier **42** 327-352 (1992)

[9] V.I. Arnol'd (Ed.) *Dynamical Systems V* (Springer, Berlin, 1994)

[10] Tadahisa Funaki *Random motion of strings and related stochastic evolution equations* Nagoya Math. J. **89** 129–193 (1983)

[11] I. Gyöngy and E. Pardoux *On quasi-linear stochastic partial differential equations* Probab. Theory Relat. Fields **94** 413–426 (1993)

[12] M.R.E. Proctor andD.W. Hughes *The false Hopf bifurcation and noise sensitivity in bifurcations with symmetry* Eur.J.Mech.,B/Fluids **10** 81–86 (1990)

[13] Miltiades Georgiou and Thomas Erneux *Pulsating laser oscillations depend on extremely-small-amplitude noise* Phys.Rev. A **45** 6636–6642 (1992)

[14] H. Haken *Synergetics* (Springer, Berlin, 1978)

[15] J.B. Walsh *An introduction to stochastic partial differential equations*, in: Ecole d'été de probabilités de St-Flour XIV P.L. Hennequin (ed.) pp266–439 (Springer, Berlin, 1986)

[16] Charles R. Doering *Nonlinear Parabolic Stochastic Differential Equations with Additive Colored Noise on $R^d \times R_+$: A Regulated Stochastic Quantization* Commun. Math. Phys. **109** 537–561 (1987)

[17] J. García-Ojalvo and J.M. Sancho *Colored noise in spatially extended systems* Phys. Rev. E **49** 2769–2778 (1994)

[18] Peter E. Kloeden and Eckhard Platen, *Numerical Solution of Stochastic Differential Equations* (Springer, Berlin,1992)

[19] Leon Lapidus and George F. Pinder *Numerical Solution of Partial Differential Equations in Science and Engineering* (Wiley, New York, 1982)

[20] G.D. Lythe, *Stochastic slow-fast dynamics* (PhD thesis, University of Cambridge, 1994)

Backward Stochastic Differential Equations and Quasilinear Partial Differential Equations

Xuerong Mao *

Department of Statistics and Modelling Science

University of Strathclyde

Glasgow G1 1XH, Scotland, U.K.

Abstract: In this paper we first review the classical Feynman-Kac formula and then introduce its generalization obtained by Pardoux-Peng via backward stochastic differential equations. It is because of the usefulness of the Feynman-Kac formula in the study of parabolic partial differential equations we see clearly how worthy to study the backward stochastic differential equations in more detail. We hence further review the work of Pardoux and Peng on backward stochastic differential equations and establish a new theorem on the existence and uniqueness of the adapted solution to a backward stochastic differential equation under a weaker condition than Lipschitz one.

Key Words: Backward stochastic differential equation, parabolic partial differential equation, adapted solution, Bihari's inequality.

1. Introduction

In 1990, Pardoux & Peng [14] initiated the study of backward stochastic differential equations motivated by optimal stochastic control (see Bensoussan [1], Bismut [3], Haussmann [9] and Kushner [10]). It is even more important that Pardoux & Peng [15], Peng [17] recently gave the probabilistic representation for the given solution of a quasilinear parabolic partial differential equation in term of the solution of the corresponding backward stochastic differential equation. In other words, they obtained a generalization of the well-known Feynman-Kac formula (cf. Freidlin [4] or Gikhman & Skorokhod [8]). In view of the powerfulness of the Feynman-Kac formula in the study of partial differential equations e.g. K.P.P. equation (cf. Freidlin [6]), one may expect that the Pardoux-Peng generalized formula will play an important role in the study of quasilinear parabolic partial differential equations. Hence from both viewpoints of the optimal stochastic control and partial differential equations, we see clearly how worthy to study the backward stochastic differential equations in more detail.

* Supported by the grant of the Faculty of Science, University of Strathclyde.

Pardoux & Peng [14] established the theory of the existence and uniqueness of the solution to a backward stochastic differential equation under the uniform Lipschitz condition. On the other hand, it is somehow too strong to require the uniform Lipschitz continuity in applications e.g. in dealing with quasilinear parabolic partial differential equations. So it is useful to find other weaker conditions under which the backward stochastic differential equation has a unique solution and this is one of the main aims of this paper.

In this paper we shall first review the classical Feynman-Kac formula and point out how useful the formula is in the study of partial differential equations. We then introduce its generalization obtained by Pardoux-Peng via backward stochastic differential equations. We therefore further review the work of Pardoux and Peng on backward stochastic differential equations and establish a more general result on the existence and uniqueness of the solution.

2.   Linear Parabolic Partial Differential Equations and Stochastic Differential Equations

Let us start with the well-known Feynman-Kac formula which shows the connection between solutions of partial differential equations and stochastic differential equations. To explain, let us consider a linear parabolic partial differential equation

$$\left.\begin{array}{c} \dfrac{\partial u(t,x)}{\partial t} + Lu(t,x) + c(t,x)u(t,x) = 0, \quad 0 \le t < 1, x \in R^d \\[2mm] u(1,x) = g(x), \quad x \in R^d \end{array}\right\} \qquad (2.1)$$

where

$$L = \frac{1}{2} \sum_{i,j=1}^{d} (\sigma\sigma^T)_{ij}(t,x)\frac{\partial^2}{\partial x_i \partial x_j} + \sum_{i=1}^{d} b_i(t,x)\frac{\partial}{\partial x_i},$$

$$\sigma = (\sigma_{ij})_{d\times d}, \qquad b = (b_1,...,b_d)^T$$

with $\sigma_{ij}(t,x)$, $b_i(t,x)$ and $c(t,x)$ are all real-valued functions defined on $[0,1] \times R^d$ and $g(x)$ on R^d. Assume $\sigma_{ij}(t,x)$, $b_i(t,x)$, $c(t,x)$ and $g(x)$ are all smooth enough so that equation (2.1) has a unique classical solution $u(t,x)$. In order to represent this solution in term of the solutions of stochastic differential equations, we introduce, for any $(t,x) \in [0,1] \times R^d$, a stochastic differential equation

$$X_s^{t,x} = x + \int_t^s b(r,X_r^{t,x})dr + \int_t^s \sigma(r,X_r^{t,x})dw_r, \quad t \le s \le 1, \qquad (2.2)$$

where $w_t, t \ge 0$ is a d-dimensional Brownian motion defined on a complete probability space $\{\Omega, \mathcal{F}_t, P\}$ with the natural filtration $\{\mathcal{F}_t\}_{t\ge 0}$ (i.e. $\mathcal{F}_t =$

$\sigma\{w_s : 0 \leq s \leq t\}$). The stochastic representation of the solution of equation (2.1) is now given by

$$u(t,x) = E\big(g(X_1^{t,x})exp[\int_t^1 c(s, X_s^{t,x})ds]\big) \tag{2.3}$$

which is called the Feynman-Kac formula. The connection between solutions of partial differential equations and stochastic differential equations is discussed in detail in Gikhman and Skorokhod [8], and in an expository article by Freidlin [4]. The above (2.3) is only one version of the formula. The Feynman-Kac formula not only shows the relationship between the two branches of mathematics but also gives a probabilistic approach to the study of partial differential equations. For the details about this probabilistic approach we refer the reader to Freidlin [5] and Freidlin & Wentzell [7] among others.

Note that (2.3) gives an explicit representation for the solution of the linear equation (2.1). How about a quasilinear parabolic partial differential equation? To be precise, let us consider the following quasilinear equation

$$\left. \begin{array}{l} \dfrac{\partial u(t,x)}{\partial t} + Lu(t,x) + c(u,x)u(t,x) = 0, \quad 0 \leq t < 1, x \in R^d \\[3mm] u(1,x) = g(x), \quad x \in R^d \end{array} \right\} \tag{2.4}$$

where c is now defined on $R \times R^d$. The Feynman-Kac formula now takes the form of

$$u(t,x) = E\bigg(g(X_1^{t,x})exp[\int_t^1 c(u(s, X_s^{t,x}), X_s^{t,x})ds]\bigg). \tag{2.5}$$

Of course this is not an explicit representation. Nevertheless it is very useful. For example, if assume

$$\underline{c}(x) \leq c(u,x) \leq \bar{c}(x)$$

and set $G_+ = \{x \in R^d : g(x) \geq 0\}$ and $G_- = R^d - G_+$, then we have the estimates

$$E\big(I_{G_+}g(X_1^{t,x})exp[\int_t^1 \underline{c}(X_s^{t,x})ds] + I_{G_-}g(X_1^{t,x})exp[\int_t^1 \bar{c}(X_s^{t,x})ds]\big)$$

$$\leq u(t,x) \leq E\big(I_{G_+}g(X_1^{t,x})exp[\int_t^1 \bar{c}(X_s^{t,x})ds]$$

$$+ I_{G_-}g(X_1^{t,x})exp[\int_t^1 \underline{c}(X_s^{t,x})ds]\big),$$

where and in the sequel I_D is the indicator function of set D. If we assume furthermore that $g(x)$ is nonnegative, i.e. $g(x) \geq 0$, then the above estimates become

$$E\big(g(X_1^{t,x})exp[\int_t^1 \underline{c}(X_s^{t,x})ds]\big)$$

$$\leq u(t,x) \leq E\big(g(X_1^{t,x})exp[\int_t^1 \bar{c}(X_s^{t,x})ds]\big). \tag{2.6}$$

Denote by $\bar{u}(t,x)$ and $\underline{u}(t,x)$ the corresponding solutions of equation (2.1) with $c(t,x)$ replaced by $\bar{c}(x)$ and $\underline{c}(x)$ respectively. Recalling (2.3), we can then rewrite (2.6) as

$$\underline{u}(t,x) \leq u(t,x) \leq \bar{u}(t,x),$$

which is a comparison result. Due to the page limit we could only show this very simple example to give the reader a feeling about the powerfulness of the Feynman-Kac formula but not in more detail. However we would like to mention that Freidlin [6] and other authors have applied the Feynman-Kac formula to obtain a lot of properties for a class of reaction-diffusion equations e.g. K.P.P equation (cf. Kolmogorov et al [11]). In other words, the Feynman-Kac formula plays a great role in the study of linear or quasilinear parabolic partial differential equations.

3. Quasilinear Parabolic Partial Differential Equations and Backward Stochastic Differential Equations

It has been seen in the previous section that although the Feynman-Kac formula does not give an explicit representation for the solution of a quasilinear parabolic partial differential equation, it has been very useful. One can then imagine that if there is a generalization of the Feynman-Kac formula which gives the explicit representation then such generalization should turn out to be even more useful. In this section we shall introduce this generalization.

To explain, let us now consider a more general quasilinear parabolic partial differential equation

$$\left.\begin{array}{c} \dfrac{\partial u(t,x)}{\partial t} + Lu(t,x) + f\big(t,x,u(t,x),(\nabla u\sigma)(t,x)\big) = 0, \quad 0 \leq t < 1, x \in R^d \\[2mm] u(1,x) = g(x), \quad x \in R^d \end{array}\right\} \tag{3.1}$$

where f is defined on $[0,1] \times R^d \times R \times R^d$ and ∇u is the gradient of u. Again we assume that equation (3.1) has a unique classical solution $u(t,x)$. Now for each $(t,x) \in [0,1] \times R^d$, let $X_s^{t,x}, t \leq s \leq 1$ be the solution of stochastic differential equation (2.2), and then solve the following backward stochastic differential equation in (Y,Z)

$$Y_s^{t,x} = g(X_1^{t,x}) + \int_s^1 f(r,X_r^{t,x},Y_r^{t,x},Z_r^{t,x})dr - \int_s^1 Z_r^{t,x}dw_r, \quad t \leq s \leq 1. \tag{3.2}$$

Then the solution of equation (3.1) can be represented as

$$u(t,x) = EY_t^{t,x} \tag{3.3}$$

which is called a generalized Feynman-Kac formula and is due to Pardoux & Peng [15]. To see this is a generalization, let

$$f(t,x,y,z) = c(t,x)y,$$

so the quasilinear equation (3.1) reduces to the linear equation (2.1) and the backward stochastic differential equation (3.2) becomes

$$Y_s^{t,x} = g(X_1^{t,x}) + \int_s^1 c(r,X_r^{t,x})Y_r^{t,x}dr - \int_s^1 Z_r^{t,x}dw_r. \tag{3.4}$$

Note that equation (3.4) has the explicit solution

$$Y_s^{t,x} = g(X_1^{t,x})exp[\int_s^1 c(r,X_r^{t,x})dr] - \int_s^1 exp[\int_s^r c(v,X_v^{t,x})dv]Z_r^{t,x}dw_r.$$

Substituting this into (3.3) gives

$$u(t,x) = EY_t^{t,x} = E\big(g(X_1^{t,x})exp[\int_s^1 c(r,X_r^{t,x})dr]\big)$$

which is the same as the classical Feynman-Kac formula (2.3). In view of the powerfulness of the Feynman-Kac formula in the study of a class of partial differential equations, one can expect that this generalized formula will play an even greater role in the study of quasilinear parabolic partial differential equations. Therefore it is important to investigate the backward stochastic differential equations in more detail.

4. Backward Stochastic Differential Equations

Backward stochastic differential equations arise from the optimal stochastic control. In fact, the equation for the adjoint process in optimal stochastic control (see Bensoussan [1], Bismut [3], Haussmann [9] and Kushner [10]) is a linear version of the following backward stochastic differential equation

$$x(t) + \int_t^1 f(s,x(s),y(s))ds + \int_t^1 [g(s,x(s)) + y(s)]dw(s) = X \tag{4.1}$$

on $0 \leq t \leq 1$. Here $\{w(t) : 0 \leq t \leq 1\}$ is a q-dimensional Brownian motion defined on the probability space $\{\Omega, \mathcal{F}_t, \{\mathcal{F}_t\}_{0 \leq t \leq 1}, P\}$ with the natural filtration $\{\mathcal{F}_t\}_{0 \leq t \leq 1}$, and X is a given $\mathcal{F}_1$-measurable R^d-valued random variable such that $E|X|^2 < \infty$. Moreover, f is a mapping from $\Omega \times [0,1] \times R^d \times R^{d \times q}$ to R^d which is assumed to be $\mathcal{P} \otimes \mathcal{B}_d \otimes \mathcal{B}_{d \times q} / \mathcal{B}_d$ measurable, where $\mathcal{P}$ denotes the σ-algebra of $\mathcal{F}_t$-progressively measurable subsets of $\Omega \times [0,1]$. Also g is a mapping from $\Omega \times [0,1] \times R^d$ to $R^{d \times q}$ which is assumed to be $\mathcal{P} \otimes \mathcal{B}_d / \mathcal{B}_{d \times q}$ measurable. In the field of control, we usually regard $y(.)$ as an adapted control and $x(.)$ as the state of the system. We are allowed to choose an adapted control $y(.)$ which drives the state $x(.)$ of the system to the given target X at time $t = 1$. This is the so-called reachability problem. So in fact we are looking for a pair of stochastic processes $\{x(t), y(t) : 0 \le t \le 1\}$ with values in $R^d \times R^{d \times q}$ which is $\mathcal{F}_t$-adapted and satisfies equation (1.1). Such a pair is called an adapted solution of the equation.

A systematic study of backward stochastic differential equations was initiated by Pardoux & Peng [14] in 1990, where they showed the existence and uniqueness of the adapted solution under the condition that $f(t, x, y)$ and $g(t, x)$ are uniformly Lipschitz continuous in (x, y) or in x respectively. In order to state their result, let us introduce some notations. In this paper, let $|x|$ denote the Euclidean norm of $x \in R^d$ and $(x, \bar{x})$ denote the inner product of $x, \bar{x} \in R^d$. An element $y \in R^{d \times q}$ will be regarded as a $d \times q$ matrix and its Euclidean norm is defined by $|y| = \sqrt{trace(yy^T)}$. Denote by $\mathcal{M}^2(0, 1; R^d)$ (resp. $\mathcal{M}^2(0, 1; R^{d \times q})$) the family of R^d-valued (resp. $R^{d \times q}$-valued) processes which are $\mathcal{F}_t$-progressively measurable and are square integrable on $\Omega \times [0,1]$ with respect to $P \times \lambda$, where λ denotes the Lebesgue measure on $[0,1]$. We also introduce the standing hypotheses:

(H1) $E|X|^2 < \infty, f(., 0, 0) \in \mathcal{M}^2(0, 1; R^d)$ and $g(., 0) \in \mathcal{M}^2(0, 1; R^{d \times q})$.

(H2) There exists a constant $c > 0$ such that

$$|f(t, x, y) - f(t, \bar{x}, \bar{y})|^2 \le c(|x - \bar{x}|^2 + |y - \bar{y}|^2) \quad a.s.$$
$$|g(t, x) - g(t, \bar{x})|^2 \le c|x - \bar{x}|^2 \quad a.s.$$

for all $x, \bar{x} \in R^d, y, \bar{y} \in R^{d \times q}$ and $0 \le t \le 1$.

We now can state the main result of Pardoux & Peng [14].

Theorem 4.1 (Pardoux & Peng [14]) *Let (H1) and (H2) hold. Then there exists a unique solution $(x(.), y(.))$ in $\mathcal{M}^2(0, 1; R^d) \times \mathcal{M}^2(0, 1; R^{d \times q})$ to equation (4.1).*

On the other hand, it is somehow too strong to require the uniform Lipschitz continuity in applications e.g. in dealing with quasilinear parabolic partial differential equations. So it is important to find some weaker conditions than the Lipschitz one under which the backward stochastic differential equation has a unique solution. In the first instance, perhaps one would like to try the local Lipschitz condition plus the linear growth condition, as these conditions guarantee the existence and uniqueness of the solution for a (forward) stochastic differential equation. To be precise, let us state these conditions as follows:

(H3) For each $n = 1, 2, ...$, there exists a constant $c_n > 0$ such that

$$|f(t, x, y) - f(t, \bar{x}, \bar{y})|^2 \le c_n(|x - \bar{x}|^2 + |y - \bar{y}|^2) \quad a.s.$$
$$|g(t, x) - g(t, \bar{x})|^2 \le c_n|x - \bar{x}|^2 \quad a.s.$$

for all $0 \le t \le 1, x, \bar{x} \in R^d, y, \bar{y} \in R^{d \times q}$ with $\max\{|x|, |\bar{x}|, |y|, |\bar{y}|\} < n$. Moreover, there exists a constant $c > 0$ such that

$$|f(t, x, y)|^2 \le c(1 + |x|^2 + |y|^2) \quad a.s.$$
$$|g(t, x)|^2 \le c(1 + |x|^2) \quad a.s.$$

for all $0 \le t \le 1$ and $x \in R^d, y \in R^{d \times q}$.

Unfortunately, it is still open whether (H3) guarantees the existence and uniqueness of the solution to the backward stochastic differential equation (4.1) or not. The difficulty here is that the technique of stopping time and localization seems not work for backward stochastic differential equations. Now the question is: Are there any weaker conditions than the Lipschitz continuity under which the backward stochastic differential equation has a unique solution? The answer is of course positive, and one of the main aims of this paper is to show these conditions.

As a matter of fact, Pardoux & Peng [16] recently considered a special case of equation (4.1), i.e. the case when $g(t, x) \equiv 0$. In other words, they considered the following backward stochastic differential equation

$$x(t) + \int_t^1 f(s, x(s), y(s))ds + \int_t^1 y(s)dw(s) = X \qquad (4.2)$$

on $0 \le t \le 1$. They gave several non-Lipschitz conditions for the existence and uniqueness of the solution to this backward stochastic differential equation. For instance, they showed that equation (4.2) has a unique solution if f is monotone in x and Lipschitzian in z, which is described as follows:

Theorem 4.2 (Pardoux & Peng [16]) *Assume that there exists a constant $c > 0$ such that*

$$\big(f(t, x, y) - f(t, \bar{x}, y), x - \bar{x}\big) \le c|x - \bar{x}|^2 \quad a.s.$$
$$|f(t, x, y) - f(t, x, \bar{y})|^2 \le c|y - \bar{y}|^2 \quad a.s.$$
$$|f(t, x, y)|^2 \le c(1 + |x|^2 + |y|^2) \quad a.s.$$

for all $0 \le t \le 1, x, \bar{x} \in R^d$ and $y, \bar{y} \in R^{d \times q}$. Assume also that for each $r > 0$, $x \to f(t, x, y)$ is continuous on $\{x \in R^d : |x| \le r\}$ uniformly with

respect to (ω, t, y). *Besides, assume that X is bounded. Then equation (4.2)
has a unique solution.*

The monotone condition has been used to replace the Lipschitz condition for the existence and uniqueness of a stochastic differential equation (cf. Krylov [12]). But in the case of backward stochastic differential equations, so far one seems need additional conditions, especially X is required to be bounded.

Although it is still open whether there exists a unique solution to the backward stochastic differential equation when $f(t, x, y)$ is locally Lipschitz continuous in (x, y), a result on the existence and uniqueness of the solution has been obtained by Peng [18] under slightly stronger condition, i.e. $f(t, x, y)$ is locally Lipschitz continuous in x but uniformly Lipschitz continuous in y.

Theorem 4.3 (Peng [18]) *Assume that for each $n = 1, 2, \cdots$, there exists a constant $c_n > 0$ such that*

$$|f(t, x, y) - f(t, \bar{x}, y)|^2 \leq c_n |x - \bar{x}|^2 \qquad a.s.$$

for all $0 \leq t \leq 1, y \in R^{d \times q}$ and $x, \bar{x} \in R^d$ with $\max\{|x|, |\bar{x}|\} < n$. Assume also that there exists a constant $c > 0$ such that

$$|f(t, x, y) - f(t, x, \bar{y})|^2 \leq c|y - \bar{y}|^2 \qquad a.s.$$

$$|f(t, x, y)|^2 \leq c(1 + |x|^2 + |y|^2) \qquad a.s.$$

for all $0 \leq t \leq 1, x \in R^d$ and $y, \bar{y} \in R^{d \times q}$. Furthermore, let X be bounded. Then equation (4.2) has a unique solution.

This result follows easily from the following useful lemma.

Lemma 4.4 (Peng [18]) *Assume that there exists a constant $c > 0$ such that $|X| \leq c$ and*

$$|f(t, x, y)|^2 \leq c(1 + |x|^2 + |y|^2) \qquad a.s.$$

for all $0 \leq t \leq 1, x \in R^d$ and $y \in R^{d \times q}$. Let $(x(t), y(t))$ be a solution of equation (4.2). Then there exists a constant K such that

$$|x(t)|^2 + E^{\mathcal{F}_t} \int_t^1 |y(s)| ds \leq K \qquad a.s.$$

for all $0 \leq t \leq 1$.

This lemma shows that, under the linear growth condition, if the final data X is bounded almost surely then x(t), a part of the solution $(x(t), y(t))$ of the backward equation, must also be bounded almost surely. This property of the backward stochastic differential equation is much different from that of the classical (forward) stochastic differential equation. For example, consider a linear one-dimensional Itô equation

$$dz(t) = bz(t)dt + \sigma z(t)dB_t, \qquad 0 \le t \le 1$$

with initial data $z(0) = z_o$, where B_t is a real-valued Brownian motion. This equation has the explicit solution

$$z(t) = z_o exp[(b - \frac{\sigma^2}{2})t + \sigma B_t].$$

So for each $t, z(t)$ may be unbounded even though the initial data z_o is bounded.

Besides the above results, Pardoux & Peng [16] also gave some other non-Lipschitz conditions. Essentially speaking, they assumed that not only $f(t, x, y)$ is continuously differentiable in (x, y) with locally bounded first order derivatives but also X satisfies certain conditions e.g. X is a bounded random variable belonging to the Wiener space and its derivatives on the Wiener space are bounded. Due to the page limit we shall not discuss their results in more detail.

A common feature in these results by Pardoux and Peng is that X needs to be bounded. However, X is generally in L^2 in applications. In what follows we shall establish a new result on the existence and uniqueness of the solution to equation (4.1) (more general than equation (4.2)). The condition to be proposed is weaker than the Lipschitz one, and X is only required to be in L^2. Let us now propose the standing hypothesis:

(H4) For all $x, \bar{x} \in R^d, y, \bar{y} \in R^{d \times q}$ and $0 \le t \le 1$,

$$|f(t, x, y) - f(t, \bar{x}, \bar{y})|^2 \le \kappa(|x - \bar{x}|^2) + c|y - \bar{y}|^2 \quad a.s.$$

$$|g(t, x) - f(t, \bar{x})|^2 \le \kappa(|x - \bar{x}|^2) \quad a.s.$$

where $c > 0$ and κ is a concave increasing function from R_+ to R_+ such that $\kappa(0) = 0, \kappa(u) > 0$ for $u > 0$ and

$$\int_{0+} \frac{du}{\kappa(u)} = \infty.$$

Let us make a few comments about this standing hypothesis before we state our main result. First of all, since κ is concave and $\kappa(0) = 0$, one can find a pair of positive constants a and b such that

$$\kappa(u) \le a + bu \qquad \text{for all } u \ge 0. \tag{4.3}$$

We therefore see that under hypotheses (H1) and (H4),

$$f(., x(.), y(.)) \in \mathcal{M}^2(0, 1; R^d) \text{ and } g(., x(.), y(.)) \in \mathcal{M}^2(0, 1; R^{d \times q})$$

whenever
$$x(.) \in \mathcal{M}^2(0,1;R^d) \text{ and } y(.) \in \mathcal{M}^2(0,1;R^{d\times q}).$$

Secondly, let us give a few examples for the function $\kappa(.)$ to see the generality of our result. In fact, let $K > 0$ and let $\delta \in (0,1)$ be sufficiently small. Define

$$\kappa_1(u) = Ku \quad \text{for } u \geq 0.$$

$$\kappa_2(u) = \begin{cases} u\log(u^{-1}) & \text{for } 0 \leq u \leq \delta, \\ \delta\log(\delta^{-1}) + \dot{\kappa}_2(\delta-)(u-\delta) & \text{for } u > \delta. \end{cases}$$

$$\kappa_3(u) = \begin{cases} u\log(u^{-1})\log\log(u^{-1}) & \text{for } 0 \leq u \leq \delta, \\ \delta\log(\delta^{-1})\log\log(\delta^{-1}) + \dot{\kappa}_3(\delta-)(u-\delta) & \text{for } u > \delta. \end{cases}$$

They are all concave non-decreasing functions satisfying

$$\int_{0+} \frac{du}{\kappa_i(u)} = \infty.$$

In particular we see that the Lipschitz condition (H2) is a special case of our proposed condition (H4). Therefore, our following result is a generalization of Theorem 4.1, i.e. the result of Pardoux & Peng [14].

Theorem 4.5 *Assume (H1) and (H4) hold. Then there exists a unique solution* $(x(.), y(.))$ *to equation (4.1) in* $\mathcal{M}^2(0,1;R^d) \times \mathcal{M}^2(0,1;R^{d\times q}).$

We shall now devote the remainder of the paper to prove this main result.

5. Lemmas

We need to prepare a number of lemmas. We first construct an approximate sequence using an iteration of the Picard type with the help of Theorem 4.1. Let $x_0(t) \equiv 0$, and let $\{x_n(t), y_n(t) : 0 \leq t \leq 1\}_{n\geq 1}$ be a sequence in $\mathcal{M}^2(0,1;R^d) \times \mathcal{M}^2(0,1;R^{d\times q})$ defined recursively by

$$x_n(t) + \int_t^1 f(s, x_{n-1}(s), y_n(s))ds + \int_t^1 [g(s, x_{n-1}(s)) + y_n(s)]dw(s) = X \tag{5.1}$$

on $0 \leq t \leq 1$. This sequence is well defined since once $x_{n-1}(.)$ is given, $f(t, x_{n-1}(t), y)$ and $g(t, x_{n-1}(t)) + y$ are both Lipschitz continuous in y hence Theorem 4.1 can be used to define $x_n(t)$ and $y_n(t)$.

Lemma 5.1 *Under hypotheses (H1) and (H4), for all* $0 \leq t \leq 1$ *and* $n \geq 1$,

$$E|x_n(t)|^2 \leq C_1 \quad \text{and} \quad E\int_0^1 |y_n(s)|^2 ds \leq C_2, \tag{5.2}$$

where C_1 and C_2 are both positive constants.

Proof. Applying Itô's formula to $|x_n(t)|^2$ we deduce that

$$|X|^2 - |x_n(t)|^2 = 2\int_t^1 \big(x_n(s), f(s, x_{n-1}(s), y_n(s))\big)\,ds$$

$$+2\int_t^1 \big(x_n(s), [g(s, x_{n-1}(s)) + y_n(s)]dw(s)\big) + \int_t^1 |g(s, x_{n-1}(s)) + y_n(s)|^2 ds.$$

Thus

$$E|x_n(t)|^2 + E\int_t^1 |y_n(s)|^2 ds$$

$$= E|X|^2 - 2E\int_t^1 \big(x_n(s), f(s, x_{n-1}(s), y_n(s))\big)\,ds$$

$$- E\int_t^1 \big(|g(s, x_{n-1}(s)|^2 + 2trace[g(s, x_{n-1}(s))y_n(s)^T]\big)ds.$$

Therefore, using the elementary inequality $2|uv| \le u^2/\alpha + \alpha v^2$ for any $\alpha > 0$, we see

$$E|x_n(t)|^2 + E\int_t^1 |y_n(s)|^2 ds$$

$$\le E|X|^2 + \frac{1}{\alpha}E\int_t^1 |x_n(s)|^2 ds + \alpha E\int_t^1 |f(s, x_{n-1}(s), y_n(s))|^2 ds$$

$$+\frac{1}{\alpha}E\int_t^1 |g(s, x_{n-1}(s))|^2 ds + \alpha E\int_t^1 |y_n(s)|^2 ds. \tag{5.3}$$

But from hypotheses (H1), (H4) and (4.3) one sees that

$$|f(s, x_{n-1}(s), y_n(s))|^2 \le 2|f(s,0,0)|^2 + 2|f(s, x_{n-1}(s), y_n(s)) - f(s,0,0)|^2$$

$$\le 2|f(s,0,0)|^2 + 2\kappa(|x_{n-1}(s)|^2) + 2c|y_n(s)|^2$$

$$\le 2|f(s,0,0)|^2 + 2a + 2b|x_{n-1}(s)|^2 + 2c|y_n(s)|^2,$$

and similarly

$$|g(s, x_{n-1}(s))|^2 \le 2|g(s,0,0)|^2 + 2a + 2b|x_{n-1}(s)|^2.$$

Substituting these into (5.3) gives

$$E|x_n(t)|^2 + E\int_t^1 |y_n(s)|^2 ds \le C_3(\alpha) + \frac{1}{\alpha}\int_t^1 E|x_n(s)|^2 ds$$

$$+ 2b(\alpha + \frac{1}{\alpha})\int_t^1 E|x_{n-1}(s)|^2 ds + \alpha(2c+1)E\int_t^1 |y_n(s)|^2 ds,$$

where

$$C_3(\alpha) = E|X|^2 + 2a(\alpha + \frac{1}{\alpha}) + 2\alpha E \int_0^1 |f(s,0,0)|^2 ds + \frac{2}{\alpha} E \int_0^1 |g(s,0,0)|^2 ds.$$

In particular, choosing $\alpha = 1/(4c+2)$ we get

$$E|x_n(t)|^2 + \frac{1}{2} E \int_t^1 |y_n(s)|^2 ds$$

$$\leq C_4 + 2(2c+1) \int_t^1 E|x_n(s)|^2 ds$$

$$+ 2b[\frac{1}{2(2c+1)} + 2(2c+1)] \int_t^1 E|x_{n-1}(s)|^2 ds$$

$$\leq C_4 + C_5 \int_t^1 \big(\sup_{1 \leq k \leq n} E|x_k(s)|^2 \big) ds, \tag{5.4}$$

where $C_4 = C_3(1/(4c+2))$ and $C_5 = 2(2c+1) + 2b[(2(2c+1))^{-1} + 2(2c+1)]$. Now let m be any integer. If $1 \leq n \leq m$, (5.4) gives (recalling $x_0(t) \equiv 0$)

$$E|x_n(t)|^2 \leq C_4 + C_5 \int_t^1 \big(\sup_{1 \leq k \leq m} E|x_k(s)|^2 \big) ds.$$

Therefore

$$\sup_{1 \leq k \leq m} E|x_k(t)|^2 \leq C_4 + C_5 \int_t^1 \big(\sup_{1 \leq k \leq m} E|x_k(s)|^2 \big) ds.$$

An application of the well-known Gronwall inequality implies

$$\sup_{1 \leq k \leq m} E|x_k(t)|^2 \leq C_4 e^{C_5(1-t)} \leq C_4 e^{C_5}.$$

Since m is arbitrary, the first inequality of (5.2) follows by setting $C_1 = C_4 e^{C_5}$. Finally it follows from (5.4) that

$$E \int_0^1 |y_n(s)|^2 ds \leq 2(C_4 + C_5 C_1) := C_2.$$

The proof is complete.

Lemma 5.2 *Under hypotheses (H1) and (H4), there exists a constant $C_6 > 0$ such that*

$$E|x_{n+m}(t) - x_n(t)|^2 \leq C_6 \int_t^1 \kappa \big(E|x_{n+m-1}(s) - x_{n-1}(s)|^2 \big) ds \tag{5.5}$$

for all $0 \leq t \leq 1$ *and* $n, m \geq 1$.

Proof. Applying Itô's formula to $|x_{n+m}(t) - x_n(t)|^2$ we have

$$-E|x_{n+m}(t) - x_n(t)|^2$$

$$= 2E \int_t^1 (x_{n+m}(s) - x_n(s),$$

$$f(s, x_{n+m-1}(s), y_{n+m}(s)) - f(s, x_{n-1}(s), y_n(s))) ds$$

$$+ E \int_t^1 |g(s, x_{n+m-1}(s)) + y_{n+m}(s) - g(s, x_{n-1}(s)) - y_n(s)|^2 ds. \quad (5.6)$$

In the same way as the proof of Lemma 5.1 one can then deduce that

$$E|x_{n+m}(t) - x_n(t)|^2 + \frac{1}{2} E \int_t^1 |y_{n+m}(s) - y_n(s)|^2 ds$$

$$\leq 2(2c + 1) \int_t^1 E|x_{n+m}(s) - x_n(s)|^2 ds$$

$$+ 2[2(2c + 1) + \frac{1}{4c + 2}] \int_t^1 \kappa(E|x_{n+m-1}(s) - x_{n-1}(s)|^2) ds. \quad (5.7)$$

Now fix $t \in [0, 1]$ arbitrarily. If $t \leq r \leq 1$, then

$$E|x_{n+m}(r) - x_n(r)|^2 \leq 2(2c + 1) \int_r^1 E|x_{n+m}(s) - x_n(s)|^2 ds$$

$$+ 2[2(2c + 1) + \frac{1}{4c + 2}] \int_t^1 \kappa(E|x_{n+m-1}(s) - x_{n-1}(s)|^2) ds.$$

In view of the Gronwall inequality we see that

$$E|x_{n+m}(t) - x_n(t)|^2 \leq 2[2(2c + 1) + \frac{1}{4c + 2}] e^{2(2c+1)(1-t)}$$

$$\times \int_t^1 \kappa(E|x_{n+m-1}(s) - x_{n-1}(s)|^2) ds.$$

So the required (5.5) follows by setting

$$C_6 = 2[2(2c + 1) + \frac{1}{4c + 2}] e^{2(2c+1)}.$$

The proof is complete.

Lemma 5.3 *Under hypotheses (H1) and (H4), there exists a constant $C_7 > 0$ such that*

$$E|x_{n+m}(t) - x_n(t)|^2 \leq C_7(1 - t)$$

for all $0 \le t \le 1$ *and* $n, m \ge 1$.

Proof. By Lemmas 5.2 and 5.1,

$$E|x_{n+m}(t) - x_n(t)|^2 \le C_6 \int_t^1 \kappa(4C_1)ds = C_6\kappa(4C_1)(1-t)$$

and the conclusion follows by letting $C_7 = C_6\kappa(4C_1)$. The proof is complete.

We now start to prepare a key lemma. To do so, let us introduce some new notations. Choose $T_1 \in [0, 1)$ such that

$$\bar{\kappa}(C_7(1-t)) \le C_7 \qquad \text{for all } T_1 \le t \le 1, \tag{5.8}$$

where $\bar{\kappa}(u) = C_6\kappa(u)$. Fix $m \ge 1$ arbitrarily and define two sequences of functions $\{\varphi_n(t)\}_{n \ge 1}$ and $\{\tilde{\varphi}_{n,m}(t)\}_{n \ge 1}$ on $t \in [0, 1]$ as follows:

$$\varphi_1(t) = C_7(1-t),$$

$$\varphi_{n+1}(t) = \int_t^1 \bar{\kappa}(\varphi_n(s))ds, \qquad n = 1, 2, \cdots,$$

$$\tilde{\varphi}_{n,m}(t) = E|x_{n+m}(t) - x_n(t)|^2, \qquad n = 1, 2, \cdots.$$

Lemma 5.4 *Under hypotheses (H1) and (H4), for any* $m \ge 1$ *and all* $n \ge 1$,

$$0 \le \tilde{\varphi}_{n,m}(t) \le \varphi_n(t) \le \varphi_{n-1}(t) \le \cdots \le \varphi_1(t) \qquad \text{whenever } t \in [T_1, 1].$$
$$\tag{5.9}$$

Proof. First of all, by Lemma 5.3,

$$\tilde{\varphi}_{1,m}(t) = E|x_{1+m}(t) - x_1(t)|^2 \le C_7(1-t) = \varphi_1(t).$$

Now by Lemma 5.2,

$$\tilde{\varphi}_{2,m}(t) = E|x_{2+m}(t) - x_2(t)|^2 \le C_6 \int_t^1 \kappa(E|x_{1+m}(s) - x_1(s)|^2)ds$$

$$= \int_t^1 \bar{\kappa}(\tilde{\varphi}_{1,m}(s))ds \le \int_t^1 \bar{\kappa}(\varphi_1(s))ds = \varphi_2(t).$$

But by (5.8) we also have

$$\varphi_2(t) = \int_t^1 \bar{\kappa}(C_7(1-s))ds \le \int_t^1 C_7 ds = C_7(1-t) = \varphi_1(t).$$

In other words, we have already showed that

$$\tilde{\varphi}_{2,m}(t) \le \varphi_2(t) \le \varphi_1(t) \qquad \text{if } t \in [T_1, 1].$$

We next assume that (5.9) holds for some $n \geq 2$. Then by Lemma 5.2 again,

$$\tilde{\varphi}_{n+1,m}(t) \leq \int_t^1 \bar{\kappa}(\tilde{\varphi}_{n,m}(s))ds \leq \int_t^1 \bar{\kappa}(\varphi_n(s))ds = \varphi_{n+1}(t)$$

$$\leq \int_t^1 \bar{\kappa}(\varphi_{n-1}(s))ds = \varphi_n(t),$$

that is, (5.9) holds for $n+1$ as well. So, by induction, (5.9) must hold for all $n \geq 1$. The proof is complete.

At last we can start to prove our main result Theorem 4.5.

6. Proof of Theorem 4.5

Existence: We first prove the existence of a solution. This will be done by four steps. In the following proof please bear in mind that the constants C_1 – C_7 as well as T_1 have already been defined in Section 5.

Step 1. We claim that

$$\sup_{T_1 \leq t \leq 1} E|x_n(t) - x_k(t)|^2 \to 0 \qquad \text{as } n, k \to \infty. \tag{6.1}$$

In fact, note that for each $n \geq 1, \varphi_n(t)$ is continuous and decreasing on $[T_1, 1]$ and for each $t, \varphi_n(t)$ is non-increasing monotonically as $n \to \infty$. Therefore we can define the function $\varphi(t)$ by $\varphi_n(t) \downarrow \varphi(t)$. It is easy to verify that $\varphi(t)$ is continuous and non-increasing on $[T_1, 1]$. By the definition of $\varphi_n(t)$ and $\varphi(t)$ we get

$$\varphi(t) = \lim_{n \to \infty} \varphi_{n+1}(t) = \lim_{n \to \infty} \int_t^1 \bar{\kappa}(\varphi_n(s))ds = \int_t^1 \bar{\kappa}(\varphi(s))ds, \quad t \in [T_1, 1].$$

Since

$$\int_{0+} \frac{du}{\bar{\kappa}(u)} = \infty,$$

the Bihari's inequality (cf. Bihari [2] or Mao [13]) implies $\varphi(t) \equiv 0$ on $t \in [T_1, 1]$. In particular, $\varphi_n(T_1) \downarrow 0$ as $n \to \infty$. So for any $\varepsilon > 0$, one can find an integer $N \geq 1$ such that $\varphi_n(T_1) < \varepsilon$ whenever $n \geq N$. Now for any $m \geq 1$ and $n \geq N$, by Lemma 5.4,

$$\sup_{T_1 \leq t \leq 1} E|x_{n+m}(t) - x_n(t)|^2 = \sup_{T_1 \leq t \leq 1} \tilde{\varphi}_{n,m}(t) \leq \sup_{T_1 \leq t \leq 1} \varphi_n(t) = \varphi_n(T_1) < \varepsilon.$$

So (6.1) must hold.

Step 2. Define

$$T_2 = \inf\{T \in [0,1] : \sup_{T \leq t \leq 1} E|x_n(t) - x_k(t)|^2 \to 0 \quad \text{as } n, k \to \infty\}. \tag{6.2}$$

Immediately we see from step 1 that $0 \leq T_2 \leq T_1 < 1$. In this step we shall show that

$$\sup_{T_2 \leq t \leq 1} E|x_n(t) - x_k(t)|^2 \to 0 \quad \text{as } n, k \to \infty. \tag{6.3}$$

Let $\varepsilon > 0$ be arbitrary. Choose $\delta \in (0, 1 - T_2)$ such that

$$C_6 \kappa(4C_1)\delta < \frac{\varepsilon}{2}. \tag{6.4}$$

Since $\kappa(0) = 0$ one can find a $\theta \in (0, \varepsilon)$ such that

$$C_6 \kappa(\theta) < \frac{\varepsilon}{2}. \tag{6.5}$$

By the definition of T_2 we observe that for a sufficiently large N

$$E|x_n(t) - x_k(t)|^2 < \theta \qquad \text{for } t \in [T_2 + \delta, 1] \text{ if } n, k \geq N. \tag{6.6}$$

Now let $n, k \geq N + 1$. By Lemmas 5.2 and 5.1 as well as inequalities (6.4) – (6.6) we can derive that if $T_2 \leq t \leq T_2 + \delta$,

$$E|x_n(t) - x_k(t)|^2 \leq C_6 \int_{T_2}^{T_2 + \delta} \kappa(E|x_{n-1}(s) - x_{k-1}(s)|^2)ds$$

$$+ C_6 \int_{T_2 + \delta}^{1} \kappa(E|x_{n-1}(s) - x_{k-1}(s)|^2)ds \leq C_6 \kappa(4C_1)\delta + C_6 \kappa(\theta) < \varepsilon.$$

This, together with (6.6) and $\theta < \varepsilon$, yields

$$\sup_{T_2 \leq t \leq 1} E|x_n(t) - x_k(t)|^2 < \varepsilon \qquad \text{whenever } n, k \geq N + 1.$$

That is, (6.3) holds.

Step 3. In this step, we shall show $T_2 = 0$. Assume $T_2 > 0$. By step 2 one can choose a sequence of numbers $\{a_k\}_{k \geq 1}$ such that $a_k \downarrow 0$ as $k \to \infty$ and

$$\sup_{T_2 \leq t \leq 1} E|x_n(t) - x_k(t)|^2 \leq a_k \qquad \text{whenever } n > k \geq 1. \tag{6.7}$$

By Lemmas 5.2 and 5.1 together with (6.7), one sees that

$$E|x_n(t) - x_k(t)|^2 \leq C_6 \int_{t}^{1} \kappa(E|x_{n-1}(s) - x_{k-1}(s)|^2)ds$$

$$\leq C_6 \kappa(a_{k-1}) + C_6 \int_{t}^{T_2} \kappa(E|x_{n-1}(s) - x_{k-1}(s)|^2)ds$$

$$\leq C_6 \kappa(a_{k-1}) + C_6 \kappa(4C_1)(T_2 - t) \tag{6.8}$$

if $0 \leq t \leq T_2$ and $n > k \geq 2$. We shall now show an assertion which is similar to Lemma 5.4. In order to state the assertion, we need introduce some new notations. Choose a positive number $\delta \in (0, T_2)$ and a positive integer $j \geq 1$ such that

$$C_6 \kappa(a_j) + C_6 \kappa(4C_1)\delta \leq 4C_1. \tag{6.9}$$

Define a sequence of functions $\{\psi_k(t)\}_{k \geq 1}$ by:

$$\psi_1(t) = C_6 \kappa(a_j) + C_6 \kappa(4C_1)(T_2 - t), \qquad T_2 - \delta \leq t \leq T_2,$$

$$\psi_{k+1}(t) = C_6 \kappa(a_{j+k}) + C_6 \int_t^{T_2} \kappa(\psi_k(s))ds, \qquad T_2 - \delta \leq t \leq T_2 \text{ and } k \geq 1.$$

Fix $m \geq 1$ arbitrarily and define a sequence of functions $\{\tilde{\psi}_{k,m}(t)\}_{k \geq 1}$ by:

$$\tilde{\psi}_{k,m}(t) = E|x_{m+j+k}(t) - x_{j+k}(t)|^2, \qquad T_2 - \delta \leq t \leq T_2.$$

We claim that

$$\tilde{\psi}_{k,m}(t) \leq \psi_k(t) \leq \psi_{k-1}(t) \leq \cdots \leq \psi_1(t), \qquad T_2 - \delta \leq t \leq T_2. \tag{6.10}$$

In fact, we first drive by (6.8) that

$$\tilde{\psi}_{1,m}(t) = E|x_{m+j+1}(t) - x_{j+1}(t)|^2 \leq C_6 \kappa(a_j) + C_6 \kappa(4C_1)(T_2 - t) = \psi_1(t),$$

and then

$$\tilde{\psi}_{2,m}(t) = E|x_{m+j+2}(t) - x_{j+2}(t)|^2$$

$$\leq C_6 \kappa(a_{j+1}) + C_6 \int_t^{T_2} \kappa(E|x_{m+j+1}(s) - x_{j+1}(s)|^2)ds$$

$$= C_6 \kappa(a_{j+1}) + C_6 \int_t^{T_2} \kappa(\tilde{\psi}_{1,m}(s))ds$$

$$\leq C_6 \kappa(a_{j+1}) + C_6 \int_t^{T_2} \kappa(\psi_1(s))ds = \psi_2(t)$$

$$\leq C_6 \kappa(a_j) + C_6 \int_t^{T_2} \kappa[C_6 \kappa(a_j) + C_6 \kappa(4C_1)(T_2 - t)]ds$$

$$\leq C_6 \kappa(a_j) + C_6 \kappa(4C_1)(T_2 - t) = \psi_1(t),$$

where (6.9) has been used. In other words, we have already showed that

$$\tilde{\psi}_{2,m}(t) \leq \psi_2(t) \leq \psi_1(t) \qquad \text{on } T_2 - \delta \leq t \leq T_2.$$

Assume now that (6.10) holds for some $k \geq 2$. Then, by (6.8)

$$\tilde{\psi}_{k+1,m}(t) = E|x_{m+j+k+1}(t) - x_{j+k+1}(t)|^2$$

$$\leq C_6 \kappa(a_{j+k}) + C_6 \int_t^{T_2} \kappa(E|x_{m+j+k}(s) - x_{j+k}(s)|^2)ds$$

$$= C_6 \kappa(a_{j+k}) + C_6 \int_t^{T_2} \kappa(\tilde{\psi}_{k,m}(s))ds$$

$$\leq C_6 \kappa(a_{j+k}) + C_6 \int_t^{T_2} \kappa(\psi_k(s))ds = \psi_{k+1}(t)$$

$$\leq C_6 \kappa(a_{j+k-1}) + C_6 \int_t^{T_2} \kappa(\psi_{k-1}(s))ds = \psi_k(t),$$

that is, (6.10) holds for $k + 1$ as well. So, by induction, (6.10) holds for all $k \geq 1$. Note that for each $k \geq 1$, $\psi_k(t)$ is continuous and decreasing on $[T_2 - \delta, T_2]$ and for each t, $\psi_k(t)$ is non-increasing monotonically as $k \to \infty$. Therefore we can define the function $\psi(t)$ by $\psi_k(t) \downarrow \psi(t)$. It is easy to verify that $\psi(t)$ is continuous and non-increasing on $[T_2 - \delta, T_2]$. By the definition of $\psi_n(t)$ and $\psi(t)$ we get

$$\psi(t) = \lim_{k\to\infty} \psi_{k+1}(t) = \lim_{k\to\infty} \left(C_6 \kappa(a_{j+k}) + C_6 \int_t^{T_2} \kappa(\psi_k(s))ds \right)$$

$$= C_6 \int_t^{T_2} \kappa(\psi(s))ds \qquad \text{on } T_2 - \delta \leq t \leq T_2.$$

Therefore by Bihari's inequality (cf Bihari [2] or Mao [13]) we see $\psi(t) \equiv 0$ for $T_2 - \delta \leq t \leq T_2$. In particular, $\psi_k(T_2 - \delta) \downarrow 0$ as $k \to \infty$. Hence, for any $\varepsilon > 0$, one can find an integer $k_o \geq 1$ such that $\psi_k(T_2 - \delta) < \varepsilon$ whenever $k \geq k_o$. It then follows from (6.10) that,

$$\sup_{T_2-\delta\leq t\leq T_2} E|x_{m+j+k}(t) - x_{j+k}(t)|^2 \leq \psi_k(T_2 - \delta) < \varepsilon \quad \text{whenever } k \geq k_o.$$

$$(6.11)$$

Since $m \geq 1$ is arbitrary, (6.11) means

$$\sup_{T_2-\delta\leq t\leq T_2} E|x_n(t) - x_k(t)|^2 \to 0 \qquad \text{as } n, k \to \infty.$$

This, together with (6.3), yields

$$\sup_{T_2-\delta\leq t\leq 1} E|x_n(t) - x_k(t)|^2 \to 0 \qquad \text{as } n, k \to \infty.$$

But this is in contradiction with the definition of T_2. So we must have $T_2 = 0$. That is, we have already showed that

$$\sup_{0\leq t\leq 1} E|x_n(t) - x_k(t)|^2 \to 0 \qquad \text{as } n, k \to \infty. \qquad (6.12)$$

Step 4. Applying (6.12) to (5.7) one sees that $\{x_n(.)\}$ is a Cauchy sequence in $\mathcal{M}^2(0,1;R^d)$ and $\{y_n(.)\}$ is a Cauchy sequence in $\mathcal{M}^2(0,1;R^{d\times q})$. Denote their limits by $x(.)$ and $y(.)$ respectively. Note from (5.1) that $\{x_n(.)\}$ also converges to $x(.)$ in $L^2(\Omega; C(0,1;R^d))$. Now letting $n \to \infty$ in (5.1) we obtain

$$x(t) + \int_t^1 f(s,x(s),y(s))ds + \int_t^1 [g(s,x(s)) + y(s)]dw(s) = X$$

on $0 \le t \le 1$. The existence of the solution has been proved.

Uniqueness: To show the uniqueness, let both $(x(.),y(.))$ and $(\bar{x}(.),\bar{y}(.))$ be the solutions of equation (4.1). Then, in the same way as the proof of Lemma 5.1 one can show that

$$E|x(t) - \bar{x}(t)|^2 + \frac{1}{2}E\int_t^1 |y(s) - \bar{y}(s)|^2 ds$$

$$\le 2\Big[2(2c+1) + \frac{1}{4c+2}\Big]\int_t^1 [E|x(s) - \bar{x}(s)|^2 + \kappa(E|x(s) - \bar{x}(s)|^2)]ds \quad (6.13)$$

for $0 \le t \le 1$. Since $\kappa(.)$ is concave function and $\kappa(0) = 0$, we have

$$\kappa(u) \ge \kappa(1)u \qquad \text{for } 0 \le u \le 1.$$

So

$$\int_{0+} \frac{du}{u + \kappa(u)} \ge \frac{\kappa(1)}{\kappa(1) + 1}\int_{0+} \frac{du}{\kappa(u)} = \infty.$$

Therefore one can apply the Bihari inequality to (6.13) to obtain

$$E|x(t) - \bar{x}(t)|^2 = 0 \qquad \text{for all } 0 \le t \le 1.$$

So $x(t) = \bar{x}(t)$ for all $0 \le t \le 1$ almost surely. It then follows from (6.13) that $y(t) = \bar{y}(t)$ for all $0 \le t \le 1$ almost surely as well. The uniqueness has been proved and the proof of the theorem is then complete.

REFERENCES

[1] Bensoussan, A., Lectures on stochastic control, in Nonlinear Filtering and Stochastic Control edited by S. K. Mitter and A. Moro, Lecture Notes in Math. 972, Springer-Verlag, 1982.

[2] Bihari, I., A generalization of a lemma of Bellman and its application to uniqueness problem of differential equations, Acta Math. Acad. Sci Hungar 7 (1956), 71-94.

[3] Bismut, J. M., Théorie probabiliste du contrôle des diffusions, Mem. Amer. Math. Soc. No. 176, 1973.

[4] Freidlin, M. I., Markov processes and differential equations, in Progress in Mathematics, Vol. 3, Plenum Press New York, 1969.

[5] Freidlin, M. I., Functional Integration and Partial Differential Equations, Princeton University Press, 1985.

[6] Freidlin, M. I., Coupled reaction-diffusion equations, the Annals of Probability, 19(1), (1991), 29-57.

[7] Freidlin, M. I. and Wentzell, A. D., Random Perturbations of Dynamical Systems, Springer-Verlag, 1984.

[8] Gikhman, I. I. and Skorokhod, A. V., Stochastic Differential Equations, Springer-Verlag, 1972.

[9] Haussmann, U. G., A Stochastic Maximum Principle for Optimal Control of Diffusions, Pitman Research Notes in Math. 151, 1986.

[10] Kushner, H. J., Necessary conditions for continuous parameter stochastic optimization problems, SIAM J. Control 10 (1972), 550-565.

[11] Kolmogorov, A., Petrovskii, I. and Piskunov, N., Etude de l'équation de la diffusion avec croissance de al matière et son application a un problème biologique, Moscow Univ. Math. Bell. 1 (1937), 1-25.

[12] Krylov, N. V., Controlled Diffusion Processes, Springer-Verlag, 1980.

[13] Mao, X., Exponential Stability of Stochastic Differential Equations, Marcel Dekker Inc., 1994.

[14] Pardoux, E. and Peng, S. G., Adapted solution of a backward stochastic differential equation, Systems and Control Letters 14 (1990), 55-61.

[15] Pardoux, E. and Peng, S. G., Backward stochastic differential equations and quasilinear parabolic partial differential equations, in Stochastic Partial Differential Equations and Their Applications (Charlotte, NC, 1991), 200-217, Lecture Notes in Control and Information Science 176, Springer, Berlin, 1992.

[16] Pardoux, E. and Peng, S. G., Some backward stochastic differential equations with non-Lipschitz coefficients, Prepublication URA 225, 94-3, Université de Provence.

[17] Peng, S. G., Probabilistic interpretation for systems of quasilinear parabolic partial differential equations, Stochastics 14 (1991), 61-74.

[18] Peng, S. G., Backward stochastic differential equation and its application in optimal control, Appl. Math. Optim. 27 (1993), 125-144.

Path Integrals and Finite Dimensional Filters

S.J.Maybank

Robotics Research Laboratory,

University of Oxford.

Department of Engineering Science,

19 Parks Road, Oxford OX1 3PJ, UK.

Abstract

A path integral representation is obtained for the optimal probability density function of the system state, conditional on the measurements. For certain non-linear systems the optimal density can be evaluated recursively using only a finite number of statistics. These systems extend the class found by Beneš, in that the drift need not be the gradient of a scalar potential.

In the one dimensional case the trajectories of the deterministic system underlying the Beneš filter fall into five classes according to their behaviour as $t \to \infty$. It is shown that an arbitrary deterministic trajectory can be approximated at small times to an accuracy of $O(t^5)$ by a trajectory for which the Beneš filter is appropriate.

1   Introduction

The Kalman-Bucy filter and its many variations are widely used throughout science and engineering for estimating the state of time varying systems. Applications to computer vision in particular are described in [2,8,9]. A typical filter contains a model of the system dynamics and of the measurement process. It is given a sequence of measurements obtained over an extended time, and produces from this sequence an estimate of the probability density function for the system state, conditional on the measurements. If the filter is optimal, then the estimated density is the exact conditional density. The best known optimal filter is the Kalman-Bucy filter. It is applicable only to the small class of systems satisfying the following conditions: i) the system evolution and the measurement process are linear, and both are subject to white Gaussian noise; ii) the prior density for the system state is Gaussian and independent of the noise processes; iii) the noise affecting the system evolution is independent of the measurement noise. There are many generalisations of the Kalman-Bucy filter to non-linear systems, the best known of which is the extended Kalman filter (EKF). These generalisations are usually suboptimal in that they do not produce the true probability density function for the system state conditional on the measurements. Further information is given in the books [14-16] which contain a detailed, applications oriented introduction to filtering.

In the non-linear case the conditional density is usually very difficult to determine. However, there do exist non-linear systems which are tractable, in that the conditional density depends on only a finite number of sufficient statistics which can be found by solving recursively a set of ordinary differential equations. The first examples were found by Beneš [4]. In this context, 'recursive' means that the conditional density ρ_t at time $t > s \geq 0$ can be computed from the single density ρ_s and the measurements obtained in the interval $(s, t]$. It is not necessary to use the measurements obtained in $[0, s]$.

In this paper the Beneš' filters are rederived. The strategy is first to find an expression for the conditional density and then to investigate the circumstances under which the expression can be evaluated easily. The class of Beneš filters is extended to include cases in which the drift velocity can differ from the gradient of a scalar potential by an arbitrary linear function.

The Beneš filters are divided into five classes, depending on the behaviour at large times of the solutions to $dx/dt = f(x)$, where f is the drift. After suitable rescalings of space and time, a trajectory $t \mapsto x(t) \in \mathbb{R}$ with $x(0) = 0$, $(dx/dt)(0) > 0$ satisfies one of i) $x(t) \sim \sqrt{2t}$; ii) $x(t) \sim t$; iii) $\lim_{t \to \infty} x(t) = a$; iv) $x(t) \sim t^2$; v) $x(t) \sim \exp(t/2)$. These trajectories are called Beneš trajectories. It is shown that at small times an initial segment of a general unperturbed trajectory can be approximated to an accuracy of $O(t^5)$ by a Beneš trajectory. In contrast, the Kalman-Bucy filter has only two classes, and an initial segment of a trajectory can be approximated to an accuracy of only $O(t^3)$.

The path integral representation of the conditional density is obtained in §2. The Beneš filter is derived in §3 and some special cases are discussed. The five types of Beneš trajectory are obtained in §4. Some concluding remarks are made in §5.

2   Stochastic Processes and Filtering

The expression for the conditional density involves an auxiliary stochastic process which is usually chosen to be an Ornstein-Uhlenbeck bridge (OU bridge). The OU process and the OU bridge are defined in §2.1. Filtering is formulated in terms of stochastic differential equations (SDEs) in §2.2. SDEs are chosen because of their versatility when making calculations. The expression for the conditional density is obtained in §2.3.

Detailed introductions to stochastic calculus can be found in [10,11,17,19, 20]. Information about filtering can be found in [5,17,18,20].

2.1 The OU process and the OU bridge

The OU process X in $\mathbb{R}^n$ is defined by the Itô SDE

$$dX_t = -MX_t \, dt + dB_t \qquad (X_0 \sim \mathcal{N}(a, \Sigma), 0 \le t) \qquad (1)$$

where M is an $n \times n$ matrix and B is a Brownian motion in $\mathbb{R}^n$ independent of X_0. The notation $X_0 \sim \mathcal{N}(a, \Sigma)$ means that X_0 is a Gaussian random variable with expectation a and covariance Σ. It follows from an application of the Itô formula that (1) has the solution

$$X_t = \exp(-Mt)X_0 + \exp(-Mt) \int_0^t \exp(Ms) \, dB_s \qquad (0 \le t) \qquad (2)$$

thus X is a Gaussian process with continuous paths. The expectation and the covariance functions of X are readily obtained from (2),

$$
\begin{aligned}
E(X_t) &= \exp(-Mt)a & (0 \le t) \qquad (3) \\
\mathrm{Cov}(X_{s_1}, X_{s_2}) &= \exp(-Ms_1)\Sigma \exp(-M^\top s_2)
\end{aligned}
$$

$$+ \int_0^{s_1} \exp(-M(s_1 - u)) \exp(-M^\top(s_2 - u)) \, du \qquad (0 \le s_1 \le s_2) \qquad (4)$$

The OU bridge $\Xi_t^{X_0 \to b}$, $t > 0$, is obtained by constraining X such that $X_t = b$. The construction of $\Xi_t^{X_0 \to b}$ involves the function Λ defined by

$$\Lambda(s, t) = -\mathrm{Cov}(X_s, X_t)\mathrm{Cov}(X_t, X_t)^{-1} \qquad (0 \le s \le t) \qquad (5)$$

Proposition 1. *Let X be the OU process defined by (1). The OU bridge $\Xi_t^{X_0 \to b}$ is Gaussian with expectation and covariance functions*

$$E(\Xi_{s,t}^{X_0 \to b}) = \exp(-Ms)(I + \exp(Ms)\Lambda(s, t) \exp(-Mt))a - \Lambda(s, t)b \quad (0 \le s \le t)$$

$$
\begin{aligned}
\mathrm{Cov}(\Xi_{s_1,t}^{X_0 \to b}, \Xi_{s_2,t}^{X_0 \to b}) &= \mathrm{Cov}(X_{s_1}, X_{s_2}) - \Lambda(s_1, t)\mathrm{Cov}(X_t, X_t)\Lambda(s_2, t)^\top \\
&\qquad\qquad (0 \le s_1 \le s_2 \le t) \quad (6)
\end{aligned}
$$

Proof. It follows from (5) that $X_s + \Lambda(s, t)X_t$ and X_t are independent random variables. The random variable X_s thus has the following orthogonal decomposition with respect to X_t,

$$X_s = (X_s + \Lambda(s, t)X_t) - \Lambda(s, t)X_t \qquad (0 \le s \le t)$$

The OU bridge $\Xi_t^{X_0 \to b}$ is defined by

$$\Xi_{s,t}^{X_0 \to b} = (X_s + \Lambda(s, t)X_t) - \Lambda(s, t)b \qquad (0 \le s \le t) \qquad (7)$$

The expectation and the covariance functions of $\Xi_t^{X_0 \to b}$ are readily obtained from (7). $\square$

Corollary 1. *The function* $\mathrm{Cov}(\Xi_{s_1,t}^{X_0 \to b}, \Xi_{s_2,t}^{X_0 \to b})$ *is independent of* b.

Corollary 2. *If the matrix* M *in the SDE (1) for* X *is symmetric and if* $\Sigma = \alpha I$, *then it follows from (4) and (5) that*

$$\Lambda(s,t) = -(\alpha \exp(-Ms) + M^{-1} \sinh(Ms))(\alpha \exp(-Mt) + M^{-1} \sinh(Mt))^{-1}$$

$$(0 \le s \le t) \qquad (8)$$

2.2   Filtering

The state evolution and the measurement process are modelled by Itô SDEs. The system state is regarded as a stochastic process in $\mathbb{R}^n$ evolving over time according to the SDE

$$dX_t = f(t, X_t)\, dt + \sigma(t, X_t)\, dB_t \qquad (0 \le t) \qquad (9)$$

where the drift $f : [0, \infty) \times \mathbb{R}^n \to \mathbb{R}^n$ and the diffusion $\sigma : [0, \infty) \times \mathbb{R}^n \to \mathbb{R}^{n \times n}$ are measurable, and B is a Brownian motion in $\mathbb{R}^n$ independent of X_0. The law of X_0 is assumed known. In many cases X_0 is deterministic, $X_0 = x$ a.s. The measurement process Y is described by the SDE

$$dY_t = g(t, X_t)\, dt + dW_t \qquad (Y_0 = 0, 0 \le t) \qquad (10)$$

where $g : [0, \infty) \times \mathbb{R}^n \to \mathbb{R}^m$ is measurable and W is a Brownian motion in $\mathbb{R}^m$, independent of B and X_0. Let Ω be the state space of $(B, W)^\top$ considered as a Brownian motion in $\mathbb{R}^{n+m}$, and let $\mathcal{F} = \{\mathcal{F}_t, 0 \le t\}$ be the filtration associated with $(B, W)^\top$. It is assumed that $(X, Y)^\top$ is adapted to $\mathcal{F}$. The equations (9), (10) are simplifications of the more general equations given in [18].

The main aim of filtering is to estimate the density of X_t, conditional on the measurements. For this reason it is appropriate to look for weak solutions to (9) and (10). To ensure that weak solutions exist, it is assumed that f, g, σ are locally bounded, measurable functions. The diffusion σ is assumed to satisfy the conditions

$$\inf_{0 \le s \le t} \{ \|\sigma \sigma^\top(s, x)\| \} > 0 \qquad (x \in \mathbb{R}^n)$$

$$\lim_{y \to x} \sup_{0 \le s \le t} \{ \|\sigma \sigma^\top(s, y) - \sigma \sigma^\top(s, x)\| \} = 0 \qquad (x \in \mathbb{R}^n)$$

$$\sup_{0 \le s \le t} \{ \|\sigma \sigma^\top(s, x)\| \} \le c_t(1 + \|x\|^2) \qquad (x \in \mathbb{R}^n) \quad (11)$$

The functions f, g are assumed to satisfy

$$\sup_{0 \le s \le t} \{ |x.k(s, x)| \} \le c_t(1 + \|x\|^2) \qquad (x \in \mathbb{R}^p) \qquad (12)$$

where $p = n$ if $k = f$ and $p = m$ if $k = g$. In (11) and (12) the variable c_t is independent of x. It is shown in [21] that if (11) and (12) hold with $k = f, g$, then the martingale problem in $\mathbb{R}^{n+m}$ for the drift $(t, x, y)^\top \mapsto (f(t, x), g(t, y))^\top$ and the diffusion $(t, x, y) \mapsto (\sigma(t, x), I)^\top$ is well-posed. It follows that (9) and (10) have a unique weak solution. [17,20]

2.3   The conditional density

The unnormalised probability density function $\rho(t, x)$ for the system state, conditional on the measurements Y_s for $0 \leq s \leq t$ is represented as a path integral. An unnormalised density is one that is defined only up to a scale factor which may depend on time; in other words, the constraint $\int \rho(t, x) \, dx = 1$ is not enforced. During the calculations terms which contribute only a scale factor to ρ are often omitted without comment.

The method is to apply the Cameron-Martin-Girsanov change of measure [17,19,20] to the SDE for $(X, Y)^\top$. The effect of the change of measure is described in general terms as follows. Let P be the probability measure on Ω associated with the Brownian motion $(B, W)^\top$. The change of measure yields firstly, a new SDE driven by Brownian motion $(\tilde{B}, \tilde{W})^\top$ with an associated probability measure $\tilde{P}$, and secondly an expression for the Radon-Nikodym derivative $M = d\tilde{P}/dP$. Let M_t be the restriction of M to $\mathcal{F}_t$, let $\mathcal{Y}_t$ be the sigma algebra generated by the random variables Y_s for $0 \leq s \leq t$ and let R be a random variable in $L^1(\Omega, \mathcal{F}_t, P)$ such that RM_t^{-1} is in $L^1(\Omega, \mathcal{F}_t, \tilde{P})$. It is shown in [18] that

$$E(R|\mathcal{Y}_t) = \frac{\tilde{E}(RM_t^{-1}|\mathcal{Y}_t)}{\tilde{E}(M_t^{-1}|\mathcal{Y}_t)} \tag{13}$$

where E is expectation with respect to P and $\tilde{E}$ is expectation with respect to $\tilde{P}$. Equation (13) is known as the Kallianpur-Striebel formula. It is observed in [12] that (13) is a version of Bayes' rule.

On setting $R = \chi_{\{X_t \in dx\}}$, (13) yields

$$\rho(t, x) \, dx = \tilde{E}(\chi_{\{X_t \in dx\}} M_t^{-1}|\mathcal{Y}_t) \tag{14}$$

The term $\tilde{E}(M_t^{-1}|\mathcal{Y}_t)$ is omitted because it is independent of x. It remains to select a change of measure such that the right-hand side of (14) takes a convenient form. This is done in the following two theorems.

Theorem 2. *Let X, Y satisfy the filter equations (9), (10), and let P be the probability measure defined by the Brownian motion $(B, W)^\top$. Let $h :$ $[0, \infty) \times \mathbb{R}^n \to \mathbb{R}^n$ be locally bounded and measurable such that (12) holds with $k = h$, $p = n$. Let $\tilde{P}$ be a new probability measure defined on the same sample space Ω as P, such that*

$$\frac{d\tilde{P}}{dP} = \exp\left(-\frac{1}{2}\int_0^t \|c(s, X_s)\|^2 \, ds - \int_0^t c(s, X_s)^\top \cdot \left(\frac{dB_s}{dW_s}\right)\right) \tag{15}$$

where $c : [0, \infty) \times \mathbb{R}^n \to \mathbb{R}^{n+m}$ *is defined by*

$$c(t, x) \equiv \begin{pmatrix} c_1(t, x) \\ c_2(t, x) \end{pmatrix} = \begin{pmatrix} \sigma(t, x)^{-1}(f(t, x) - h(t, x)) \\ g(t, x) \end{pmatrix} \tag{16}$$

Then $\rho(t, x)$ *is given by*

$$\rho(t, x)\, dx =$$
$$\tilde{E}\left(\chi_{\{\xi_t \in dx\}} \exp\left(\frac{1}{2} \int_0^t \|c(s, \xi_s)\|^2\, ds + \int_0^t c(s, \xi_s)^\top \cdot \begin{pmatrix} dB_s \\ dW_s \end{pmatrix} \right) \right) \tag{17}$$

where $\tilde{B}$ *is a standard Brownian motion on* $\mathbb{R}^n$ *and* ξ *satisfies the SDE*

$$d\xi_t = h(t, \xi_t)\, dt + \sigma(t, \xi_t)\, d\tilde{B}_t \qquad (\xi_0 = X_0, 0 \le t) \tag{18}$$

Proof. It follows from the Cameron-Martin-Girsanov change of measure that $(X, Y)^\top$ is a weak solution of the SDE

$$\begin{pmatrix} dX_t \\ dY_t \end{pmatrix} = \begin{pmatrix} h(t, X_t) \\ 0 \end{pmatrix} dt + \begin{pmatrix} \sigma(t, X_t)\, d\tilde{B}_t \\ d\tilde{W}_t \end{pmatrix} \qquad (0 \le t) \tag{19}$$

where $(\tilde{B}, \tilde{W})^\top$ is a Brownian motion in $\mathbb{R}^{n+m}$ with associated measure $\tilde{P}$, as defined by (15). The SDE (18) for ξ is the first component of (19), with ξ written in place of X. The result follows from (14) and (19) on observing that

$$\tilde{E}(\chi_{\{\xi_t \in dx\}} M_t^{-1} \mid \mathcal{Y}_t) = \tilde{E}(\chi_{\{\xi_t \in dx\}} M_t^{-1})$$

$\square$

The freedom to choose h is of central importance in the applications of Theorem 2. In the next theorem the expression (17) for $\rho(t, x)$ is rewritten.

Theorem 3. *Let* X, Y *satisfy the filter equations (9), (10). Let* $r : [0, \infty) \times \mathbb{R}^n \to \mathbb{R}^n$ *be defined by*

$$r(t, x) = (\sigma \sigma^\top)^{-1}(f(t, x) - h(t, x)) \tag{20}$$

and let r *also be the gradient of a* $C^{1,2}$ *scalar potential* Φ, *i.e.* $r_i = \partial \Phi / \partial x_i$ *for* $1 \le i \le n$. *Let* $V : [0, \infty) \times \mathbb{R}^n \to \mathbb{R}$ *be defined by*

$$V(t, x) = f^\top (\sigma \sigma^\top)^{-1} f - h^\top (\sigma \sigma^\top)^{-1} h + \sum_{i,j=1}^n \frac{\partial r_i}{\partial x_j} (\sigma \sigma^\top)_{ij} + 2 \frac{\partial \Phi}{\partial t} + \|g\|^2 \tag{21}$$

Let ξ *be the stochastic process in* $\mathbb{R}^n$ *defined by*

$$d\xi_t = h(t, \xi_t)\, dt + \sigma(t, \xi_t)\, d\tilde{B}_t \qquad (\xi_0 = X_0, 0 \le t) \tag{22}$$

where $\tilde{B}$ *is a Brownian motion in* $\mathbb{R}^n$. *Then* $\rho(t, x)$ *is given by*

$$\rho(t, x)\, dx = \exp(\Phi(t, x))$$

$$\tilde{E}\left(\chi_{\{\xi_t \in dx\}} \exp\left(-\Phi(0, \xi_0) - \frac{1}{2}\int_0^t V(s, \xi_s)\, ds + \int_0^t g(s, \xi_s)^{\mathsf{T}}.dY_s\right)\right) \qquad (23)$$

Proof. Let $c : [0, \infty) \times \mathbb{R}^n \to \mathbb{R}^{n+m}$ be defined by (16). Let $e(t, \xi_t)$ be the exponent on the right-hand side of (17),

$$e(t, \xi_t) = \frac{1}{2}\int_0^t \|c(s, \xi_s)\|^2\, ds + \int_0^t c(s, \xi_s)^{\mathsf{T}}.\left(\frac{dB_s}{dW_s}\right) \qquad (24)$$

The exponent $e(t, \xi_t)$ is reduced to a convenient form. It follows from (9) and (10) that

$$\left(\frac{dB_t}{dW_t}\right) = \left(\begin{matrix} \sigma(t, X_t)^{-1}(dX_t - f(t, X_t)\, dt) \\ dY_t - g(t, X_t)\, dt \end{matrix}\right)$$

thus

$$\int_0^t c(s, \xi_s)^{\mathsf{T}}.\left(\frac{dB_s}{dW_s}\right) = \int_0^t c_1(s, \xi_s)^{\mathsf{T}}.dB_s + \int_0^t c_2(s, \xi_s)^{\mathsf{T}}.dW_s$$

$$= \int_0^t r(s, \xi_s)^{\mathsf{T}}.d\xi_s - \int_0^t r(s, \xi_s)^{\mathsf{T}}.f(s, \xi_s)\, ds + \int_0^t g(s, \xi_s)^{\mathsf{T}}.dY_s - \int_0^t \|g(s, \xi_s)\|^2\, ds \qquad (25)$$

It follows from (21), (24) and (25) that

$$\begin{aligned} e(t, \xi_t) &= -\frac{1}{2}\int_0^t \left[V(s, \xi_s) - 2\frac{\partial \Phi}{\partial s}(s, \xi_s) - \sum_{i,j=1}^n \frac{\partial r_i}{\partial x_j}(\sigma\sigma^{\mathsf{T}})_{ij} \right] ds \\ &\quad + \int_0^t r(s, \xi_s)^{\mathsf{T}}.d\xi_s + \int_0^t g(s, \xi_s)^{\mathsf{T}}.dY_s \end{aligned} \qquad (26)$$

An application of the Itô formula [10] to $\Phi(t, \xi_t)$ yields

$$\begin{aligned} d(\Phi(t, \xi_t)) &= \frac{\partial \Phi}{\partial t}(t, \xi_t)\, dt + \sum_{i=1}^n \frac{\partial \Phi}{\partial x_i}(t, \xi_t)\, d\xi_i + \frac{1}{2}\sum_{i,j=1}^n \frac{\partial^2 \Phi}{\partial x_i \partial x_j}(t, \xi_t)\, d\xi_i d\xi_j \\ &= \frac{\partial \Phi}{\partial t}(t, \xi_t)\, dt + r(t, \xi_t)^{\mathsf{T}}.d\xi_t + \frac{1}{2}\sum_{i,j=1}^n \frac{\partial r_i}{\partial x_j}(t, \xi_t)(\sigma\sigma^{\mathsf{T}})_{ij}\, dt \end{aligned}$$

thus

$$\begin{aligned} \int_0^t r(s, \xi_s)^{\mathsf{T}}.d\xi_s &= \Phi(t, \xi_t) - \Phi(0, \xi_0) - \int_0^t \frac{\partial \Phi}{\partial s}(s, \xi_s)\, ds \\ &\quad - \frac{1}{2}\sum_{i,j=1}^n \int_0^t \frac{\partial r_i}{\partial x_j}(s, \xi_s)(\sigma\sigma^{\mathsf{T}})_{ij}\, ds \end{aligned} \qquad (27)$$

It follows from (26) and (27) that

$$e(t, \xi_t) = -\frac{1}{2}\int_0^t V(s, \xi_s)\, ds + \Phi(t, \xi_t) - \Phi(0, \xi_0) + \int_0^t g(s, \xi_s)^{\mathsf{T}}.dY_s \qquad (28)$$

The result follows. $\qquad\qquad\square$

The bridge process $\eta_t^{\xi_0 \to x}$ is used to remove the term $\chi_{\{\xi_t \in dx\}}$ from (23), to yield

$$\rho(t, x) = \exp(\Phi(t, x))$$
$$\tilde{E}\left(\exp\left(-\Phi(0, \xi_0) - \frac{1}{2}\int_0^t V(s, \eta_{s,t}^{\xi_0 \to x})\, ds + \int_0^t g(s, \eta_{s,t}^{\xi_0 \to x})^\top . dY_s\right)\right) \qquad (29)$$

Other representations of $\rho(t, x)$ as a path integral are given in [5], Chapter 4.

3 Finite Dimensional Filters

A finite dimensional filter is one in which the conditional density depends on only a finite number of statistics and the evolution of the statistics over time is governed by a set of ordinary differential equations. The best known example is the Kalman-Bucy filter. The finite dimensional filters are important because they are computationally tractable. The characterisation of the systems with exact finite dimensional filters appears to be difficult. Brockett [6] and Brockett & Clark [7] suggest a theoretical framework for this problem based on Lie algebras and the Duncan-Mortensen-Zakaï equation [25]. Among the successes of this approach are rigorous proofs that certain simple SDEs define filters which are truly infinite dimensional [12,18]. Although the Lie algebra approach is of great interest, it lies outside the scope of this paper.

The path integral representation of $\rho(t, x)$ is used in §3.1 to define a new class of finite dimensional filters. In addition to the Beneš filters, the class includes examples in which the drift velocity of the system state differs by a general linear function from the gradient of a scalar potential. The relation of the Beneš filter to the Kalman-Bucy filter is discussed in §3.2. Some simple examples of Beneš filters are described in §3.3.

3.1 Beneš filters

The Beneš filters are obtained by constraining the filter equations such that $\rho(t, x)$, as given by (23), takes a particularly simple form. The next theorem is a preliminary.

Theorem 4. *With the notation of Theorem 3, let $p_\xi(t, x)$ be the probability density function for ξ_t and let $\eta_t^{\xi_0 \to b}$ be the bridge process for ξ. If the process $L(t, \eta_t^{\xi_0 \to x})$ defined by*

$$L(t, \eta_t^{\xi_0 \to x}) = -\Phi(0, \xi_0) - \frac{1}{2}\int_0^t V(s, \eta_{s,t}^{\xi_0 \to x})\, ds + \int_0^t g(s, \eta_{s,t}^{\xi_0 \to x})^\top . dY_s \qquad (30)$$

is Gaussian with expectation $\nu(t, x)$ and variance $S(t, x)$, then the conditional density ρ for the system state is given by

$$\rho(t, x) = \exp(\Phi(t, x))\, p_\xi(t, x) \exp\left(\nu(t, x) + \frac{1}{2}S(t, x)\right) \qquad (31)$$

Proof. The result follows from (29). □

The Beneš filters are defined in the following theorem.

Theorem 5. *With the notation of Theorems 3 and 4, let ξ be an OU process defined by*

$$d\xi_t = -M\xi_t\, dt + d\tilde{B}_t \qquad\qquad (\xi_0 = a \text{ a.s.}, \ 0 \le t) \qquad (32)$$

Let $g(t, x) = \Pi(t)x$ where $\Pi(t)$ is an $m \times n$ matrix, and let V be linear in x,

$$V(t, x) = v_1(t)^{\mathsf{T}}.x + v_0(t) \qquad\qquad (33)$$

Then the unnormalised conditional density for the system state is

$$\rho(t, x) = \exp(\Phi(t, x)) \exp\left(-\frac{1}{2}(x - c(t))^{\mathsf{T}} D(t)^{-1}(x - c(t))\right) \exp(\mu(t, x)) \quad (34)$$

where $\mu(t, x)$, $c(t)$, $D(t)$ are defined by

$$
\begin{aligned}
\mu(t, x) &= \frac{1}{2}\int_0^t v_1(s)^{\mathsf{T}}\Lambda(s, t)x\, ds - \int_0^t (dY_s)^{\mathsf{T}}\Pi(s)\Lambda(s, t)x\, ds \\
c(t) &= \exp(-Mt)a \\
D(t) &= \int_0^t \exp(-M(t - u))\exp(-M^{\mathsf{T}}(t - u))\, du
\end{aligned}
\qquad (35)
$$

The function Λ in the definition of $\mu(t, x)$ is

$$\Lambda(s, t) = -\mathrm{Cov}(\xi_s, \xi_t)\mathrm{Cov}(\xi_t, \xi_t)^{-1} \qquad (0 \le s \le t) \qquad (36)$$

Proof. The term $v_0(t)$ in (33) is set equal to zero without loss of generality, because it contributes only a scale factor to ρ. The exponent $L(t, \eta_t^{a \to x})$ defined by (30) reduces to

$$L(t, \eta_t^{a \to x}) = -\frac{1}{2}\int_0^t v_1(s)^{\mathsf{T}}.\eta_{s,t}^{a \to x}\, ds + \int_0^t (dY_s)^{\mathsf{T}}\Pi(t)\eta_{s,t}^{a \to x}$$

Let $m(s)$ be the expectation of the OU_n bridge $\eta_t^{a \to x}$,

$$m(s) = E(\eta_{s,t}^{a \to x})$$

$$= \exp(-Ms)(I + \exp(Ms)\Lambda(s, t)\exp(-Mt))a - \Lambda(s, t)x \quad (0 \le s \le t) \quad (37)$$

The expression for $m(s)$ is taken from (6). The process $\eta_t^{a \to x}$ is Gaussian, thus $L(t, \eta_t^{a \to x})$ is Gaussian. The expectation of $L(t, \eta_t^{a \to x})$ is

$$E(L(t, \eta_t^{a \to x})) = -\frac{1}{2}\int_0^t v_1(s)^{\mathsf{T}}.m(s)\, ds + \int_0^t (dY_s)^{\mathsf{T}}\Pi(s)m(s) \qquad (38)$$

The covariance $S(t,x)$ of $L(t,\eta_t^{a\to x})$ and the difference $E(L(t,\eta_t^{a\to x}))$ − $\mu(t,x)$ are both independent of x. The density $\rho(t,x)$ is unnormalised, thus the exponent on the right-hand side of (31) can be and is replaced by $\mu(t,x)$.

The density $p_\xi(t,x)$ of ξ_t is Gaussian, with expectation $c(t)$ and covariance $D(t)$,

$$p_\xi(t,x) = (2\pi)^{-n/2}(\det(D(t)))^{-1/2}\exp\left(-\frac{1}{2}(x-c(t))^{\mathsf{T}}D(t)^{-1}(x-c(t))\right)$$

The expressions for $c(t)$, $D(t)$ given in (35) are obtained from (3), (4). The result, (34), now follows.   □

It is shown in [12], Section 5.2 that the state space $\mathbb{R}^n$ can be reparameterised such that the Beneš filter reduces to the Kalman-Bucy filter. The advantage of the development of the Beneš filter, as given in Theorem 5, over such a reduction is that the different terms appearing in the expression for the conditional density are directly related to physical quantities. This facilitates the implementation of the filter and also suggests ways in which the filter can be adapted to systems outside the class for which it generates the exact unnormalised conditional density.

3.2   The Kalman-Bucy filter

The Kalman-Bucy filter is obtained as a special case of the Beneš filter. The filter equations are

$$\begin{aligned}
dX_t &= FX_t\,dt + dB_t & (0 \le t)\\
dY_t &= \Pi X_t\,dt + dW_t & (0 \le t)
\end{aligned}$$

where F is a constant $n \times n$ matrix and Π is a constant $m \times n$ matrix. Let the function h of Theorem 3 be linear in x, $h(x) = -Mx$. It follows from (21) that

$$V(t,x) = x^{\mathsf{T}}(F^{\mathsf{T}}F - M^{\mathsf{T}}M + \Pi^{\mathsf{T}}\Pi)x$$

on omitting terms independent of x. It is necessary to choose M such that

$$F^{\mathsf{T}}F - M^{\mathsf{T}}M + \Pi^{\mathsf{T}}\Pi = 0 \tag{39}$$

and such that $M + F$ is symmetric. The symmetry condition ensures that $r(x) = (M+F)x$ is the gradient of a scalar potential. Let $M = Q - F$, with Q symmetric. Equation (39) yields

$$Q^2 - F^{\mathsf{T}}Q - QF - \Pi^{\mathsf{T}}\Pi = 0 \tag{40}$$

Equation (40) is an algebraic Riccati equation. It has a unique, positive definite solution for Q under mild conditions on F, Π [22]. In particular, such a solution exists if $\Pi^{\mathsf{T}}\Pi$ has maximal rank n.

3.3   Special cases

The expressions for D and Λ given in Theorem 5 are easy to evaluate if M is symmetric.

Proposition 6. *Let the matrix M in the equation (32) defining the OU process ξ be symmetric and let R be an orthogonal matrix such that $RMR^{\mathsf{T}} = L$ where L is diagonal. Then the functions D, Λ of Theorem 5 are given by*

$$D(t) = R^{\mathsf{T}} L^{-1} \sinh(Lt) \exp(-Lt) R \qquad (0 \le t)$$
$$\Lambda(s,t) = -R^{\mathsf{T}} \sinh(Ls) \sinh(Lt)^{-1} R \qquad (0 \le s \le t) \qquad (41)$$

Proof. The result follows from (35) and (36).   □

The hypotheses of Theorem 5 entail severe restrictions on the drift f and the measurement function g. The following proposition is typical.

Proposition 7. *With the notation of Theorems 3,4,5, let $\sigma(t,x) = I$ and let $g(t,x) = \Pi x$ where $\Pi : \mathbb{R}^n \to \mathbb{R}^m$ is linear and independent of time. Let $V(t,x) = v_1^{\mathsf{T}}.x + v_0$ where v_1, v_0 are constant. If the drift velocity f is independent of t and if $h(x) = -Mx$ where M is a constant $n \times n$ matrix, then f satisfies the Riccati equation*

$$\|f\|^2 + \sum_{i=1}^{n} \frac{\partial f}{\partial x_i} = Q(x) \qquad (42)$$

where

$$Q(x) = x^{\mathsf{T}} q_2 x + q_1^{\mathsf{T}}.x + q_0$$

The coefficients q_0, q_1, q_2 are constant and $q_2 + \Pi^{\mathsf{T}}\Pi$ is a positive matrix.

Proof. Equation (21) and the hypotheses of the proposition yield

$$v_1^{\mathsf{T}}.x + v_0 = V(x)$$
$$= \|f\|^2 - \|h\|^2 + \sum_{i=1}^{n} \left(\frac{\partial f_i}{\partial x_i} - \frac{\partial h_i}{\partial x_i} \right) + \|\Pi x\|^2 \qquad (43)$$

The result follows on observing that (43) reduces to

$$\|f\|^2 + \sum_{i=1}^{n} \frac{\partial f_i}{\partial x_i} = v_1^{\mathsf{T}}.x + v_0 + x^{\mathsf{T}}(M^{\mathsf{T}}M - \Pi^{\mathsf{T}}\Pi)x - \text{trace}(M) \qquad (44)$$

□

The hypotheses of Proposition 7 are insufficient to ensure that ρ is finite dimensional. It is also necessary that $f - h$ should be the gradient of a scalar potential,

$$f_i(x) + \sum_{j=1}^{n} M_{ij}x_j = \frac{\partial \Phi}{\partial x_i} \qquad (45)$$

On using (45) to eliminate f from (44) it follows that

$$\|\nabla\Phi\|^2 - 2x^\top M^\top(\nabla\Phi) + \nabla^2\Phi = -x^\top\Pi^\top\Pi x + v_1^\top.x + v_0$$

In the one dimensional case M is just a scalar, $M = \beta$. A potential Φ can always be defined by

$$\Phi(x) = \int^x f(u)\,du + \frac{1}{2}\beta x^2 \tag{46}$$

and (42) reduces to a Riccati equation [3],

$$f^2 + \frac{df}{dx} = q_2 x^2 + q_1 x + q_0 \qquad (q_2 \geq -\Pi^2) \tag{47}$$

4 Beneš Trajectories in One Dimension

The advantage of the Beneš filter over the Kalman-Bucy filter is that it can be applied to a wider range of systems. As indicated in §1, a Beneš trajectory in one dimension is defined to be a solution $t \mapsto x(t)$ to the ODE $dx/dt = f(x)$, where f satisfies the Riccati equation (47). In this section it is shown that there are five possible types of behaviour for the Beneš trajectories at large times. It is also shown that at small times an arbitrary smooth trajectory can be approximated to $O(t^5)$ by a Beneš trajectory.

It is assumed throughout this section that $\sigma(t, x) \equiv 1$, $g(t, x) = x$ and that f satisfies the Riccati equation (47). It is also assumed that $x(0) = 0$ and that f is normalised such that $f(0) > 0$. The sign of $f(0)$ can be changed if necessary by applying the transformation $x \mapsto -x$.

The solutions to (47) need not be global, in that f may have one or more singularities. In the neighbourhood of a singularity, condition (12) for the existence of a unique weak solution to the filter equations fails. Nevertheless, the Beneš trajectories are of interest in practical applications of filtering. They provide new, tractable filters with non-linear drift. Any singularities of f can be safely ignored provided there is only a very small probability that the system state approaches them.

Some general properties of Beneš trajectories are obtained in §4.1. In particular it is shown that the normalised Beneš trajectories are strictly monotonic increasing with time. The trajectories are obtained explicitly in §4.2 for the case in which $q_1 = q_2 = 0$ on the right-hand side of (47). The trajectories for $q_1 \neq 0$ and $q_2 \neq 0$ are discussed in §4.3. It is shown in §4.4 that an arbitrary trajectory can be approximated to an accuracy of $O(t^5)$ by a Beneš trajectory.

4.1 General results

A single integration of the ODE $dx/dt = f(x)$ yields

$$\int^x f(u)^{-1}\,du = t \tag{48}$$

from which the system state x can in principle be obtained as a function of t. It is convenient to write $G(t)$ for the system state $x(t)$,

$$dG/dt = f(G(t)) \tag{49}$$

This is done to avoid overuse of the symbol x. It follows from (47) and (49) that

$$
\begin{aligned}
\frac{d^2 G}{dt^2} &= \frac{df}{dx}(G)\frac{dG}{dt} \\
&= \left(-\left(\frac{dG}{dt}\right)^2 + Q(G) \right) \frac{dG}{dt}
\end{aligned} \tag{50}
$$

where $Q(G) = q_0 + q_1 G + q_2 G^2$. A rearrangement of (50) yields

$$\frac{d^2 G}{dt^2} + \left(\frac{dG}{dt}\right)^3 = \frac{dG}{dt}Q(G) \tag{51}$$

In the Kalman-Bucy filter, $f(x) = a_0 + a_1 x$, where a_0, a_1 are constants. The integral in (48) is evaluated to yield

$$
\begin{aligned}
G(t) &= a_1^{-1} a_0(\exp(a_1 t) - 1) \qquad & (G(0) = 0, a_1 \neq 0) \\
G(t) &= a_0 t \qquad & (G(0) = 0, a_1 = 0)
\end{aligned} \tag{52}
$$

There are only two possible behaviours of G at large times: either G increases linearly with time or G increases exponentially with time.

It is well known [3] that any solution to the Riccati equation (47) is of the form

$$f = \phi^{-1}\frac{d\phi}{dx} \tag{53}$$

where ϕ is a solution of the linear ODE

$$\frac{d^2 \phi}{dx^2} - Q(x)\phi = 0 \tag{54}$$

It follows from (53) and (54) that f is real analytic in any region not including the zeros of ϕ or the poles of $d\phi/dx$. If $q_2 \neq 0$, then ϕ is a parabolic cylinder function. If $q_2 = 0$, $q_1 \neq 0$, then ϕ is an Airy function. If $q_2 = q_1 = 0$ and $q_0 \neq 0$, then ϕ is an exponential function, possibly with a complex exponent. If $Q \equiv 0$, then ϕ is linear. Further information about the various solutions to (54) is given in [1].

Proposition 8. *Let f have a zero in $[0, \infty)$, and let $a > 0$ be the least such zero. Let f be bounded on $[0, a]$. Then G is strictly monotonically increasing with t, $G(t) < a$ for all $t \geq 0$ and $\lim_{t \to \infty} G(t) = a$.*

Proof. Let $\epsilon > 0$ be a small positive number. Then f is bounded away from zero on $[0, a - \epsilon]$ thus there exists a time $t_0 < \infty$ such that $G(t_0) = a - \epsilon$, and G is strictly monotonically increasing for $0 \leq t \leq t_0$.

Suppose, if possible, that $f'(a) = 0$. Then it follows from (53) and (54) that $\phi'(a) = \phi''(a) = 0$, thus ϕ is identically zero. The case $f'(a) = 0$ is thus excluded, because f is not properly defined. In addition, $f'(a) > -\infty$, because f satisfies the Riccati equation (47) at a. It follows that $0 > f'(a) > -\infty$. After rescaling f it can be and is assumed that $f'(a) = -1$. It follows that

$$
\begin{aligned}
f(x) &= f(a) + (x - a)f'(a) + O((x - a)^2) \\
&= -(x - a) + O((x - a)^2)
\end{aligned}
$$

Let $\alpha_1 < 1 < \alpha_2$ be positive numbers such that

$$\alpha_1(a - x) \leq f(x) \leq \alpha_2(a - x) \qquad (a - \epsilon \leq x \leq a) \qquad (55)$$

Equations (48) and (55) yield

$$\alpha_1^{-1} \int_{a-\epsilon}^{x} (a - u)^{-1}\, du \geq t - t_0 \geq \alpha_2 \int_{a-\epsilon}^{x} (a - u)^{-1}\, du$$

from which it follows that

$$\epsilon \exp(-\alpha_2(t - t_0)) \leq a - x \leq \epsilon \exp(-\alpha_1(t - t_0))$$

It follows that $\lim_{t \to \infty} G(t) = a$ and that G does not reach a in a finite time. Finally, G is strictly monotonically increasing for all t, because f is strictly positive on $[0, a)$. □

Proposition 9. *Let f be bounded on each finite interval $[0, N]$ for $N > 0$. Each trajectory $t \mapsto G(t)$, $t \geq 0$ is a monotonic function of t. If f has no zero in $[0, \infty)$, then $\lim_{t \to \infty} G(t) = \infty$.*

Proof. The result is proved in Proposition 8 for the case in which f has a zero in $[0, \infty)$. If f has no zero in $[0, \infty)$, then it is strictly bounded away from zero on any finite interval $[0, N]$. Thus the trajectory G reaches N in a finite time t_0 and G is strictly monotonically increasing for $0 \leq t \leq t_0$. The result follows. □

Propositions 8 and 9 are special cases of more general results for the Riccati equation [3].

4.2   Constant Q

The case $Q(x) \equiv q_0$ can be resolved completely by elementary integration. It is convenient to rescale space and time. Let $\tilde{G}$ be defined by $\tilde{G} = aG(a^{-1}t)$.

Then it follows that

$$
\frac{dG}{dt} = \frac{d\tilde{G}}{dt}(at)
$$

$$
\frac{d^2G}{dt^2} = a\frac{d^2\tilde{G}}{dt^2}(at) \tag{56}
$$

It follows from (51), (56) and the assumption $Q \equiv q_0$ that

$$
a\frac{d^2\tilde{G}}{dt^2}(at) + \left(\frac{d\tilde{G}}{dt}(at)\right)^3 = q_0\frac{d\tilde{G}}{dt}(at) \tag{57}
$$

Let $s = at$. Then it follows from (57) that

$$
\frac{d^2\tilde{G}}{ds^2} + \left(\frac{d\tilde{G}}{ds}\right)^3 = a^{-2}q_0\frac{d\tilde{G}}{ds}
$$

Thus it suffices to consider the three cases $q_0 = 0, \pm 1$. The assumptions $G(0) = 0$, $(dG/dt)(0) > 0$ made at the beginning of this section continue to hold.

Proposition 10. *If $Q(x) \equiv 1$, then $\lim_{t\to\infty} t^{-1}G(t) = 1$. If $Q(x) \equiv -1$, then $\lim_{t\to\infty} G(t)$ is finite.*

Proof. If $q_2 = 0, q_1 = 0$ and $q_0 = 1$, then (51) reduces to

$$
\frac{d^2G}{dt^2} + \left(\frac{dG}{dt}\right)^3 = \frac{dG}{dt} \tag{58}
$$

If $d^2G/dt^2 \equiv 0$, then (58) has the solution $G(t) = t$. Otherwise, define the function $t \mapsto u(t)$ by $u = dG/dt$. Then (58) yields

$$
\frac{du}{dt} = u - u^3
$$

from which it follows that

$$
u(t) = \frac{A\exp(t)}{\sqrt{1 + A^2\exp(2t)}} \tag{59}
$$

where $0 < A < 1$ is a disposable constant. On integrating (59) it follows that

$$
G(t) = \ln\left(\frac{A\exp(t) + \sqrt{1 + A^2\exp(2t)}}{A + \sqrt{1 + A^2}}\right) \tag{60}
$$

The result $\lim_{t\to\infty} t^{-1}G(t) = 1$ follows.

If $q_2 = 0, q_1 = 0$ and $q_0 = -1$, then a similar calculation yields

$$
G(t) = \sin^{-1}(A) - \sin^{-1}(A\exp(-t)) \tag{61}
$$

where $A > 0$ is a disposable constant. The trajectory (61) is bounded such that $\lim_{t \to \infty} G(t) = \sin^{-1}(A)$.   □

Proposition 11. *Let* $Q(x) \equiv 0$. *Then* $f(x) = (x + a)^{-1}$ *for some constant* a. *If* $a > 0$, *then* $\lim_{t \to \infty} t^{-1/2} G(t) = \sqrt{2}$.

Proof. The function $f(x) = (x + a)^{-1}$ is obtained by solving (42) with $Q \equiv 0$. If $a > 0$, then $dx/dt = f(x)$ has a solution in the interval $t \geq 0$. The solution satisfies $x^2/2 + ax = t$ from which it follows that

$$G(t) = -a + \sqrt{a^2 + 2t}$$

□

4.3   Large times

If $f(x) > 0$ for $x \geq 0$, then a Beneš trajectory $t \mapsto G(t)$ such that $G(0) = 0$ attains large values at large times, $\lim_{t \to \infty} G(t) = \infty$. In this subsection the rate of growth of a Beneš trajectory at large times is investigated. The case $q_2 = q_1 = 0$ has already been described in §4.2. It remains to consider the case $q_2 = 0, q_1 \neq 0$ and the case $q_2 \neq 0$.

If $q_2 = 0$, $q_1 \neq 0$, then

$$f(x)^2 + \frac{df}{dx} = q_1 x + q_0$$

After a rescaling, as described in §4.2, and a translation of the x axis, it suffices to examine the case $q_1 = 1$, $q_0 = 0$. Then f is a ratio, $f = \phi^{-1} d\phi/dx$ where ϕ is a solution of

$$\frac{d^2 \phi}{dx^2} - x\phi = 0 \tag{62}$$

The solutions of (62) are known as Airy functions [1]. There are two linearly independent Airy functions Ai, Bi with contrasting behaviours at large $x > 0$.

$$\mathrm{Ai}(x) \sim \frac{1}{2} \pi^{-1/2} x^{-1/4} \exp\left(-\frac{2}{3} x^{3/2}\right)(1 + O(x^{-3/2}))$$

$$\mathrm{Bi}(x) \sim \pi^{-1/2} x^{-1/4} \exp\left(\frac{2}{3} x^{3/2}\right)(1 + O(x^{-3/2})) \tag{63}$$

In any linear combination $r_1 \mathrm{Ai} + r_2 \mathrm{Bi}$ with $r_2 \neq 0$ the function Bi predominates over Ai for $x > 0$ and large. It is thus sufficient to consider two cases, $f(x) = \mathrm{Ai}(x)^{-1} d\mathrm{Ai}/dx$ and $f(x) = \mathrm{Bi}(x)^{-1} d\mathrm{Bi}/dx$. The first case is ruled out by the condition $f(0) > 0$, because $\mathrm{Ai}(x) > 0$, $d\mathrm{Ai}/dx < 0$ for $x \geq 0$. In the second case $\mathrm{Bi}(x) > 0$, $d\mathrm{Bi}/dx > 0$ for $x \geq 0$. The hypothesis $f(x) > 0$ for $x \geq 0$ is satisfied and the growth of f is given at large x by

$$f(x) = \mathrm{Bi}(x)^{-1} \frac{d\mathrm{Bi}}{dx} \sim x^{1/2}$$

from which it follows that $G(t) \sim t^2$.

If $q_2 \neq 0$, then after rescaling x, t and applying a constant translation to the x coordinate it suffices to consider the two cases

$$\frac{d^2\phi}{dx^2} - \left(\frac{1}{4}x^2 + a\right)\phi = 0 \tag{64}$$

$$\frac{d^2\phi}{dx^2} + \left(\frac{1}{4}x^2 - a\right)\phi = 0 \tag{65}$$

The solutions ϕ to (64) or (65) are known as parabolic cylinder functions [1]. The solutions to (65) are oscillatory if $a < 0$. If $a \geq 0$, then the solutions to (65) are oscillatory for $|x| > 2\sqrt{a}$. Thus (65) does not yield velocities f that are non-zero in $[0, \infty)$. The equation (64) has two linearly independent solutions U, V with contrasting behaviours for large $x > 0$,

$$U(a, x) = \exp\left(-\frac{1}{4}x^2\right) x^{-a-1/2}(1 + O(x^{-2}))$$

$$V(a, x) = \sqrt{\frac{2}{\pi}} \exp\left(\frac{1}{4}x^2\right) x^{a-1/2}(1 + O(x^{-2})) \tag{66}$$

If $a < 0$, then U, V are oscillatory for $|x| < 2\sqrt{|a|}$. Thus it suffices to examine the case $a \geq 0$. In the range $x \geq 0$ the function V dominates any linear combination $r_1 U + r_2 V$ at large x provided $r_2 \neq 0$. The case $f(x) = U(x)^{-1} dU/dx$ is ruled out, because $f(x) \sim -x/2 < 0$ if x is large. There remains only the case

$$f(x) = V(a, x)^{-1}\frac{dV}{dx} \sim \frac{1}{2}x + O(x^{-1})$$

from which it follows that $G(t) \sim \exp(t/2)$.

The behaviours of the Beneš trajectories at large times are summarised.

Theorem 12. *Let $t \mapsto G(t)$ be a Beneš trajectory such that $G(0) = 0$, $(dG/dt)(0) > 0$. Then after the appropriate rescalings of space and time the trajectory exhibits one of the following behaviours as $t \to \infty$.*

$$\begin{array}{lll}
i) & \lim_{t \to \infty} G(t) = a & iv) \quad G(t) \sim t^2 \\
ii) & G(t) \sim \sqrt{2t} & v) \quad G(t) \sim \exp(t/2) \\
iii) & G(t) \sim t &
\end{array}$$

In all cases G is strictly monotonic increasing for $t \geq 0$.   □

Examples of the five types of trajectory are shown in Figure 1, drawn with the aid of Mathematica [23]. The trajectories are chosen such that $(dG/dt)(0) = 1/2$. At time $t = 5$ the trajectories in Figure 1 are in order (i) to (v) as shown.

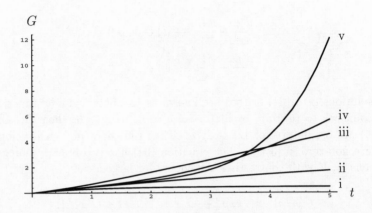

Figure 1. *The five types of trajectory*

4.4　Small times

Let H be a general trajectory such that

$$H(t) = h_1 t + h_2 t^2 + h_3 t^3 + h_4 t^4 + O(t^5) \qquad (h_1 > 0) \qquad (67)$$

At small times H can be approximated to an accuracy of $O(t^3)$ by a trajectory for which f is linear. It suffices to choose a_0, a_1 in (52) such that

$$
\begin{aligned}
a_0 &= h_1 \\
a_1 &= 2h_1^{-1} h_2
\end{aligned}
$$

Let $f = \phi^{-1} d\phi/dx$ where ϕ is as in (53). The inverse drift velocity f^{-1} has an expansion as a power series in x away from the zeros of $d\phi/dx$ and the poles of ϕ. It follows that G also has an expansion as a power series in t. Let this expansion be

$$G(t) = \sum_{i=1}^{\infty} g_n t^n \qquad (g_1 > 0)$$

On substituting the power series for G into (51) it follows that

$$2g_2 + 6g_3 t + 12 g_4 t^2 + (g_1 + 2g_2 t + 3g_3 t^2)^3 =$$

$$q_2 g_1^3 t^2 + q_1 (g_1^2 t + 3g_1 g_2 t^2) + q_0 (g_1 + 2g_2 t + 3g_3 t^2) + O(t^3) \qquad (68)$$

The equation (68) holds for $t \geq 0$ if and only if

$$
\begin{aligned}
q_0 &= g_1^2 + 2g_1^{-1}g_2 \\
q_1 &= 6g_1^{-2}g_3 + 4g_2 - 4g_1^{-3}g_2^2 \\
q_2 &= 12g_1^{-5}g_2^3 - 24g_1^{-4}g_2g_3 + 12g_1^{-3}g_4 + 6g_1^{-1}g_3
\end{aligned}
\tag{69}
$$

It follows that the g_i for $1 \leq i \leq 4$ can be assigned arbitrary values, subject only to the conditions $g_1 > 0$ and $q_2 \geq -\Pi^2$. The g_i for $i \geq 5$ are uniquely determined by a set of equations of the form

$$
n(n-1)g_n t^{n-2} = c_n(g_i, 1 \leq i \leq n-1)t^{n-2} \qquad (n \geq 5)
$$

where c_n is a complicated polynomial in the g_i for $1 \leq i \leq n-1$. Thus the coefficients g_i, $1 \leq i \leq 4$ once chosen, determine G uniquely. These observations have the following consequence.

Proposition 13. *Let the coefficients h_i of H as defined in (67) satisfy*

$$
q_2(h_i, 1 \leq i \leq 4) > -\Pi^2
$$

Then there exists a Beneš filter for which the deterministic trajectory G approximates H to an accuracy of $O(t^5)$.

Proof. It suffices to set $g_i = h_i$ for $1 \leq i \leq 4$. The remaining coefficients g_i of G are then determined uniquely by the condition that G is a solution to (51). □

5   Conclusion

The probability density function for the system state conditional on the measurements is represented as a path integral. For certain non-linear systems the conditional density depends on only a finite number of statistics which can be calculated recursively by solving a set of linked ordinary differential equations. These systems were first described by Beneš in [4]. The class of Beneš systems is extended to include certain non-linear systems in which the drift is the sum of a general linear function and the gradient of a scalar potential.

The Beneš filters appear to be the only non-linear, finite dimensional, recursive filters with a straightforward construction. In the case of a one dimensional state space there is strong evidence that the Beneš filters are the only finite dimensional, recursive filters [12]. Additional examples are known for higher dimensional state spaces, but their construction is contrived. The filter due to Wong [12] is a good example.

Path integral representations of the optimal density have not so far been used widely for the design of practical filters. This situation is likely to change in future. The path integral approach has the following advantages over more traditional approaches to filter design.

i) It is straightforward to find finite dimensional, recursive approximations to the optimal density.

ii) The optimal density is approximated directly, rather than through an initial simplification of the filter equations. This gives a much greater insight into the effects of the different choices made in constructing the approximation.

iii) It is straightforward to construct non-Gaussian approximations.

The above advantages are illustrated by the examples in [13].

Acknowledgements: Thanks are due to Roger Brockett for the references [4,6,7] and for information about finite dimensional filters. Part of this work was carried out during the Computer Vision Programme at the Isaac Newton Institute, Cambridge, UK (July-December 1993). The author thanks the Institute for its hospitality. The work is funded by the Royal Society through an Industrial Fellowship held by the author at the Department of Engineering Science, Oxford, UK. The author is employed by the GEC-Marconi Hirst Research Centre, Borehamwood, UK.

References

[1] Abramovitz, M. & Stegun, I.A. 1965 *Handbook of Mathematical Functions*. New York: Dover.

[2] Ayache, N. 1991 *Artificial Vision for Mobile Robots: stereo vision and multisensory perception*. Cambridge, Mass: MIT Press.

[3] Bellman, R. 1969 *Stability Theory of Differential Equations*. New York: Dover.

[4] Beneš, V.E. 1981 Exact finite-dimensional filters for certain diffusions with non-linear drift. *Stochastics* **5**, 65-92.

[5] Bensoussan, A. 1992 *Stochastic Control of Partially Observable Systems*. Cambridge: Cambridge University Press.

[6] Brockett, R.W. 1980 Remarks on finite dimensional nonlinear estimation. In *Analyse des Systemes (Bordeaux, 11-16 September 1978)*, Astérisque **75-76**. Société Mathématique de France.

[7] Brockett, R.W. & Clark, J.M.C. 1980 The geometry of the conditional density equation. In *Analysis and Optimisation of Stochastic Systems*, eds. Jacobs, O.L.R., Davis, M.H.A., Dempster, M.A.H., Harris, C.J. & Parks, P.C. The Institute of Mathematics and Its Applications Conference Series, Academic Press.

[8] Broida, T.J. & Chellappa, R. 1991 Estimating the kinematics and structure of a rigid object from a sequence of monocular images. *IEEE Transactions on Pattern Analysis and Machine Intelligence* **13**, 497-513.

[9] Faugeras, O.D. 1993 *Three Dimensional Computer Vision: a geometric approach*. Cambridge, Mass: MIT Press.

[10] Karatzas, I. & Shreve, S.E. 1988 *Brownian Motion and Stochastic Calculus*. Graduate Texts in Mathematics **113**. New York: Springer-Verlag.

[11] Karlin, S. & Taylor, H.M. 1981 *A Second Course in Stochastic Processes*. Academic Press.

[12] Marcus, S.I. 1984 Algebraic and geometric methods in nonlinear filtering. *SIAM J. on Control and Optimization* **22**, 817-844.

[13] Maybank, S.J. 1994 Recursive approximation to the optimal filter. Submitted for publication.

[14] Maybeck, P.S. 1979 *Stochastic Models, Estimation and Control - Volume 1*. Mathematics in Science and Engineering Series, vol 141-1. San Diego, CA, USA: Academic Press.

[15] Maybeck, P.S. 1979 *Stochastic Models, Estimation and Control - Volume 2*. Mathematics in Science and Engineering Series, vol 141-2. New York, NY, USA: Academic Press.

[16] Maybeck, P.S. 1982 *Stochastic Models, Estimation and Control - Volum 3*. Mathematics in Science and Engineering Series, vol 141-3. London, UK: Academic Press.

[17] Øksendal, B. 1992 *Stochastic Differential Equations: an introduction with applications*. Berlin: Springer-Verlag (third edition).

[18] Pardoux, E. 1991 Filtrage non-linéaire et équations aux dérivées partielles stochastiques associeés. In *Ecole d'Eté de Probabilités de Saint-Flour XIX*, Lecture Notes in Mathematics **1464**. Berlin: Springer-Verlag.

[19] Revuz, D. & Yor, M. 1991 *Continuous Martingales and Brownian Motion*. Grundlehren der mathematischen Wissenschaften **293**. Berlin: Springer-Verlag.

[20] Rogers, L.C.G. & Williams, D.W. 1987 *Diffusions, Markov Processes and Martingales, Volume 2: Itô calculus*. Chichester: John Wiley and Sons (reprinted 1990).

[21] Stroock, D.W. & Varadhan, S.R.S 1979 *Multidimensional Diffusion Processes*. Grundlehren der mathematischen Wissenschaften **233**. Springer-Verlag: Berlin.

[22] Willems, J.C. 1971 Least squares stationary optimal control and the algebraic Riccati equation. *IEEE Transactions on Automatic Control* **16**, 621-634.

[23] Wolfram, S. 1991 *Mathematica: a system for doing mathematics by computer*. Redwood City, CA, USA: Addison Wesley (2nd edition).

[24] Yor, M. 1992 *Some Aspects of Brownian Motion, Part 1: some special functions*. Basel: Birkhäuser Verlag.

[25] Zakaï, M. 1969 On the optimal filtering of diffusion processes. *Z. Wahrscheinlichkeitstheorie verw. Geb.* **11**, 230-243.

A skew-product representation for the generator of a two sex population model

JOACHIM A. REBHOLZ
Department of Mathematics
University of California at Berkeley
Berkeley, Ca 94720

Abstract

We consider a reproducing two sex population model, whose dynamics is governed by the generator of a time-changed planar Brownian motion. This time-changed process enjoys a skew-product representation and is in duality to a pure jump process on an integer grid. Using duality we can estimate the growth of the moments of the studied process.

1 Introduction and Motivation

The main object of interest is the infinitesimal generator

$$A = \frac{1}{2}x_1 x_2 (\sigma_1^2 \frac{\partial^2 f}{\partial x_1^2} + 2\rho\sigma_1\sigma_2 \frac{\partial^2 f}{\partial x_1 \partial x_2} + \sigma_2^2 \frac{\partial^2 f}{\partial x_2^2}),$$

which arises as a limit of particle systems as follows: Two populations, say males M and females F, reproduce in a box (i.e. one point state space) and give birth to offspring at rates equal to the product of the number of individuals of the two populations. Starting with more and more individuals and using a scaling argument similar to that in Feller's branching diffusion approximation, we arrive at the limiting generator from above. We have to use (unbiased) offspring distributions with $\mathbb{E}M = \mathbb{E}F = 1$, where σ_1, σ_2 and ρ denote the variance in M, variance in F and the correlation between M and F, respectively. For certain correlation coefficients the offspring distribution is very simple. Let us give three examples:

$\rho = 1$

$$\boxed{M}\,\boxed{F} \to \begin{cases} \boxed{M}\,\boxed{M}\,\boxed{F}\,\boxed{F} & \text{with prob. } \tfrac{1}{2} \\ \varnothing & \text{with prob. } \tfrac{1}{2} \end{cases} \tag{1}$$

$\rho = -1$

$$\boxed{M}\,\boxed{F} \to \begin{cases} \boxed{M}\,\boxed{M} & \text{with prob. } \tfrac{1}{2} \\ \boxed{F}\,\boxed{F} & \text{with prob. } \tfrac{1}{2} \end{cases} \tag{2}$$

$\rho = 0$

$$\boxed{M}\,\boxed{F} \to \begin{cases} \boxed{M}\,\boxed{M}\,\boxed{F} & \text{with prob. } \tfrac{1}{4} \\ \boxed{M}\,\boxed{F}\,\boxed{F} & \text{with prob. } \tfrac{1}{4} \\ \boxed{M} & \text{with prob. } \tfrac{1}{4} \\ \boxed{F} & \text{with prob. } \tfrac{1}{4} \end{cases} \tag{3}$$

We note that by choosing biased offspring distributions, one can produce deterministic models, which are studied in [2]. In certain cases it is possible to give explicit representations, see [3].

The generator A is a time-change of a standard two-dimensional Brownian motion. A priori, there is no reason why this time-changed process should posses a skew-product representation like ordinary two-dimensional Brownian motion. We will give a heuristic argument to show that a transform of A should have such a representation and then prove it rigorously. We observe the phenomenon that the radial and angular part switch roles in the skew-product representation of the time-changed process. We will also show that the generator A is in duality to a jump process on an integer grid. This will allow us to extract some information about the growth of the moments of A.

To motivate what follows, let us consider the usual skew-product representation of standard two-dimensional Brownian motion (B_1, B_2) in terms of generators. Express the generator of (B_1, B_2), $\frac{1}{2}\Delta$, in polar coordinates (r, θ) as

$$\frac{1}{2}\Delta = \left(\frac{1}{2}\frac{d^2}{dr^2} + \frac{1}{2r}\frac{d}{dr}\right) + \frac{1}{r^2}\left(\frac{1}{2}\frac{d^2}{d\theta^2}\right)$$

where we have not used partial derivatives notation to emphasize the interpretation of this representation: Start with the piece $\frac{1}{2}\frac{d^2}{d\theta^2}$, the generator of a linear Brownian motion. It is multiplied by $1/r^2$, which means the Brownian motion is run at a different speed, namely at rate of $1/r^2$. Now we add the part obtained so far by a Trotter product argument to that of the generator of the radial part r, which is a genuine Markov process in its own right.

We now make the following heuristic calculation for zero correlation:

$$A = \frac{1}{2}x_1 x_2\left(\frac{\partial^2 f}{\partial x_1^2} + \frac{\partial^2 f}{\partial x_2^2}\right) \;=\; \frac{1}{2}r^2 \cos\theta \sin\theta\left(\frac{d^2}{dr^2} + \frac{1}{r}\frac{d}{dr} + \frac{1}{r^2}\frac{d^2}{d\theta^2}\right) \tag{4}$$

$$= \; \frac{1}{2}\cos\theta \sin\theta \frac{d^2}{d\theta^2} + \frac{1}{2}\cos\theta \sin\theta\left(r^2\frac{d^2}{dr^2} + r\frac{d}{dr}\right) \tag{5}$$

This means that the time-changed process still satisfies a skew-product representation. Note that this need not be true in general. Also observe that the roles of r and θ have switched: The process in θ is now an autonomous Markov process in its own right and the process r is a time-changed θ process. To make these ideas precise, this suggests we look for a skew-product representation in terms of $\theta = \cot(\frac{x_1}{x_2})$ and $r^2 = x_1^2 + x_2^2$. This is carried out in the next section.

2   The skew-product representation

If (X_1, X_2) is a solution to the $(A, \delta_{x_1} \times \delta_{x_2})$ martingale problem on the state space $[0, \infty) \times [0, \infty)$, then we can view (X_1, X_2) alternatively as a solution to the following two-dimensional SDE:

$$dX_1(t) = \sqrt{X_1(t)X_2(t)}\sigma_1 dB_1(t), \ X_1(0) = x_1 \text{ a.s.} \tag{6}$$

$$dX_2(t) = \sqrt{X_1(t)X_2(t)}\sigma_2(\rho dB_1(t) + \sqrt{1 - \rho^2}dB_2(t)), \ X_2(0) = x_2 \text{ a.s.} \tag{7}$$

where (B_1, B_2) is a pair of independent standard Brownian motions.

Proposition 2.1 *Suppose (X_1, X_2) is a solution to the $(A, \delta_{x_1} \times \delta_{x_2})$ martingale problem. Then $X(t) = \frac{X_1(t)}{X_2(t)}$ is a solution to the SDE*

$$dX(t) = (\sigma_2^2 X(t)^2 - \rho\sigma_1\sigma_2 X(t))dt$$
$$+((\sigma_1 X(t)^{\frac{1}{2}} - \rho\sigma_2 X(t)^{\frac{3}{2}})^2 + \sigma_2(1 - \rho^2)X(t)^3)^{\frac{1}{2}}dB(t), \tag{8}$$
$$X(0) = \frac{x_1}{x_2} \text{ a.s.}$$

with a standard Brownian motion B. Pathwise uniqueness holds for this SDE and X is strong Markov up to the explosion time.

Proof. We have by Itô's Lemma,

$$dg(X_1(t), X_2(t)) \tag{9}$$
$$= g_{x_1}(X_1(t), X_2(t))\sqrt{X_1(t)X_2(t)}\sigma_1 dB_1(t)$$
$$+ g_{x_2}(X_1(t), X_2(t))\sqrt{X_1(t)X_2(t)}\sigma_2(\rho dB_1(t) + \sqrt{1 - \rho^2}dB_2(t))$$
$$+ \frac{1}{2}(\sigma_1^2 g_{x_1 x_1}(X_1(t), X_2(t)) + 2\rho\sigma_1\sigma_2 g_{x_1 x_2}(X_1(t), X_2(t))$$
$$+ \rho^2\sigma_2^2 g_{x_2 x_2}(X_1(t), X_2(t)))X_1(t)X_2(t)dt$$

for C^2 functions g. Letting $g(x_1, x_2) = \frac{x_1}{x_2}$, we can write the right hand side as

$$\sigma(g(X_1(t), X_2(t)))dB(t) + b(g(X_1(t), X_2(t)))dt$$

with B some standard Brownian motion and

$$\sigma^2(x) = (\sigma_1\sqrt{x} - \rho\sigma_2\sqrt{x^3})^2 + \sigma_2(1 - \rho^2)x^3$$
$$b(x) = x^2 - \rho\sigma_1\sigma_2 x$$

This follows immediately from above, using the explicit form of g. Furthermore, pathwise uniqueness holds for the SDE

$$
\begin{aligned}
dX(t) &= (\sigma_2^2 X(t)^2 - \rho\sigma_1\sigma_2 X(t))dt && (10)\\
&\quad +((\sigma_1 X(t)^{\frac{1}{2}} - \rho\sigma_2 X(t)^{\frac{3}{2}})^2 + \sigma_2(1 - \rho^2)X(t)^3)^{\frac{1}{2}}dB(t),\\
X(0) &= \frac{x_1}{x_2} \text{ a.s.}
\end{aligned}
$$

by the version allowing for explosions of the Yamada-Watanabe theorem, see [4]. The strong Markov property is a consequence of pathwise uniqueness. ■

We will give a skew-product representation for a transform (Y_1, Y_2) of (X_1, X_2).

Set

$$Y_1 = \frac{X_1}{X_2} \tag{11}$$

and

$$Y_2 = f(X_1, X_2) \tag{12}$$

for a function f to be determined.
We have by Itô's Lemma,

$$
\begin{aligned}
dY_1(t) &= (\sigma_1(\frac{X_1(t)}{X_2(t)})^{\frac{1}{2}} - \rho\sigma_2(\frac{X_1(t)}{X_2(t)})^{\frac{3}{2}})dB_1(t) + \sigma_2(1 - \rho^2)(\frac{X_1(t)}{X_2(t)})^3 dB_2(t)\\
&\quad + (\sigma_2^2(\frac{X_1(t)}{X_2(t)})^2 - \rho\sigma_1\sigma_2\frac{X_1(t)}{X_2(t)})dt\\
&= dM_1(t) + dD_1(t)
\end{aligned}
$$

and

$$
\begin{aligned}
dY_2(t) &= (\sigma_1 f_{x_1}(X_1(t), X_2(t)) + \rho\sigma_2 f_{x_2}(X_1(t), X_2(t)))\sqrt{X_1(t)X_2(t)}dB_1(t)\\
&\quad + \sigma_2\sqrt{1 - \rho^2}f_{x_2}(X_1(t), X_2(t))\sqrt{X_1(t)X_2(t)}dB_2(t)\\
&\quad + \frac{1}{2}(\sigma_1^2 f_{x_1 x_1}(X_1(t), X_2(t)) + 2\rho\sigma_1\sigma_2 f_{x_1 x_2}(X_1(t), X_2(t))\\
&\quad + \rho^2\sigma_2^2 f_{x_2 x_2}(X_1(t), X_2(t)))X_1(t)X_2(t)dt\\
&= dM_2(t) + dD_2(t)
\end{aligned}
$$

where M_i (D_i) denotes the martingale (drift) part of Y_i. For a skew-product representation to hold, we have to find a diffusion Z with generator $\frac{1}{2}\sigma^2(y_2)\frac{\partial^2}{\partial y_2^2} + b(y_2)\frac{\partial}{\partial y_2}$, say, and a rate function $g(y_1)$ such that

$$\text{generator}(Y_1, Y_2) = \text{generator}(Y_1) + g(y_1)\text{generator}(Z) \tag{13}$$

where $g(y_1)$generator(Z) plays the role of generator(Y_2).

This is equivalent to the following three conditions:

(1) $\quad\quad [M_1, M_2](t) = 0$
(2) $\quad\quad d[M_2](t) = g(Y_1)\sigma^2(Y_2)dt$
(3) $\quad\quad dD_2(t) = g(Y_1)b(Y_2)dt$

So, we have three conditions for the four unknown functions f, g, σ and b. It turns out that we are able to find these functions with the desired properties, suggested by the heuristic calculation from above.

Condition (1):
   A straightforward computation shows that this condition is equivalent to

$$\frac{x_1}{x_2}(\sigma_1(\sigma_1 x_2 - \rho\sigma_2 x_1)f_{x_1} + \sigma_2(\rho\sigma_1 x_2 - \sigma_2 x_1)f_{x_2}) = 0$$

All solutions of this linear p.d.e. are given by C^1 functions of $\sigma_2^2 x_1^2 - 2\rho\sigma_1\sigma_2 x_1 x_2 + \sigma_1^2 x_2^2$, see [5, p.62]. We choose

$$f(x_1, x_2) = \sigma_2^2 x_1^2 - 2\rho\sigma_1\sigma_2 x_1 x_2 + \sigma_1^2 x_2^2 \tag{14}$$

and thus

$$Y_2(t) = \sigma_2^2 X_1^2(t) - 2\rho\sigma_1\sigma_2 X_1(t)X_2(t) + \sigma_1^2 X_2^2(t) \tag{15}$$

Condition (3):
   Having determined f, we get easily

$$dD_2(t) = 2(1 - \rho^2)\sigma_1^2\sigma_2^2 X_1(t)X_2(t)dt.$$

So, we have to write $2(1 - \rho^2)\sigma_1^2\sigma_2^2 x_1 x_2$ as $g(\frac{x_1}{x_2})b(\sigma_2^2 x_1^2 - 2\rho\sigma_1\sigma_2 x_1 x_2 + \sigma_1^2 x_2^2)$ for some g and σ. But it is readily verified that

$$2(1 - \rho^2)\sigma_1^2\sigma_2^2 x_1 x_2 = (1 + \rho)\sigma_1\sigma_2 h(\frac{x_1}{x_2})(\sigma_2^2 x_1^2 - 2\rho\sigma_1\sigma_2 x_1 x_2 + \sigma_1^2 x_2^2)$$

with $h(x) = \frac{2(1-\rho)\sigma_1\sigma_2 x}{\sigma_2^2 x_1^2 - 2\rho\sigma_1\sigma_2 x_1 x_2 + \sigma_1^2 x_2^2} \in (0, 1)$. Hence we can set,

$$b(y_2) \;=\; y_2 \tag{16}$$
$$g(y_1) \;=\; (1 + \rho)\sigma_1\sigma_2 h(y_1) \tag{17}$$

Condition (2):
   Finally we have to determine σ. Again skipping some computations we obtain,

$$d[M_2](t) \;=\; 4(1 - \rho^2)\sigma_1^2\sigma_2^2 f(X_1(t), X_2(t))X_1(t)X_2(t)dt$$
$$\;=\; 2f(X_1(t), X_2(t))^2 g(\frac{X_1(t)}{X_2(t)})dt$$

Therefore we can put

$$\sigma^2(y_2) = 2y_2^2 \tag{18}$$

Putting everything together, we have obtained the following theorem.

Theorem 2.2 *If (X_1, X_2) denotes the solution to the martingale problem for the generator*

$$A = \frac{1}{2}x_1 x_2 (\sigma_1^2 \frac{\partial^2 f}{\partial x_1^2} + 2\rho\sigma_1\sigma_2 \frac{\partial^2 f}{\partial x_1 \partial x_2} + \sigma_2^2 \frac{\partial^2 f}{\partial x_2^2}),$$

then

$$(Y_1, Y_2) = (\frac{X_1}{X_2}, \sigma_2^2 X_1^2 - 2\rho\sigma_1\sigma_2 X_1 X_2 + \sigma_1^2 X_2^2)$$

is a strong Markov process whose generator has the following skew-product representation

$$(\sigma_2^2 y_1^2 - \rho\sigma_1\sigma_2 y_1)\frac{\partial}{\partial y_1} + \frac{1}{2}((\sigma_1\sqrt{y_1} - \rho\sigma_2\sqrt{y_1^3})^2 + \sigma_2(1 - \rho^2)y_1^3)\frac{\partial^2}{\partial y_1^2}$$

$$+ g(y_1)(\frac{1}{2}y_2^2 \frac{\partial^2}{\partial y_2^2} + y_2 \frac{\partial}{\partial y_2}) \tag{19}$$

where

$$g = (1 + \rho)\sigma_1\sigma_2 h$$

and with h given by

$$h(x) = \frac{2(1 - \rho)\sigma_1\sigma_2 x}{\sigma_2^2 x_1^2 - 2\rho\sigma_1\sigma_2 x_1 x_2 + \sigma_1^2 x_2^2}$$

Remark 2.3 *We note that the martingale problem for the diffusion Z with generator $\frac{1}{2}y_2^2 \frac{\partial^2}{\partial y_2^2} + y_2 \frac{\partial}{\partial y_2}$ is well-posed (again, this follows from the Yamada-Watanabe Theorem). Observe also that Z is a lognormal process, i.e. Z is the exponential of a linear Brownian motion with drift, and has therefore moments of all orders. The boundedness of the time-change rate function g implies the same for Y_2 and hence for (X_1, X_2).*

By our description of $(X_1(t), X_2(t))$ as a time-changed two-dimensional correlated Brownian motion run until it hits the coordinate axes, some properties are easily read off from those of Brownian motion. We have the following lemma.

Lemma 2.4   *(i) For $t \to \infty$ we have $X_1(t)X_2(t) \to 0$ a.s.*

(ii) The diffusion on natural scale $Y(t) = \tan^{-1}(\frac{X_1(t)}{X_2(t)})$ can be written as a time-changed linear Brownian motion as follows

$$\tan^{-1}(Y(t)) = \beta(H(t)) \tag{20}$$

where β is a linear Brownian motion and $H(t) = \int_0^t \frac{\tan(Y(t))}{1+\tan^2(Y(t))}ds$.

(iii) We have

$$\mathbb{P}^{x_1,x_2}(X_1 \text{ reaches } 0 \text{ before } X_2) = 1 - \frac{2}{\pi}\tan^{-1}(\frac{x_1}{x_2}) \tag{21}$$

Remark 2.5 *Part (i) of the above lemma implies in particular that $Y_1 = \frac{X_1}{X_2}$ may explode in a finite time. Part (ii) is a simple application of one-dimensional diffusion theory and part (iii) follows from (ii) and the fact that Brownian motion on an interval, here $[0, \pi/2]$, hits its end points with probability proportional to where it is started.*

3   Duality to a pure jump process

Here we give a dual process to the process (X_1, X_2) with generator $A_1 = A = \frac{1}{2}x_1x_2(\sigma_1^2\frac{\partial^2 f}{\partial x_1^2} + \rho\sigma_1\sigma_2\frac{\partial^2 f}{\partial x_1 \partial x_2} + \sigma_2^2\frac{\partial^2 f}{\partial x_2^2})$. Duality is usually applied to show uniqueness in the martingale problem by transforming the uniqueness problem into an existence problem, which is usually easier to handle. Yet this approach would fail in our example due to the fact that the moments of (X_1, X_2) grow too fast, at least in the zero correlation case. One usually applies Carleman's condition which gives a sufficient condition for uniqueness of a distribution in terms of growth of the moments, see e.g. [1, p.116]. Nevertheless, we pursue this method since it is of independent interest and since it gives us information on exactly how fast the moments grow. We first show that A_1 is in fact in duality to a pure jump process on an integer lattice. We use the necessary results about the method of duality from [1, Chap. 4].

Definition 3.1 *Let (E_1, r_1) and (E_2, r_2) be separable metric spaces. Let $A_1 \subset \mathcal{B}(E_1) \times \mathcal{B}(E_1)$, $A_2 \subset \mathcal{B}(E_2) \times \mathcal{B}(E_2)$, $F E_1 \times E_2$-measurable, α E_1-measurable, β E_2- measurable, $\mu_1 \in M_1(E_1)$ and $\mu_2 \in M_1(E_2)$. Then the martingale problems for (A_1, μ_1) and (A_2, μ_2) are called Feynman-Kac dual with respect to (F, α, β) if for each solution X of the martingale problem for (A_1, μ_1) and each solution Y for (A_2, μ_2), $\int_0^t |\alpha(X(s))|ds < \infty$ a.s., $\int_0^t |\beta(Y(s))|ds < \infty$ a.s.,*

$$\int \mathbb{E}[|F(X(t), y)| \exp(\int_0^t \alpha(X(s))ds)|]\mu_2(dy) < \infty, \tag{22}$$

$$\int \mathbb{E}[|F(x, Y(t))\exp(\int_0^t \beta(Y(s))ds)|]\mu_1(dx) < \infty, \tag{23}$$

and

$$\int \mathbb{E}[F(X(t), y)\exp(\int_0^t \alpha(X(s))ds)]\mu_2(dy) \tag{24}$$
$$= \int \mathbb{E}[F(x, Y(t))\exp(\int_0^t \beta(Y(s))ds)]\mu_1(dx)$$

for every $t \geq 0$*. Note that if* X *and* Y *are defined on the same probability space and are independent, then (24) can be written*

$$\mathbb{E}[F(X(t), Y(0))\exp(\int_0^t \alpha(X(s))ds)] = \mathbb{E}[F(X(0), Y(t))\exp(\int_0^t \beta(Y(s))ds)]. \tag{25}$$

We consider the following set up:

$$A_1 f(x_1, x_2) = \frac{1}{2}x_1 x_2(\sigma_1^2\frac{\partial^2 f}{\partial x_1^2} + \rho\sigma_1\sigma_2\frac{\partial^2 f}{\partial x_1\partial x_2} + \sigma_2^2\frac{\partial^2 f}{\partial x_2^2}) \tag{26}$$

$$\mu_1 = \delta_{(x_1, x_2)} \tag{27}$$

$$E_1 = [0, \infty) \times [0, \infty) \tag{28}$$

and

$$A_2 f(y_1, y_2) = \frac{1}{2}\sigma_1 y_1(y_1 - 1)(f(y_1 - 1, y_2 + 1) - f(y_1, y_2)) \tag{29}$$
$$+ \frac{1}{2}\sigma_2 y_2(y_2 - 1)(f(y_1 + 1, y_2 - 1) - f(y_1, y_2))$$

$$\mu_2 = \delta_{(y_1, y_2)} \tag{30}$$

$$E_2 = \{0, 1, 2, \ldots\} \times \{0, 1, 2, \ldots\} \tag{31}$$

This means A_2 governs a pure jump process on an integer lattice moving only on the line segment $\{(y_1', y_2') \in \mathbb{N} \times \mathbb{N} \mid y_1' + y_2' = y_1 + y_2\}$ passing through the starting point (y_1, y_2). The process associated with A_2 is, as a jump process Feller, and so the (A_2, μ_2) martingale problem is well-posed. Denoting by (Y_1, Y_2) its unique solution, observe that (Y_1, Y_2) is bounded.

Now we define the triple of functions (F, α, β) setting up the duality between A_1 and A_2.

Let

$$F(x_1, x_2; y_1, y_2) = x_1^{y_1}x_2^{y_2} \tag{32}$$

$$\alpha(x_1, x_2) = 0 \tag{33}$$

$$\beta(y_1, y_2) = \frac{1}{2}(\sigma_1 y_1(y_1 - 1) + \rho\sigma_1\sigma_2 y_1 y_2 + \sigma_2 y_2(y_2 - 1)) \tag{34}$$

In order to show the proposed duality we have to state a theorem, see [1, p. 192], which allows us to check the duality conditions 24.

Theorem 3.2 *Let X and Y be independent measurable processes in E_1 and E_2, respectively. Let $F, g, h \in M(E_1 \times E_2)$, $\alpha \in M(E_1)$ and $\beta \in M(E_2)$. Suppose that for each $T > 0$ there exists an integrable random variable Γ_T and a constant C_T such that*

$$\sup_{r,s,t \leq T} (|\alpha(X(r))| + 1)|F(X(s), Y(t))| \leq \Gamma_T, \tag{35}$$

$$\sup_{r,s,t \leq T} (|\beta(Y(r))| + 1)|F(X(s), Y(t))| \leq \Gamma_T, \tag{36}$$

$$\sup_{r,s,t \leq T} (|\alpha(X(r))| + 1)|g(X(s), Y(t))| \leq \Gamma_T, \tag{37}$$

$$\sup_{r,s,t \leq T} (|\beta(Y(r))| + 1)|h(X(s), Y(t))| \leq \Gamma_T, \tag{38}$$

$$\tag{39}$$

and

$$\int_0^T |\alpha(X(u))|du + \int_0^T |\beta(Y(u))|du \leq C_T. \tag{40}$$

Suppose that

$$F(X(t), y) - \int_0^t g(X(s), y)ds \tag{41}$$

is an $\mathcal{F}_\sqcup^X$-martingale for each y, and

$$F(x, Y(t)) - \int_0^t h(x, Y(s))ds \tag{42}$$

is an $\mathcal{F}_\sqcup^Y$-martingale for each x. Then for almost every $t \geq 0$,

$$\mathbb{E}[F(X(t), Y(0)) \exp(\int_0^t \alpha(X(s))ds)] \tag{43}$$

$$-\mathbb{E}[F(X(0), Y(t)) \exp(\int_0^t \beta(Y(s))ds)]$$

$$= \mathbb{E}[\int_0^t (g(X(s), Y(t-s)) - h(X(s), Y(t-s))$$

$$+(\alpha(X(s)) - \beta(Y(t-s)))F(X(s), Y(t-s)))$$

$$\times \exp(\int_0^s \alpha(X(u))du + \int_0^{t-s} \beta(Y(u))du)]$$

In particular, the condition $g(x, y) + \alpha(x)F(x, y) = h(x, y) + \beta(y)F(x, y)$ together with the boundedness conditions establishes duality.

Theorem 3.3 *The martingale problems for (A_1, μ_1) and (A_2, μ_2) are dual with respect to $(F, 0, \beta)$.*

Proof. We apply the preceding theorem with

$$g(x_1, x_2; y_1, y_2) = \frac{1}{2} x_1 x_2 (\sigma_1 y_1 (y_1 - 1) x_1^{y_1 - 2} x_2^{y_2} + \rho \sigma_1 \sigma_2 y_1 y_2 x_1^{y_1 - 1} x_2^{y_2 - 1}$$
$$+ \sigma_2 y_2 (y_2 - 1) x_1^{y_1} x_2^{y_2 - 2})$$

and

$$h(x_1, x_2; y_1, y_2) = \frac{1}{2} \sigma_1 y_1 (y_1 - 1)(x_1^{y_1 - 1} x_2^{y_2 + 1} - x_1^{y_1} x_2^{y_2})$$
$$+ \frac{1}{2} \sigma_2 y_2 (y_2 - 1)(x_1^{y_1 + 1} x_2^{y_2 - 1} - x_1^{y_1} x_2^{y_2})$$

Then a simple calculation shows that $g = h + \beta F$, so that we only have to check the boundedness conditions. The ones involving Y are clearly met, since Y itself is bounded. The ones involving (X_1, X_2) are easily seen to be implied by the single condition

$$\mathbb{E}[\sup_{0 \leq s \leq t} (X_1(s) X_2(s))^n] < \infty \text{ for all } t \geq 0 \text{ and all positive integers } n$$

As $(X_1(t) X_2(t), t \geq 0)$ is a continuous martingale by Lemma 2.4, this condition reduces by the submartingale inequalities to (X_1, X_2) having finite moments. But in Remark 2.3 we have seen that is the case. ∎

Duality being established, we can estimate the moments of (X_1, X_2). For convenience let us assume $\sigma_1 = \sigma_2 = 1$.

Corollary 3.4 *We have the following upper bound for the moments of (X_1, X_2), if $(n, m) = (Y_1(0), Y_2(0))$,*

$$\mathbb{E}[X_1(t)^n X_2(t)^m] \leq \max(x_1^{n+m}, 1) \max(x_2^{n+m}, 1) \exp(\frac{1}{2}((n + m)^2 - (n + m))t) \tag{44}$$

In case of $\rho = 0$, we have also a lower bound,

$$\mathbb{E}[X_1(t)^n X_2(t)^m] \geq \max(x_1^{n+m}, 1) \max(x_2^{n+m}, 1) \exp((\frac{1}{4}(n + m)^2 - \frac{1}{2}(n + m))t) \tag{45}$$

Proof. By duality, we have

$$\mathbb{E}[X_1(t)^n X_2(t)^m] \tag{46}$$
$$= \mathbb{E}[x_1^{Y_1(t)} x_2^{Y_2(t)} \exp(\frac{1}{2} \int_0^t (Y_1(s)(Y_1(s) - 1)$$
$$+ \rho Y_1(s) Y_2(s) + Y_2(s)(Y_2(s) - 1))ds)]$$

where $(n, m) = (Y_1(0), Y_2(0))$. Note that $Y_1(s) + Y_2(s)$ is constantly equal to $n + m$. For all $x, y_1, y_2 \geq 0$, we have the inequalities

$$y_1^2 + \rho y_1 y_2 + y_2^2 \leq (y_1 + y_2)^2$$
$$x^{y_i} \leq \max(x^{y_1 + y_2}, 1)$$
$$y_1^2 + y_2^2 \geq \frac{1}{2}(y_1 + y_2)^2$$

so that the claims follow.   ∎

Acknowledgements. This work is based on the first chapter of my dissertation [3], which was completed under the direction of Prof. Steven Evans whom I would to thank for his advice and support during my stay at Berkeley.

References

[1] S. Ethier and T. Kurtz, *Markov Processes : Characterization and Convergence*, 1986, Wiley, New York.

[2] J.D. Murray, *Mathematical Biology*, 1989, Springer Verlag, New York.

[3] J.A. Rebholz, *Planar Diffusions with Applications to Mathematical Finance*, 1994, Ph.D. Dissertation, University of California at Berkeley.

[4] L.C. Rogers and D. Williams, 1987, Wiley, New York.

[5] E.C. Zachmanoglou and D.W. Thoe, *Introduction to partial differential equations with applications*, 1986, Dover Publications, Inc., New York.

A NONLINEAR HYPERBOLIC SPDE: APPROXIMATIONS AND SUPPORT[*]

Carles Rovira and *Marta Sanz-Solé*

Facultat de Matemàtiques

Universitat de Barcelona

Gran Via, 585

08007 Barcelona–Spain

1 Introduction

In [14], Stroock and Varadhan gave a probabilistic proof of Nirenberg's maximum principle. This proof is based on the so-called support theorem for diffusions. Consider the stochastic differential equation

$$(1.1) \qquad X_t = X_0 + \int_0^t [\sigma(s, X_s) \circ dW_s + b(s, X_s) \, ds], \quad t \in [0, \tau],$$

defined on some probability space $(\Omega, \mathcal{F}, P)$ with deterministic initial condition X_0. Here $\sigma : [0, \tau] \times \mathbb{R}^m \longrightarrow \mathbb{R}^m \otimes \mathbb{R}^d$, $b : [0, \tau] \times \mathbb{R}^m \longrightarrow \mathbb{R}^m$ are measurable functions, $W = \{W_t, t \in [0, \tau]\}$ is a d-dimensional standard Wiener process and the stochastic integral is a Stratonovich one.

The support theorem provides a characterization for the topological support of the probability $P \circ X^{-1}$ on $\mathcal{C}([0, \tau]; \mathbb{R}^m)$, that means, a precise description of the smallest closed subset F of $\mathcal{C}([0, \tau]; \mathbb{R}^m)$, endowed with the topology of uniform convergence, such that $(P \circ X^{-1})(F) = 1$. Let $\mathcal{H}_0$ be the set of smooth paths in $\mathbb{R}^d$. For any $h \in \mathcal{H}_0$ we consider the solution to the differential equation

$$(1.2) \qquad \Phi_t^h = X_0 + \int_0^t [\sigma(s, \Phi_s^h) \, dh_s + b(s, \Phi_s^h) \, ds], \quad t \in [0, \tau].$$

Assuming $\sigma \in \mathcal{C}_b^{1,2}$ and b uniformly Lipschitz in the second variable and bounded, Stroock and Varadhan proved that the support of X, $\mathrm{supp}(P \circ X^{-1})$, is the closure in $\mathcal{C}([0, \tau]; \mathbb{R}^m)$ of the set $\{\Phi^h, h \in \mathcal{H}_0\}$ (see [14], [15]).

---

[*]Partially supported by a grant of the DGICYT n° PB 90–0452.

The support theorem also provides a useful tool in the qualitative theory of stochastic systems (see [2]) and can be of interest in proving irreducibility of transition probabilities.

In the last years there have been several contributions to this subject in several directions. Stroock-Varadhan's result has been extended to more general spaces, for instance to Hölder and Besov spaces ([1], [4], [9], [11], [12]). Moreover, alternative approaches have also been considered (see, for instance [1], [7], [8], [10], [13]) and applied to some interesting classes of Wiener functionals such as the solutions to stochastic partial differential equations ([3], [13]). These two research directions are not independent. As we will next describe, some methods seem to be more appropriate in cases where no Itô formula is available and, at the same time, they work in a rather abstract framework allowing worthy generalizations.

The following proposition synthesizes one possible method to state support theorems. In the sequel $(\Omega, \mathcal{F}, P)$ denotes the canonical space associated with a d-dimensional multiparameter Wiener process $W = \{W_z, z \in [0, \tau]^k\}$ (in the next section k will be 2) and $\mathcal{H}$ the corresponding Cameron-Martin space.

Proposition 1.1 *Let $F : \Omega \longrightarrow \mathbb{B}$ be a random vector taking its values in some separable Banach space $(\mathbb{B}, \| \quad \|)$ and $\mathcal{H}_0 \subset \mathcal{H}$.*

(a)   Let $\xi_1 : \mathcal{H}_0 \longrightarrow \mathbb{B}$ be measurable and assume that there exists a sequence of random variables $H_n : \Omega \longrightarrow \mathcal{H}_0$ such that for any $\epsilon > 0$,

$$(1.3) \qquad \lim_{n \to \infty} P\{\|F(\omega) - \xi_1(H_n(\omega))\| > \epsilon\} = 0.$$

Then,

$$(1.4) \qquad \mathrm{supp}(P \circ F^{-1}) \subset \overline{\xi_1(\mathcal{H}_0)}.$$

(b)   Let $\xi_2 : \mathcal{H}_0 \longrightarrow \mathbb{B}$ be measurable and suppose that for each $h \in \mathcal{H}_0$ there exists a sequence of measurable transformations $T_n^h : \Omega \longrightarrow \Omega$ such that $P \circ (T_n^h)^{-1}$ is absolutely continuous with respect to P for any $n \geq 1$. If

$$(1.5) \qquad \limsup_n P\{\|F(T_n^h(\omega)) - \xi_2(h)\| < \epsilon\} > 0, \quad \epsilon > 0,$$

then

$$(1.6) \qquad \overline{\xi_2(\mathcal{H}_0)} \subset \mathrm{supp}(P \circ F^{-1}).$$

The notation $\overline{(\quad)}$ means the closure in the norm $\| \; \|$.

Proof: Condition (1.3) yields the convergence in distribution of the sequence $\{\xi_1(H_n), n \geq 1\}$ to F. Therefore, Portmanteau's theorem yields

$$1 = \limsup_n \left(P \circ (\xi_1(H_n))^{-1}\right)\overline{\left(\xi_1(\mathcal{H}_0)\right)} \leq \left(P \circ F^{-1}\right)\overline{\left(\xi_1(\mathcal{H}_0)\right)},$$

and consequently, by the definition of support, (1.4) is established .

Assume now (1.5). There exists some $n > 0$ with $P\{\|F(T_n^h(\omega)) - \xi_2(h)\| < \epsilon\} > 0$. Since $P \circ (T_n^h)^{-1}$ is absolutely continuous with respect to P, it holds $P\{\|F(\omega) - \xi_2(h)\| < \epsilon\} > 0$ and therefore, $\xi_2(h) \in \mathrm{supp}(P \circ F^{-1})$. Then, (1.6) follows, because $h \in \mathcal{H}_0$ is arbitrary and $\mathrm{supp}(P \circ F^{-1})$ is a closed set.   ∎

In most of the applications $\mathcal{H}_0$ will be a set of smooth paths in $\mathbb{R}^d$ and $\{H_n(\omega), n \geq 1\}$ a sequence of smooth approximations of the Wiener process. Then, the statement (1.3) is an approximation result for the Wiener functional F. For instance, consider the sequence of dyadic grids on $[0, \tau]$ and set

$$W_n(t) = W_{i2^{-n}} + 2^n (W_{(i+1)2^{-n}} - W_{i2^{-n}})(t - i2^{-n}),$$

$t \in [i2^{-n}, (i+1)2^{-n}), \ i = 0, \ldots, [2^n\tau] - 1, \ n \geq 1$. Then $\{W_n, n \geq 1\}$ is a sequence of $\mathcal{H}$-valued random variables. The convergence of $\{\Phi^{W_n}, n \geq 1\}$ given by (1.2), to the process $\{X_t, t \in [0, \tau]\}$ defined by (1.1), in probability in $\mathcal{C}([0, \tau]; \mathbb{R}^m)$, stated first in [14], yields the inclusion supp$(P \circ X^{-1}) \subset \overline{\{\Phi^h, h \in \mathcal{H}\}}$. A proof of the converse inclusion following the ideas presented in part (b) of the previous proposition, which are different from those of Stroock and Varadhan, is given in [12].

In the next section we will apply Proposition 1.1 to establish a support theorem in Hölder norm for the solution to a hyperbolic spde (see equation (2.1)).

2   Characterization of the support

In this section we consider the hyperbolic stochastic partial differential equation

$$(2.1) \quad \frac{\partial^2 X_{s,t}}{\partial s \partial t} = a_3(X, s, t)\dot{W}_{s,t} + a_4(X, s, t) + a_1(s, t)\frac{\partial X_{s,t}}{\partial t} + a_2(s, t)\frac{\partial X_{s,t}}{\partial s},$$

with initial condition $X_{s,t} = X_0$ if $(s,t) \in T$, $s \cdot t = 0$, $T = [0,1]^2$. The coefficients are measurable functions $a_i : T \longmapsto \mathbb{R}$, $i = 1, 2$, and $a_i : \mathcal{C} \times T \longrightarrow \mathbb{R}$, $i = 3, 4$, where $\mathcal{C} = \mathcal{C}(T; \mathbb{R})$. $\{\dot{W}_{s,t}, (s,t) \in T\}$ is a white noise on T and X_0 is a $\mathcal{F}_{0,0}$-measurable random variable, where $\mathcal{F}_{s,t}, (s,t) \in T$, is the completion of $\sigma\{W_{u,v}, (u,v) \leq (s,t)\}$. These kind of equations appear, for example, in the problem of constructing a Wiener sheet on manifolds.

One can give a rigorous meaning to (2.1) as follows. Consider the Green function $\gamma_{s,t}(u,v)$ associated with the second order differential operator

$$\mathcal{L}f(s,t) = \frac{\partial^2 f(s,t)}{\partial s \partial t} - a_1(s,t)\frac{\partial f(s,t)}{\partial t} - a_2(s,t)\frac{\partial f(s,t)}{\partial s},$$

which is introduced and studied in the appendix (see (3.1)). Then, by a solution of (2.1) we mean a stochastic process $X = \{X_{s,t}, (s,t) \in T\}$ adapted to the filtration $\{\mathcal{F}_{s,t}, (s,t) \in T\}$, satisfying

$$(2.2) \quad X_{s,t} = X_0 + \int_{R_{s,t}} \gamma_{s,t}(u,v)\Big(a_4(X, u, v)du\,dv + a_3(X, u, v)W(du, dv)\Big),$$

where, for $(s,t) \in T$, $R_{s,t}$ denotes the rectangle $[0, (s,t)]$.

Our first purpose is to prove existence and uniqueness of solution for (2.1). Consider the following set (H) of hypotheses:

(H1) a_3 and a_4 are globally Lipschitz functions in $f \in \mathcal{C}$ and satisfy a linear growth condition. That is, there exists a constant $C > 0$ such that, for all $f, f' \in \mathcal{C}$, $(s, t) \in T$

$$|a_3(f, s, t) - a_3(f', s, t)| + |a_4(f, s, t) - a_4(f', s, t)| \leq C|f(s, t) - f'(s, t)|,$$

$$|a_3(f, s, t)| + |a_4(f, s, t)| \leq C(1 + |f(s, t)|).$$

(H2) a_1 and a_2 are derivable, bounded and have bounded derivatives.

Proposition 2.1 *Assume that the coefficients satisfy hypotheses (H) and $X_0 \in L^{2p}$ for some $p \geq 1$. Then, there exists a unique continuous solution $X = \{X_{s,t}, (s, t) \in T\}$ of (2.2) which is bounded in L^{2p}.*

Proof. The method is standard, once we know that the Green function $\gamma_{s,t}(u, v)$ is uniformly bounded (see (3.8)) . Consider the Picard approximations defined by

$$X_{s,t}^{(0)} = X_0,$$
$$X_{s,t}^{(n+1)} = X_0 + \int_{R_{s,t}} \gamma_{s,t}(u, v)\Big(a_4(X^{(n)}, u, v)du\, dv + a_3(X^{(n)}, u, v)W(du, dv)\Big).$$

Clearly, the processes $X^{(n)}$, $n \geq 0$, are well–defined. Burkholder's and Hölder's inequalities, (3.8) and hypotheses (H1) yield

$$E(|X_{s,t}^{(n+1)} - X_{s,t}^{(n)}|^{2p}) \leq \frac{C^{n+1}}{(n!)^2}, \quad n \geq 0.$$

Thus, $X_{s,t}^{(n)}$ converges in L^{2p} and the convergence is uniform in T. The process defined by $X_{s,t} = L^2 - \lim_{n \to \infty} X_{s,t}^{(n)}$ is a solution of (2.2) and is L^{2p}–bounded. Let X, Y be two processes satisfying $\sup_{(s,t) \in T} \{E\{|X_{s,t}|^{2p}\} + E\{|Y_{s,t}|^{2p}\}\} < +\infty$ and (2.2). Then,

$$E\{|X_{s,t} - Y_{s,t}|^2\} \leq C \int_{R_{s,t}} E(|X_{u,v} - Y_{u,v}|^2)du\, dv, \quad (s, t) \in T,$$

and, consequently $E(|X_{s,t} - Y_{s,t}|^2) = 0$. Hence, uniqueness is proved.   ∎

Remark.   Following a different approach based on two-parameter stochastic calculus, in [5] an existence and uniqueness of solution to equation (2.1) has also been established. Although their notion of solution is apparently different from that proposed in this paper, they can be shown to be equivalent.

Using the classical Kolmogorov's continuity criterion the next result can be stated.

Proposition 2.2 *Suppose $X_0 \in L^{2p}$ for all $p \geq 1$. Then, almost surely, the paths of X, solution to (2.2), are α–Hölder continuous for any $\alpha \in \left(0, \frac{1}{2}\right)$.*

We want now to apply the ideas developed in the introduction to establish a characterization for the topological support of the law of the process $\{X_z, z \in T\}$, $z = (s, t)$, solution to (2.2) in the case where the coefficients a_i, $i = 3, 4$ are of the form $a_i(f, z) = a_i(f_z)$, $f \in \mathcal{C}$, $z \in T$. So, instead of (2.2) we will consider the equation

$$(2.3) \qquad X_z = X_0 + \int_{R_z} \gamma_z(\eta) \big(a_4(X_\eta) \, d\eta + a_3(X_\eta) \, dW_\eta \big).$$

Furthermore, assumptions (H) will be replaced by (H′), which are stronger:

(H′1) a_3 and a_4 are bounded Lipschitz functions defined on $\mathbb{R}$. Moreover, a_3 has third order bounded derivatives.

(H′2) a_1 and a_2 are bounded, derivable functions defined on T with bounded derivatives.

If $a_i \equiv 0$, $i = 1, 2$ this problem has been studied in [13]; in this case the corresponding Green function is $\gamma_z(\eta) = \mathbb{1}_{R_z}(\eta)$. In the more general situation considered here, the additional difficulties are solved using the results on the Green function proved in the appendix.

Fix $h \in \mathcal{H}$, and consider the differential equation

$$(2.4) \qquad S_z^h = X_0 + \int_{R_z} \gamma_z(\eta) \big(a_4(S_\eta^h) + a_3(S_\eta^h) \, \dot{h}_\eta \big) \, d\eta, \quad z \in T.$$

Our aim is to prove the following result.

Theorem 2.3 *Assume that the coefficients of the equation (2.3) satisfy the hypothesis (H′). Then* $\mathrm{supp}\,(P \circ X^{-1})$ *is the closure of the set* $\{S^h, h \in \mathcal{H}\}$ *in the topology of the α-Hölder convergence, for any $\alpha \in (0, \frac{1}{2})$.*

As in other examples of spde's (see, for instance, [13] and [3]), Theorem 2.3 will follow from a result on approximation in Hölder norm for the solution of (2.3).

Let us introduce some notation. Given two points $z = (s, t)$ and $z' = (s', t')$ in T, we denote by $z \otimes z'$ the point (s, t'). For any fixed positive integer n, consider the dyadic grid on T, $\mathcal{P}^n = \{z_{i,j} = (i2^{-n}, j2^{-n}), \ 0 \le i, j \le 2^n - 1\}$ and let $\Delta_{i,j} = (z_{i,j}, z_{i+1,j+1}], 0 \le i, j \le 2^n - 1$. Given $z \in T$, let $z_n = z_{i-1,j-1}$ if $z \in \Delta_{i,j}$ with $i \cdot j \ne 0$ and $z_n = 0$, otherwise. For $z \in \Delta_{i,j}$, set $\tilde{z}_n = z_{i,j}$. Let $f : T \longrightarrow \mathbb{R}$ and Δ a rectangle in T, then $f(\Delta)$ denotes the increment of f in the sense of distribution functions and $\dot{f}$ means the mixed derivative $\frac{\partial^2 f}{\partial s \partial t}$, whenever it exists. All constants will be denoted by C.

Define a sequence of approximations of the Brownian sheet as follows:

$$\omega_z^n = 0, \quad \text{if} \quad z \in \Delta_{i,j} \quad \text{with} \quad i \cdot j = 0,$$

$$\omega_z^n - \omega_{z \otimes \tilde{z}_n}^n - \omega_{\tilde{z}_n \otimes z}^n + \omega_{\tilde{z}_n}^n = 2^{2n} W(\Delta_{i-1,j-1}) \Big(s - \frac{i}{2^n} \Big) \Big(t - \frac{j}{2^n} \Big),$$

if $z \in \Delta_{i,j}$ with $i \cdot j \neq 0$. For any $n \geq 1$, $\omega \mapsto \omega^n$ defines a sequence of $\mathcal{H}$-valued random variables $\{H_n; n \geq 1\}$. Notice

$$\dot{\omega}_z^n = \begin{cases} 0, & \text{if } z \in \Delta_{i,j} \text{ with } i \cdot j = 0, \\ 2^{2n} W(\Delta_{i-1,j-1}), & \text{if } z \in \Delta_{i,j} \text{ with } i \cdot j \neq 0. \end{cases}$$

Moreover, for any $n \geq 1$ and $p \in [1, \infty)$

$$(2.5) \qquad \sup_{z \in T} E(|\dot{\omega}_z^n|^{2p}) \leq C\, 2^{2np}.$$

For any fixed $h \in \mathcal{H}$, consider the adapted processes $\{Y_z, z \in T\}$, $\{Y_z^n, z \in T\}$ solutions to

$$(2.6)$$
$$Y_z^n = Y_0 + \int_{R_z} \left(\gamma_z(\eta) F(Y_\eta^n) dW_\eta + \gamma_z(\eta)(G(Y_\eta^n)\dot{h}_\eta + H(Y_\eta^n)\dot{\omega}_\eta^n + B(Y_\eta^n)) d\eta \right),$$

$$(2.7) \qquad \begin{aligned} Y_z &= Y_0 + \int_{R_z} \Big\{ \gamma_z(\eta)(F + H)(Y_\eta) dW_\eta + \\ & \gamma_z(\eta)\Big(G(Y_\eta)\dot{h}_\eta + B(Y_\eta) + H'(Y_\eta)F(Y_\eta) + \tfrac{1}{4}H'(Y_\eta)H(Y_\eta)\Big) d\eta \Big\}, \end{aligned}$$

respectively, where Y_0 is a $\mathcal{F}_{0,0}$-adapted random variable and the coefficients satisfy the hypothesis (H''):

(H''1) F, G, H and B are bounded Lipschitz functions defined on $\mathbb{R}$.

(H''2) F is of class $\mathcal{C}^1$ with bounded derivatives, H is of class $\mathcal{C}^3$ with bounded derivatives up to the third order.

Under (H''), existence and uniqueness of solutions to (2.6) and (2.7), respectively, can be easily obtained. Our first aim is to prove an approximation result.

Theorem 2.4 *For any* $p \in [1, \infty)$ *and* $\alpha \in (0, \tfrac{1}{2})$, $\lim_{n \to \infty} E(\|Y_z^n - Y_z\|_\alpha^p) = 0$, *where* $\| \quad \|_\alpha$ *denotes the* α-*Hölder norm in the space of real-valued functions defined on* T.

The proof of the preceding theorem is a consequence of the next two propositions (see, for instance, Lemma A.1 [3]).

Proposition 2.5 *For any* $z, z' \in T$, $p \in [1, \infty)$,

$$(2.8) \qquad \sup_n E(|Y_z^n - Y_{z'}^n|^{2p}) + E(|Y_z - Y_{z'}|^{2p}) \leq C\, d(z, z')^p,$$

where d *denotes the Euclidean distance on* T.

Proposition 2.6 *For any* $p \in [1, \infty)$, $n \geq 1$,

$$(2.9) \qquad \sup_{z \in T} E(|Y_z^n - Y_z|^{2p}) \leq C\, 2^{-np}.$$

Before checking Propositions 2.5 and 2.6 we prove some preliminary results.

Lemma 2.7 *Let* $\left\{ A_{i,j} = \sum_{\substack{0 \le k \le i-1 \\ 0 \le \ell \le j-1}} a_{k,\ell} b_{k,\ell}, \ i = 0, \ldots, i_0 - 1; \ j = 0, \ldots, j_0 - 1 \right\}$
be a stochastic process and $\{\mathcal{F}_{i,j}, \ i, j \ge 1\}$ *a filtration satisfying the usual
conditions. Assume that* $a_{k,\ell}$ *is a* $\mathcal{F}_{k,\ell+2} \vee \mathcal{F}_{k+2,\ell}$*-measurable random variable and* $b_{k,\ell}$ *is a centered* $\mathcal{F}_{k+2,l+2}$*-measurable random variable independent
of* $\mathcal{F}_{k,\ell}^* := \mathcal{F}_{k,j_0} \vee \mathcal{F}_{i_0,\ell}, \ k = 0, \ldots, i_0 - 2, \ \ell = 0, \ldots, j_0 - 2$. *Then, for any*
$p \in [1, \infty)$

$$(2.10) \qquad E(|A_{i,j}|^{2p}) \le C(i \cdot j)^{p-1} \sum_{\substack{0 \le k \le i-1 \\ 0 \le \ell \le j-1}} E(|a_{k,\ell}|^{2p}) E(|b_{k,\ell}|^{2p}).$$

Proof. Set $I = \{0, 1, \ldots, i-1\}$, $J = \{0, 1, \ldots, j-1\}$. We can consider the
decomposition $A_{i,j} = \sum_{\rho=1}^{4} A_{i,j}^{(\rho)}$, where

$$A_{i,j}^{(1)} = \sum_{\substack{(k,\ell) \in I \times J \\ k=2, \ell=\dot{2}}} a_{k,\ell} b_{k,\ell}, \qquad A_{i,j}^{(2)} = \sum_{\substack{(k,\ell) \in I \times J \\ k=2, \ell=\dot{2}+1}} a_{k,\ell} b_{k,\ell},$$

$$A_{i,j}^{(3)} = \sum_{\substack{(k,\ell) \in I \times J \\ k=\dot{2}+1, \ell=\dot{2}}} a_{k,\ell} b_{k,\ell}, \qquad A_{i,j}^{(4)} = \sum_{\substack{(k,\ell) \in I \times J \\ k=\dot{2}+1, \ell=\dot{2}+1}} a_{k,\ell} b_{k,\ell}.$$

We will establish (2.10) for each $A_{ij}^{(\rho)}$, $\rho = 1, \ldots, 4$. Since the arguments are
similar for each term we will only give the details for $\rho = 3$. So, we will prove
$M_{i,j}^{(3)} = \sum_{k=0}^{i-1} \sum_{\ell=0}^{j-1} a_{2k+1,2\ell} \ b_{2k+1,2\ell}$, satisfies (2.10) for any pair of positive
integers (i,j) with $2i - 1 \le i_0 - 2$ and $2j - 2 \le j_0 - 2$.
Let $\mathcal{G}_{i,j} = \mathcal{F}_{2i+1,2j}$. The process $\{M_{i,j}^{(3)}, \ \mathcal{G}_{i,j}, (i,j) \in \mathbb{Z}^+, \ 2i-1 \le i_0-2, \ 2j-2 \le
j_0 - 2\}$ is a two–parameter discrete martingale. Indeed, it is clear that $M_{i,j}^{(3)}$
is $\mathcal{G}_{i,j}$–measurable. Moreover,

$$E\left(M_{i+1,j}^{(3)} - M_{i,j}^{(3)} / \mathcal{G}_{i,j}\right) = E\left(\sum_{\ell=0}^{j-1} a_{2i+1,2\ell} \ b_{2i+1,2\ell} / \mathcal{F}_{2i+1,2j}\right)$$

$$= \sum_{\ell=0}^{j-1} E\left\{ E\left\{ a_{2i+1,2\ell} \ b_{2i+1,2\ell} / \mathcal{F}_{2i+1,2\ell}^* \right\} / \mathcal{F}_{2i+1,2j}\right\} = 0.$$

Analogously, $E\left(M_{i,j+1}^{(3)} - M_{i,j}^{(3)} / \mathcal{G}_{i,j}\right) = 0$ and consequently, the martingale
property is established. Since $a_{k,\ell}$ and $b_{k,\ell}$ are independent, Burkholder's and
Hölder's inequality yield

$$E\left(|M_{i,j}^{(3)}|^{2p}\right) \le C(i \ j)^{p-1} \sum_{k=0}^{i-1} \sum_{\ell=0}^{j-1} E\left(|a_{k,\ell}|^{2p}\right) E\left(|b_{k,\ell}|^{2p}\right). \qquad \blacksquare$$

Set $\mathcal{F}_\eta^* := \mathcal{F}_{\eta n + (0, \frac{2}{2^n})} \vee \mathcal{F}_{\eta n + (\frac{2}{2^n}, 0)}$. Lemma 2.7 will be of interest in order to
obtain majorizations of L^p-norms for integrals, with respect to $\{\omega_z^n, z \in T\}$, of
$\mathcal{F}_\eta^*$-adapted processes. The following statement is an abstract formulation of
a situation that we will often encounter along the paper.

Corollary 2.8 *Fix* $z, z' \in T$, $z' \geq z_{1,1}$, $z' = (s,t)$, $z = (s+b,t)$, $b > 0$. *Let* g *be a deterministic function defined on* T *and* $\{F_\eta, \eta \in T\}$ *an* $\mathcal{F}_\eta^*$ *–adapted process. Then, for any* $p \in [1, \infty)$

$$(2.11) \quad E\Big(\big|\int_{R_z - R_{z'}} g(\eta) F(\eta) \dot{\omega}_\eta^n \, d\eta\big|^{2p}\Big) \leq C\, b^p \sup_{\eta > z_{1,1}} |g(\eta)|^{2p} \sup_{\eta > z_{1,1}} E(|F_\eta|^{2p}).$$

Proof. Suppose $z' \in \Delta_{i_0, j_0}$, $z \in \Delta_{i_1, j_0}$, with $i_1 > i_0 + 1$. The cases $i_1 = i_0 + 1$ and $i_1 = i_0$ are studied by similar arguments.

Set $M^n(z, z') = \int_{R_z - R_{z'}} g(\eta) F(\eta) \dot{\omega}_\eta^n \, d\eta$. Clearly, $M^n(z, z') = \sum_{k=1}^3 M_k^n(z, z')$, where

$$M_1^n(z, z') = \int_{R_{(i_0+1)2^{-n}, t} - R_{z'}} g(\eta) F(\eta) \dot{\omega}_\eta^n \, d\eta,$$

$$M_2^n(z, z') = \int_{R_{i_1 2^{-n}, t} - R_{(i_0+1)2^{-n}, t}} g(\eta) F(\eta) \dot{\omega}_\eta^n \, d\eta,$$

$$M_3^n(z, z') = \int_{R_z - R_{i_1 2^{-n}, t}} g(\eta) F(\eta) \dot{\omega}_\eta^n \, d\eta.$$

Lemma 2.7 yields

$$E(|M_1^n(z, z')|^{2p}) = E\big(\big| \sum_{1 \leq \beta \leq 2^{n-1}} 2^{2n} W(\Delta_{i_0-1, \beta-1})$$
$$\int_{\Delta_{i_0, \beta} \cap (R_{(i_0+1)2^{-n}, t} - R_{z'})} g(\eta) F(\eta) d\eta \big|^{2p}\big)$$
$$\leq C\, 2^{n(p-1)} \sum_{1 \leq \beta \leq 2^{n-1}} 2^{4np} E(|W(\Delta_{i_0-1, \beta-1})|^{2p})$$
$$E\big(\big| \int_{\Delta_{i_0, \beta} \cap (R_{(i_0+1)2^{-n}, t} - R_{z'})} g(\eta) F(\eta) d\eta \big|^{2p}\big)$$
$$\leq C b^p \sup_{\eta > z_{1,1}} |g(\eta)|^{2p} \sup_{\eta > z_{1,1}} E(|F_\eta|^{2p}).$$

Analogously,

$$E(|M_2^n(z, z')|^{2p}) = E\big(\big| \sum_{\substack{i_0+1 \leq \alpha \leq i_1 \\ 0 \leq \beta \leq 2^{n-1}}} 2^{2n} W(\Delta_{\alpha-1, \beta-1})$$
$$\int_{\Delta_{\alpha, \beta} \cap (R_{i_1 2^{-n}, t} - R_{(i_0+1)2^{-n}, t})} g(\eta) F(\eta) d\eta \big|^{2p}\big)$$
$$\leq C[(i_1 - (i_0+1))2^n]^{p-1} \sum_{\substack{i_0+1 \leq \alpha \leq i_1 \\ 0 \leq \beta \leq 2^{n-1}}} E(|W(\Delta_{\alpha, \beta})|^{2p}) \sup_{\eta > z_{1,1}} |g(\eta)|^{2p}$$
$$\sup_{\eta > z_{1,1}} E(|F_\eta|^{2p}) \leq C b^p \sup_{\eta > z_{1,1}} |g(\eta)|^{2p} \sup_{\eta > z_{1,1}} E(|F_\eta|^{2p}).$$

The same bound can be obtained for $E(|M_3^n(z, z')|^{2p})$. Hence (2.11) is proved. Notice that the roles of the coordinates s and t can be exchanged. ∎

Lemma 2.9 *For any* $p \in [1, \infty)$ *and* $n \geq 0$,

$$(2.12) \qquad \sup_{z > z_{1,1}} E(|Y^n(z_n, z)|^{2p}) \leq C\, 2^{-2np}.$$

Proof. For any $z > z_{1,1}$, we decompose the rectangle R_z into four rectangles, $R_z = \bigcup_{j=1}^{n} \tilde{R}_j(z)$, with $\tilde{R}_1(z) = R_{z_n}$, $\tilde{R}_2(z) = (z_n \otimes 0,\ z \otimes z_n]$, $\tilde{R}_3(z) = (z_n, z]$, $\tilde{R}_4(z) = (0 \otimes z_n,\ z_n \otimes z]$. Set

$$h_z^{(1)}(\eta) = \gamma_z(\eta) - \gamma_{z \otimes z_n}(\eta) - \gamma_{z_n \otimes z}(\eta) + \gamma_{z_n}(\eta),$$
$$h_z^{(2)}(\eta) = \gamma_z(\eta) - \gamma_{z \otimes z_n}(\eta), \quad h_z^{(3)} = \gamma_z(\eta),$$
$$h_z^{(4)}(\eta) = \gamma_z(\eta) - \gamma_{z_n \otimes z}(\eta).$$

Proposition 3.3 yields $\sup_\eta |h_z^{(1)}(\eta)| \le C\ 2^{-2n}$ and $\sup_\eta |h_z^{(2)}(\eta)| + \sup_\eta |h_z^{(4)}(\eta)| \le C\ 2^{-n}$ (see (3.11) and (3.10), respectively). We can write

$$(2.13) \qquad Y^n(z_n, z] = \sum_{i=1}^{4} (T_{i,F}^n(z) + T_{i,G}^n(z) + T_{i,H}^n(z) + T_{i,B}^n(z)),$$

where

$$T_{i,F}^n(z) = \int_{\tilde{R}_i(z)} h_z^{(i)}(\eta)\ F(Y_\eta^n)dW_\eta, \quad T_{i,G}^n(z) = \int_{\tilde{R}_i(z)} h_z^{(i)}(\eta)\ G(Y_\eta^n)\ \dot{h}_\eta d\eta,$$
$$T_{i,H}^n(z) = \int_{\tilde{R}_i(z)} h_z^{(i)}(\eta)\ H(Y_\eta^n)\ \dot{\omega}_\eta d\eta, \quad T_{i,B}^n(z) = \int_{\tilde{R}_i(z)} h_z^{(i)}(\eta)\ B(Y_\eta^n)\ d\eta.$$

The estimates of increments of the Green function mentioned before, (2.5), Burkholder's, Hölder's and Schwarz's inequalities yield

$$(2.14) \qquad \begin{aligned} &E\left(|T_{i,F}^n(z)|^{2p} + |T_{i,G}^n(z)|^{2p} + |T_{i,H}^n(z)|^{2p} + |T_{i,B}^n(z)|^{2p}\right) \\ &\le C(1 + \|h\|_{\mathcal{H}}^{2p})\, 2^{-2np}, \end{aligned}$$

for any $i = 1, \cdots, 4$. We recall that $\|h\|_{\mathcal{H}}^2 = \int_T |\dot{h}_z|^2\, dz$. Indeed, consider the term $T_{2,H}^n(z)$, the other cases are studied using the same method. We have,

$$E(|T_{2,H}^n(z)|^{2p}) \le C\, 2^{-n(2p-1)} \int_{\tilde{R}_2(z)} |h_z^{(2)}(\eta)|^{2p} E(|\dot{\omega}_\eta|^{2p})\, d\eta \le C 2^{-2np}.$$

The identity (2.13) together with (2.14) yield (2.12). $\blacksquare$

Proof of Proposition 2.5. Assume first $z' = (s,t)$, $z = (s+b, t)$. Set $Y_z^n - Y_{z'}^n = \sum_{i=1}^4 M_i(z, z')$, where

$$\begin{aligned}
M_1(z, z') &= \int_{(z' \otimes 0, z]} \gamma_z(\eta) F(Y_\eta^n)dW_\eta + \int_{(z' \otimes 0, z]} \gamma_z(\eta) G(Y_\eta^n)\dot{h}_\eta d\eta, \\
M_2(z, z') &= \int_{R_{z'}} (\gamma_z(\eta) - \gamma_{z'}(\eta)) F(Y_\eta^n)dW_\eta + \int_{(z' \otimes 0, z]} \gamma_z(\eta) B(Y_\eta^n)d\eta \\
&\quad + \int_{R_{z'}} (\gamma_z(\eta) - \gamma_{z'}(\eta))[G(Y_\eta^n)\dot{h}_\eta + B(Y_\eta^n)]d\eta, \\
M_3(z, z') &= \int_{R_{z'}} (\gamma_z(\eta) - \gamma_{z'}(\eta)) H(Y_\eta^n)\dot{\omega}_\eta^n d\eta, \\
M_4(z, z') &= \int_{(z' \otimes 0, z]} \gamma_z(\eta) H(Y_\eta^n)\dot{\omega}_\eta^n\, d\eta.
\end{aligned}$$

Burkholder's, Hölder's and Schwarz's inequalities and properties (3.8), (3.10) yield $E(|M_1(z,z')|^{2p}) \leq C\,b^p$, $E(|M_2(z,z')|^{2p}) \leq C\,b^{2p}$. By Taylor's formula,

$$(2.15) \quad H(Y_\eta^n) = H'(Y_{\eta,\eta_n}^n)(Y_\eta^n - Y_{\eta_n \otimes \eta}^n - Y_{\eta \otimes \eta_n}^n + Y_{\eta_n}^n) + H(Y_{\eta_n \otimes \eta}^n + Y_{\eta \otimes \eta_n}^n - Y_{\eta_n}^n),$$

where Y_{η,η_n}^n denotes a point depending on Y_η^n and $Y_{\eta_n \otimes \eta}^n + Y_{\eta \otimes \eta_n}^n - Y_{\eta_n}^n$. Then, $M_3(z,z') = M_{3,1}(z,z') + M_{3,2}(z,z')$, with

$$M_{3,1}(z,z') = \int_{R_{z'}} (\gamma_z(\eta) - \gamma_{z'}(\eta))H'(Y_{\eta,\eta_n}^n)Y^n(\eta_n,\eta]\dot{\omega}_\eta^n d\eta,$$

$$M_{3,2}(z,z') = \int_{R_{z'}} (\gamma_z(\eta) - \gamma_{z'}(\eta))H(Y_{\eta_n \otimes \eta}^n + Y_{\eta \otimes \eta_n}^n - Y_{\eta_n}^n)\dot{\omega}_\eta^n d\eta.$$

Lemma 2.9, (2.5) and (3.10) ensure $E(|M_{3,1}(z,z')|^{2p}) \leq C\,b^{2p}$. Furthermore, Corollary 2.8 yields $E(|M_{3,2}(z,z')|^{2p}) \leq C\,b^{2p}$. So, we have

$$E(|M_3(z,z')|^{2p}) \leq C\,b^{2p}.$$

Finally, we use again (2.15) and write $M_4(z,z') = M_{4,1}(z,z') + M_{4,2}(z,z')$, where

$$M_{4,1}(z,z') = \int_{(z' \otimes 0,z]} \gamma_z(\eta)H'(Y_{\eta,\eta_n}^n)Y^n(\eta_n,\eta]\dot{\omega}_\eta^n d\eta,$$

$$M_{4,2}(z,z') = \int_{(z' \otimes 0,z]} \gamma_z(\eta)H(Y_{\eta_n \otimes \eta}^n + Y_{\eta \otimes \eta_n}^n - Y_{\eta_n}^n)\dot{\omega}_\eta^n d\eta.$$

Then, $E(|M_{4,1}(z,z')|^{2p}) \leq C\,b^{2p}$ and Corollary 2.8 yields $E(|M_{4,2}(z,z')|^{2p}) \leq C\,b^p$. Therefore,

$$E(|M_4(z,z')|^{2p}) \leq C\,b^p.$$

Consequently, we have obtained $E(|Y_{s+b,t}^n - Y_{s,t}^n|^{2p}) \leq C\,b^p$. Analogously, $E(|Y_{s,t+b}^n - Y_{s,t}^n|^{2p}) \leq C\,b^p$. Therefore,

$$\sup_n E(|Y_z^n - Y_{z'}^n|^{2p}) \leq C\,d(z,z')^p.$$

Notice that the equation defining Y_z can be considered as a particular case of that defining Y_z^n. Thus, (2.8) is completely proved. ∎

Proof of Proposition 2.6. Set

$$\psi(x) := H'(x)F(x) + \frac{1}{4}H'(x)H(x).$$

Notice that ψ is Lipschitz.
Let

$$I_1^n(z) = \int_{R_z} \gamma_z(\eta)(F(Y_\eta^n) - F(Y_\eta) + H(Y_\eta^n) - H(Y_\eta))dW_\eta,$$

$$I_2^n(z) = \int_{R_z} \gamma_z(\eta)(G(Y_\eta^n) - G(Y_\eta))\dot{h}_\eta\, d\eta,$$

$$I_3^n(z) = \int_{R_z} \gamma_z(\eta)(B(Y_\eta^n) - B(Y_\eta) + \psi(Y_\eta^n) - \psi(Y_\eta))d\eta,$$

$$I_4^n(z) = \int_{R_z} \gamma_z(\eta)H(Y_\eta^n)\dot{\omega}_\eta^n d\eta - \int_{R_z} \gamma_z(\eta)H(Y_\eta^n)dW_\eta - \int_{R_z} \gamma_z(\eta)\psi(Y_\eta^n)d\eta.$$

Then, $Y_z^n - Y_z = \sum_{i=1}^4 I_i^n(z)$. Burkholder's, Hölder's and Schwarz's inequalities yield

$$E(|I_i^n(z)|^{2p}) \leq C \int_{R_z} E(|Y_\eta^n - Y_\eta|^{2p})\, d\eta \, , \ i = 1, 3,$$

$$E(|I_2^n(z)|^{2p}) \leq C \|h\|_{\mathcal{H}}^{2p} \int_{R_z} E(|Y_\eta^n - Y_\eta|^{2p})\, d\eta \, .$$

Hence, by Gronwall's lemma,

$$E(|Y_z^n - Y_z|^{2p}) \leq C \sup_{z \in T} E(|I_4^n(z)|^{2p}).$$

In Proposition 2.12 we establish the following property

$$(2.16) \qquad \sup_{z \in T} E(|I_4^n(z)|^{2p}) \leq C 2^{-np},$$

therefore the proof of the proposition is complete. ∎

Property (2.16) is the key of the approximation result. Its proof is rather long and technical. Some auxiliary estimates are presented separately in the next two lemmas.

Lemma 2.10 *For any $z \in T$, $p \in [1, \infty)$, $n \geq 1$,*

$$(2.17)$$
$$E\Big(\Big| \int_{R_z} \gamma_z(\eta_n) H'(Y_{\eta_n}^n)\Big(F(Y_{\eta_n}^n) \int_{(\eta_n, \eta]} dW_\xi + H(Y_{\eta_n}^n) \int_{(\eta_n, \eta]} \dot{\omega}_\xi^n\, d\xi\Big)\dot{\omega}_\eta^n\, d\eta$$
$$- \int_{R_z} \gamma_z(\eta_n)\psi(Y_{\eta_n}^n) d\eta \Big|^{2p}\Big) \leq C\, 2^{-2np} \, .$$

Proof. Suppose $z \in \Delta_{i_0, j_0}$ and set

$$T_1^{(n)}(z) = \int_{R_z - R_{z_{i_0, j_0}}} \gamma_z(\eta_n) H'(Y_{\eta_n}^n)\Big(F(Y_{\eta_n}^n) \int_{(\eta_n, \eta]} dW_\xi$$
$$+ H(Y_{\eta_n}^n) \int_{(\eta_n, \eta]} \dot{\omega}_\xi^n d\xi\Big)\dot{\omega}_\eta^n d\eta \, ,$$

$$T_2^{(n)}(z) = \int_{R_{z_{i_0, j_0}}} \gamma_z(\eta_n) H'(Y_{\eta_n}^n)\Big(F(Y_{\eta_n}^n) \int_{(\eta_n, \eta]} dW_\xi$$
$$+ H(Y_{\eta_n}^n) \int_{(\eta_n, \eta]} \dot{\omega}_\xi^n d\xi\Big)\dot{\omega}_\eta^n d\eta - \int_{R_{z_{i_0, j_0}}} \gamma_z(\eta_n)\psi(Y_{\eta_n}^n)\, d\eta,$$

$$T_3^{(n)}(z) = -\int_{R_z - R_{z_{i_0, j_0}}} \gamma_z(\eta_n)\psi(Y_{\eta_n}^n)\, d\eta \, .$$

One can easily check, using Hölder's inequality,

$$E(|T_i^{(n)}(z)|^{2p}) \leq 2^{-2np}, \quad i = 1, 3.$$

Thus, since the left-hand side of (2.17) is bounded by $C \sum_{i=1}^3 E(|T_i^{(n)}(z)|^{2p})$, we only need to prove

$$(2.18) \qquad E(|T_2^{(n)}(z)|^{2p}) \leq C\, 2^{-2np} \, .$$

For any $\eta \in \Delta_{\alpha,\beta}$ with $\alpha \cdot \beta \neq 0$ we decompose the rectangle $(\eta_n, \eta]$ into four rectangles. More precisely, $(\eta_n, \eta] = \bigcup_{j=1}^4 R_j(\eta)$, with

$$R_1(\eta) = \Delta_{\alpha-1,\beta-1}, \quad R_2(\eta) = (z_{\alpha,\beta-1}, \eta \otimes z_{\alpha,\beta}],$$
$$R_3(\eta) = (z_{\alpha,\beta}, \eta], \quad R_4(\eta) = (z_{\alpha-1,\beta}, z_{\alpha,\beta} \otimes \eta].$$

For $z = z_{i_0,j_0}$, $z' \in \Delta_{i_0,j_0}$ and $r = 1, \dots, 4$ let

$$U_r^n = \int_{R_z} \gamma_{z'}(\eta_n) H'(Y_{\eta_n}^n) F(Y_{\eta_n}^n) \Big(\int_{R_r(\eta)} dW_\xi \Big) \dot{\omega}_\eta^n \, d\eta,$$
$$V_r^n = \int_{R_z} \gamma_{z'}(\eta_n) H'(Y_{\eta_n}^n) H(Y_{\eta_n}^n) \Big(\int_{R_r(\eta)} \dot{\omega}_\xi^n \, d\xi \Big) \dot{\omega}_\eta^n \, d\eta.$$

Then, the proof of (2.18) reduces to showing the following inequalities

$$E \Big(|U_1^n - \int_{R_z} \gamma_{z'}(\eta_n) H'(Y_{\eta_n}^n) F(Y_{\eta_n}^n) \, d\eta|^{2p} \Big) \leq C \, 2^{-2np},$$
$$E \Big(|V_3^n - \frac{1}{4} \int_{R_z} \gamma_{z'}(\eta_n) H'(Y_{\eta_n}^n) H(Y_{\eta_n}^n) \, d\eta|^{2p} \Big) \leq C \, 2^{-2np},$$
$$E(|U_r^n|^{2p}) \leq C \, 2^{-2np}, \quad r = 2,3,4; \quad E(|V_r^n|^{2p}) \leq C \, 2^{-2np}, \quad r = 1,2,4.$$

They are obtained following very closely the arguments presented in the proof of Proposition 4.5 [13], taking into account the boundedness of $\gamma_z(\eta)$. The details are left to the reader. ∎

Lemma 2.11 *For any $z \in T$, $p \in [1, \infty)$, $n \geq 1$,*

$$(2.19) \quad E \Big(| \int_{R_z} \gamma_z(\eta) H'(Y_{\eta_n}^n) Y^n(\eta_n, \eta] \dot{\omega}_\eta^n \, d\eta - \int_{R_z} \gamma_z(\eta) \psi(Y_\eta^n) d\eta |^{2p} \Big) \leq C \, 2^{-np}.$$

Proof. Using the notation introduced in (2.13), set $T_{1,2,4}^n(\eta) := \sum_{i=1,2,4} (T_{i,F}^n(\eta) + T_{i,G}^n(\eta) + T_{i,H}^n(\eta) + T_{i,B}^n(\eta))$. Then, using the identity (2.13) we can write

$$E \Big(| \int_{R_z} \gamma_z(\eta) H'(Y_{\eta_n}^n) Y^n(\eta_n, \eta] \dot{\omega}_\eta^n \, d\eta - \int_{R_z} \gamma_z(\eta) \psi(Y_\eta^n) \, d\eta |^{2p} \Big)$$
$$\leq C \sum_{j=1} U_j^n(z),$$

where

$$U_1^n(z) = E \Big(| \int_{R_z} \gamma_z(\eta) H'(Y_{\eta_n}^n) T_{1,2,4}^n(\eta) \dot{\omega}_\eta^n d\eta |^{2p} \Big),$$
$$U_2^n(z) = E \Big(| \int_{R_z} \gamma_z(\eta) H'(Y_{\eta_n}^n) T_{3,G}^n(\eta) \dot{\omega}_\eta^n d\eta |^{2p} \Big),$$
$$U_3^n(z) = E \Big(| \int_{R_z} \gamma_z(\eta) H'(Y_{\eta_n}^n) T_{3,B}^n(\eta) \dot{\omega}_\eta^n d\eta |^{2p} \Big),$$
$$U_4^n(z) = E \Big(| \int_{R_z} \gamma_z(\eta) H'(Y_{\eta_n}^n) (T_{3,F}^n(\eta) + T_{3,H}^n(\eta)) \dot{\omega}_\eta^n d\eta$$
$$- \int_{R_z} \gamma_z(\eta) \psi(Y_\eta^n) d\eta |^{2p} \Big).$$

In Lemma 2.9 we have checked $E(|T^n_{1,2,4}(\eta)|^{2p}) \leq C\,2^{-2np}$. Obviously $T^n_{1,2,4}(\eta)$ is $\mathcal{F}^*_\eta$-measurable (see Corollary 2.8 for the notation). Then, Corollary 2.8 yields $U^n_1(z) \leq C\,2^{-2np}$. Moreover,

$$U^n_2(z) \leq C \Big\{ E\Big(|\int_{R_z} \gamma_z(\eta) H'(Y^n_{\eta_n}) G(Y^n_{\eta_n}) \Big(\int_{(\eta_n,\eta]} \gamma_\eta(\xi) \dot{h}_\xi\, d\xi \Big) \dot{\omega}^n_\eta\, d\eta|^{2p} \Big)$$
$$+ E\Big(|\int_{R_z} \gamma_z(\eta) H'(Y^n_{\eta_n}) \Big(\int_{(\eta_n,\eta]} \gamma_\eta(\xi) (G(Y^n_{\eta_n}) - G(Y^n_\xi)) \dot{h}_\xi\, d\xi \Big) \dot{\omega}^n_\eta\, d\eta|^{2p} \Big) \Big\}$$
$$\leq C(1 + \|h\|^{2p}_{\mathcal{H}})\, 2^{-np}\,,$$

where we have applied Corollary 2.8, (3.11) and Proposition 2.5.

Using $E(|T^n_{3,B}(\eta)|^{4p}) \leq C\,2^{-8np}$, Höölder's inequality and (2.5), we obtain $U^n_3(z) \leq C\,2^{-2np}$. So, the proof of (2.19) reduces to showing $U^n_4(z) \leq C\,2^{-np}$. Set

$$U^n_{4,1}(z) = E\Big(|\int_{R_z} (\gamma_z(\eta) - \gamma_z(\eta_n)) H'(Y^n_{\eta_n}) \Big[\int_{(\eta_n,\eta]} \gamma_\eta(\xi) F(Y^n_\xi) dW_\xi$$
$$+ \int_{(\eta_n,\eta]} \gamma_\eta(\xi) H(Y^n_\xi) \dot{\omega}^n_\xi d\xi \Big] \dot{\omega}^n_\eta\, d\eta|^{2p} \Big),$$
$$U^n_{4,2}(z) = E\Big(|\int_{R_z} \gamma_z(\eta_n) H'(Y^n_{\eta_n}) \Big[\int_{(\eta_n,\eta]} \gamma_\eta(\xi) (F(Y^n_\xi) - F(Y^n_{\eta_n})) dW_\xi$$
$$+ \int_{(\eta_n,\eta]} \gamma_\eta(\xi) (H(Y^n_\xi) - H(Y^n_{\eta_n})) \dot{\omega}^n_\xi d\xi \Big] \dot{\omega}^n_\eta\, d\eta|^{2p} \Big),$$
$$U^n_{4,3}(z) = E\Big(|\int_{R_z} \gamma_z(\eta_n) H'(Y^n_{\eta_n}) \Big(F(Y^n_{\eta_n}) \int_{(\eta_n,\eta]} dW_\xi + H(Y^n_{\eta_n}) \int_{(\eta_n,\eta]} \dot{\omega}^n_\xi\, d\xi \Big)$$
$$\times \dot{\omega}^n_\eta d\eta - \int_{R_z} \gamma_z(\eta_n) \psi(Y^n_{\eta_n}) d\eta|^{2p} \Big),$$
$$U^n_{4,4}(z) = E\Big(|\int_{R_z} (\gamma_z(\eta_n) \psi(Y^n_{\eta_n}) - \gamma_z(\eta) \psi(Y^n_\eta)) d\eta|^{2p} \Big).$$

For each $i = 1, \cdots, 4$, $U^n_{4,i}(z) \leq C\,2^{-np}$. For $i = 3$ this has been proved in Lemma 2.10. Let us, for instance, check this property for $i = 2$. Hölder's and Burkholder's inequalities, (2.5) and Proposition 2.5 yield

$$U^n_{4,2} \leq C\,2^{2np} \sup_{\eta > z_{1,1}} \Big\{ E\Big(|\int_{(\eta_n,\eta]} \gamma_\eta(\xi) (F(Y^n_\xi) - F(Y^n_{\eta_n})) dW_\xi|^{4p} \Big)$$
$$+ E\Big(|\int_{(\eta_n,\eta]} \gamma_\eta(\xi) (H(Y^n_\xi) - H(Y^n_{\eta_n})) \dot{\omega}^n_\xi\, d\xi|^{4p} \Big) \Big\}^{\frac{1}{2}} \leq C\,2^{-np}.$$

The other cases are proved analogously. The proof is now complete, because $U^n_4(z) \leq C \sum^4_{i=1} U^n_{4,i}(z)$. ∎

Proposition 2.12 *For any* $p \in [1,\infty)$, $n \geq 1$,

$$\sup_{z \in T} E(|I^n_4(z)|^{2p}) \leq C2^{-np},$$

where

$$I^n_4(z) = \int_{R_z} \gamma_z(\eta) H(Y^n_\eta) \dot{\omega}^n_\eta d\eta - \int_{R_z} \gamma_z(\eta) H(Y^n_\eta) dW_\eta - \int_{R_z} \gamma_z(\eta) \psi(Y^n_\eta) d\eta\,.$$

Proof. Consider the following decomposition:

$$I_4^n(z) \leq C(\tilde{J}_1^n(z) + \tilde{J}_2^n(z) + \tilde{J}_3^n(z) + \tilde{J}_4^n(z))$$

with

$$\tilde{J}_1^n(z) = E\Big(\big| \int_{R_z} \gamma_z(\eta)(H(Y_\eta^n) - H(Y_{\eta_n}^n))\dot{\omega}_\eta^n d\eta - \int_{R_z} \gamma_z(\eta)\psi(Y_\eta^n)d\eta \big|^{2p}\Big),$$

$$\tilde{J}_2^n(z) = E\Big(\big| \int_{R_z} (\gamma_z(\eta) - \gamma_z(\eta_n))H(Y_{\eta_n}^n)\dot{\omega}_\eta^n d\eta \big|^{2p}\Big),$$

$$\tilde{J}_3^n(z) = E\Big(\big| \int_{R_z} \gamma_z(\eta_n)H(Y_{\eta_n}^n)\dot{\omega}_\eta^n d\eta - \int_{R_z} \gamma_z(\eta_n)H(Y_{\eta_n}^n)dW_\eta \big|^{2p}\Big),$$

$$\tilde{J}_4^n(z) = E\Big(\big| \int_{R_z} \big((\gamma_z(\eta_n) - \gamma_z(\eta))H(Y_{\eta_n}^n) + \gamma_z(\eta)(H(Y_{\eta_n}^n)$$

$$-H(Y_\eta^n))\big)dW_\eta \big|^{2p}\Big).$$

Corollary 2.8 yields $\tilde{J}_2^n(z) \leq C\, 2^{-2np}$. Moreover, Burkholder's inequality, (3.8), (3.9) and Proposition 2.5 ensures $\tilde{J}_4^n(z) \leq C\, 2^{-np}$. We now show

$$(2.20) \qquad\qquad \tilde{J}_3^n(z) \leq C\, 2^{-np}.$$

Let $z \in \Delta_{i_0,j_0}$, then,

$$\int_{R_z} \gamma_z(\eta_n)H(Y_{\eta_n}^n)\dot{\omega}_\eta^n d\eta - \int_{R_z} \gamma_z(\eta_n)H(Y_{\eta_n}^n)dW_\eta$$
$$= \int_{R_z - R_{z_{i_0,j_0}}} \gamma_z(\eta_n)H(Y_{\eta_n}^n)(\dot{\omega}_\eta^n d\eta - dW_\eta)$$
$$+ \int_{R_{z_{i_0,j_0}}} \gamma_z(\eta_n)H(Y_{\eta_n}^n)(\dot{\omega}_\eta^n d\eta - dW_\eta).$$

Burkholder's inequality yields

$$E\Big(\big| \int_{R_z - R_{z_{i_0,j_0}}} \gamma_z(\eta_n)H(Y_{\eta_n}^n)(\dot{\omega}_\eta^n d\eta - dW_\eta)\big|^{2p}\Big) \leq C\, 2^{-np}.$$

Moreover,

$$\int_{R_{z_{i_0,j_0}}} \gamma_z(\eta_n) H(Y_{\eta_n}^n)(\dot{\omega}_\eta^n d\eta - dW_\eta)$$

$$= \sum_{\substack{1 \le \alpha \le i_0-2 \\ 1 \le \beta \le j_0-2}} W(\Delta_{\alpha,\beta})(\gamma_z(z_{\alpha,\beta}) H(Y_{z_{\alpha,\beta}}^n) - \gamma_z(z_{\alpha-1,\beta-1}) H(Y_{z_{\alpha-1,\beta-1}}^n))$$

$$- \sum_{1 \le \beta \le j_0-1} W(\Delta_{i_0-1,\beta})\gamma_z(z_{i_0-2,\beta-1}) H(Y_{z_{i_0-2,\beta-1}}^n)$$

$$- \sum_{1 \le \alpha \le i_0-2} [W(\Delta_{\alpha,j_0-1})\gamma_z(z_{\alpha-1,j_0-2}) H(Y_{z_{\alpha-1,j_0-2}}^n)$$

$$- W(\Delta_{\alpha,0})(\gamma_z(z_{\alpha,0}) - \gamma_z(0)) H(Y_0^n)]$$

$$+ \sum_{1 \le \beta \le j_0-2} W(\Delta_{0,\beta})(\gamma_z(z_{0,\beta}) - \gamma_z(0)) H(Y_0^n)$$

$$- (W(\Delta_{0,j_0-1}) + W(\Delta_{i_0-1,0}))\gamma_z(0) H(Y_0^n) .$$

Burkholder's inequality, Proposition 2.5, (3.8) and (3.9) allow us to bound all the terms of the previous equality by $C\,2^{-np}$. So, (2.20) holds true. Finally, we deal with $\tilde{J}_1^n(z)$. Taylor's formula yields

$$H(Y_\eta^n) - H(Y_{\eta_n}^n) = H'(Y_{\eta_n}^n)(Y_\eta^n - Y_{\eta_n}^n) + \frac{1}{2} H''(Y_{\eta_n}^n)(Y_\eta^n - Y_{\eta_n}^n)^2$$
$$+ \frac{1}{6} H'''(\overline{Y}_{\eta,\eta_n}^n)(Y_\eta^n - Y_{\eta_n}^n)^3,$$

where $\overline{Y}_{\eta,\eta_n}^n$ denotes a point between Y_η^n and $Y_{\eta_n}^n$. Then, $\tilde{J}_1^n(z) \le C\sum_{j=1}^5 J_j^n(z)$, where

$$J_1^n(z) = E\Big(\Big| \int_{R_z} \gamma_z(\eta) H'(Y_{\eta_n}^n) Y^n(\eta_n,\eta]\dot{\omega}_\eta^n d\eta - \int_{R_z} \gamma_z(\eta)\psi(Y_\eta^n) d\eta\Big|^{2p}\Big),$$

$$J_2^n(z) = E\Big(\Big| \int_{R_z} \gamma_z(\eta) H'(Y_{\eta_n}^n)[(Y_{\eta\otimes\eta_n}^n - Y_{\eta_n}^n) + (Y_{\eta_n\otimes\eta}^n - Y_{\eta_n}^n)]\dot{\omega}_\eta^n d\eta\Big|^{2p}\Big),$$

$$J_3^n(z) = E\Big(\Big| \int_{R_z} \gamma_z(\eta) H''(Y_{\eta_n}^n) G_3^n(\eta) \dot{\omega}_\eta^n d\eta\Big|^{2p}\Big),$$

$$J_4^n(z) = E\Big(\Big| \int_{R_z} \gamma_z(\eta) H''(Y_{\eta_n}^n) G_4^n(\eta) \dot{\omega}_\eta^n d\eta\Big|^{2p}\Big),$$

$$J_5^n(z) = E\Big(\Big| \int_{R_z} \frac{1}{6}\gamma_z(\eta) H'''(\overline{Y}_{\eta,\eta_n}^n)(Y_\eta^n - Y_{\eta_n}^n)^3 \dot{\omega}_\eta^n d\eta\Big|^{2p}\Big),$$

with

$$G_3^n(\eta) = \frac{1}{2}Y^n(\eta_n,\eta]^2 + Y^n(\eta_n,\eta][(Y_{\eta\otimes\eta_n}^n - Y_{\eta_n}^n) + (Y_{\eta_n\otimes\eta}^n - Y_{\eta_n}^n)],$$
$$G_4^n(\eta) = \frac{1}{2}((Y_{\eta\otimes\eta_n}^n - Y_{\eta_n}^n) + (Y_{\eta_n\otimes\eta}^n - Y_{\eta_n}^n))^2.$$

By Lemma 2.11, $J_1^n(z) \le C\,2^{-np}$. Corollary 2.8 and Proposition 2.5 yield $J_2^n(z) \le C2^{-2np}$ and $J_4^n(z) \le C2^{-2np}$. Finally, Proposition 2.5 and Lemma

2.9 ensure $J_3^n(z) \leq C\, 2^{-np}$ and $J_5^n(z) \leq C\, 2^{-np}$. Thus, $\tilde{J}_1^n(z) \leq C\, 2^{-np}$. This completes the proof of the proposition. ∎

Fix $h \in \mathcal{H}$, $a \in \mathbb{R}$ and consider the differential equation

$$
\begin{aligned}
(2.21) \quad S_z^a(h) \;&= X_0 + \int_{R_z} \gamma_z(\eta)\Big(a_4(S_\eta^a(h)) - a\, a_3'(S_\eta^a(h)) a_3(S_\eta^a(h))\Big) d\eta \\
&+ \int_{R_z} \gamma_z(\eta)\, a_3(S_\eta^a(h))\, \dot{h}_\eta\, d\eta \, ,
\end{aligned}
$$

$z \in T$. Notice that (2.21) is a particular case of equation (2.6).
Let ρ^a be the closure with respect to the α–Hölder topology ($\alpha \in (0, \frac{1}{2})$) of $\{S^a(h),\ h \in \mathcal{H}\}$. Fix $h \in \mathcal{H}$ and set $T_n^h(w) = w - w^n + h$. An extension of the classical Girsanov theorem for two–parameter processes (see [6]) states that $P \circ (T_n^h)^{-1}$ is absolutely continuous with respect to P, for any $n \geq 1$.
The processes $\{S_z^a(w^n),\ z \in T\}$ and $\{X_z(w - w^n + h),\ z \in T\}$ are particular cases of solutions of (2.6). Indeed, let $X^n(w) := X(w - w^n + h)$, $h \in \mathcal{H}$. Then, approximating stochastic integrals by Riemann sums, it is easy to see that X^n satisfies the following stochastic differential equation

$$
\begin{aligned}
X_z^n \;&= X_0 + \int_{R_z} \gamma_z(\eta)[a_4(X_\eta^n) + a_3(X_\eta^n)\dot{h}_\eta - a_3(X_\eta^n)\dot{\omega}_\eta^n]\, d\eta \\
&+ \int_{R_z} \gamma_z(\eta) a_3(X_\eta^n)\, dW_\eta \, .
\end{aligned}
$$

We can now prove the support theorem.

Proof of Theorem 2.3 Our aim is to check the following convergences for all $\alpha \in (0, \frac{1}{2})$, $\varepsilon > 0$,

$$
(2.22) \qquad \lim_{n \to \infty} P(w:\ \|X(w) - S^{\frac{1}{4}}(w^n)\|_\alpha > \varepsilon) = 0
$$

and

$$
(2.23) \qquad \lim_{n \to \infty} P(w:\ \|X(w - w^n + h) - S^{\frac{3}{4}}(h)\|_\alpha > \varepsilon) = 0,\ h \in \mathcal{H},
$$

which, according to Proposition 1.1 ensure the inclusions $\rho^{\frac{3}{4}} \subset \text{supp}(P \circ X^{-1}) \subset \rho^{\frac{1}{4}}$. Finally, we will show that $\rho^a = \rho^0$ for any $a \in \mathbb{R}$. Consequently, $\text{supp}\,(P \circ X^{-1}) = \rho^0$.
Choose in (2.6) and (2.7) $Y_0 = X_0$ and the following set of coefficients: $F = G = 0$, $H = a_3$ and $B = a_4 - \frac{1}{4}\, a_3'\, a_3$. Then, $Y_z^n = S_z^{\frac{1}{4}}(w^n)$ and $Y_z = X_z$. Next choose $F = G = a_3$, $H = -a_3$ and $B = a_4$. then, $Y_z^n = X_z(w - w^n + h)$ and $Y_z = S_z^{\frac{3}{4}}(h)$. Hence, the convergences (2.22) and (2.23) follow from Theorem 2.4.
Finally, we prove that, for all $a \in \mathbb{R}$,

$$
(2.24) \qquad \{S^0(h),\ h \in \mathcal{H}\} = \{S^a(h),\ h \in \mathcal{H}\}.
$$

Fix $h \in \mathcal{H}$ and let $S^a(h)$ be the solution of (2.21). We define $g \in \mathcal{H}$ as follows: $\dot{g} := \dot{h} - a\, a'_3(S^a(h))$. Then, the process $S^0(g)$ satisfies

$$S_z^0(g) = X_0 + \int_{R_z} \gamma_z(\eta) \Big[a_4(S_\eta^0(g)) + a_3(S_\eta^0(g)) \big(\dot{h}_\eta - a\, a'_3(S_\eta^a(h)) \big) \Big] d\eta\,.$$

The uniqueness of the solution for (2.21) implies $S_z^0(g) = S_z^a(h)$. So,

$$\{ S^a(h),\ h \in \mathcal{H} \} \subset \{ S^0(h),\ h \in \mathcal{H} \}.$$

Similar arguments show the converse inclusion. Thus, (2.24) holds true and the proof of the theorem is finished. ∎

3 Appendix

In this section we will show the existence of a function $\gamma_{s,t}(u,v)$, $(0,0) \leq (u,v) \leq (s,t)$, $(s,t) \in T$, satisfying

$$(3.1) \qquad \gamma_{s,t}(u,v) = 1 + \int_u^s a_1(r,v)\gamma_{s,t}(r,v)dr + \int_v^t a_2(u,w)\gamma_{s,t}(u,w)dw\,.$$

This is the Green function associated with the second order differential operator $\mathcal{L}$ introduced in Section 2. We also check some smooth properties of this function. As a by-product we obtain the following properties:

$$(\mathrm{P}) \qquad \frac{\partial^2 \gamma_{s,t}}{\partial u\, \partial v} + \frac{\partial(a_1 \gamma_{s,t})}{\partial v} + \frac{\partial(a_2 \gamma_{s,t})}{\partial u} = 0\,,$$

$$\frac{\partial \gamma_{s,t}}{\partial u} = -a_1(u,v)\gamma_{s,t},\ \text{if } v = t, \qquad \frac{\partial \gamma_{s,t}}{\partial v} = -a_2(u,v)\gamma_{s,t},\ \text{if } u = s,$$
$$\gamma_{s,t} = 1,\ \text{if } u = s,\ v = t\,.$$

Set $H_0(s,t;u,v) \equiv 1$ and
$$(3.2)$$
$$H_{n+1}(s,t;u,v) = \int_u^s a_1(r,v)H_n(s,t;r,v)dr + \int_v^t a_2(u,w)H_n(s,t;u,w)dw\,,$$

for $n \geq 0$. Define

$$(3.3) \qquad \gamma_{s,t}(u,v) = \sum_{n=0}^{\infty} H_n(s,t;u,v)\,,\ (u,v) \leq (s,t).$$

Proposition 3.1 *The series in (3.3) defining the function γ is absolutely convergent and satisfies (3.1). In addition, for any $(s,t) \in T$, $\gamma_{s,t}(\cdot)$ has uniformly bounded derivatives of first order and mixed second order partial derivative on $\{(u,v) : 0 < u < s,\ 0 < v < t\}$.*

Proof. First, we will check that for all $n \geq 0$, $(u, v) \leq (s, t)$

$$(3.4) \qquad |H_n(s, t; u, v)| \leq C^n \sum_{\alpha=0}^{n} \binom{n}{\alpha} \frac{(s - u)^\alpha}{\alpha!} \frac{(t - v)^{n-\alpha}}{(n - \alpha)!} .$$

For $n = 0$, it is obvious. By induction,

$$\begin{aligned}
|H_{n+1}(s, t; u, v)| &= | \int_u^s a_1(r, v) H_n(s, t; r, v) dr \\
&\quad + \int_v^t a_2(u, w) H_n(s, t; u, w) \, dw| \\
&\leq C^{n+1} \sum_{\alpha=0}^{n+1} \binom{n+1}{\alpha} \frac{(s - u)^\alpha}{\alpha!} \frac{(t - v)^{n+1-\alpha}}{(n + 1 - \alpha)!} .
\end{aligned}$$

Hence (3.4) holds and, consequently, $|H_n(s, t; u, v)| \leq (2C)^n \max_{\alpha \in \{0, \cdots, n\}}$ $\left(\frac{1}{\alpha!}, \frac{1}{(n-\alpha)!} \right)$. Notice that $\max_{\alpha \in \{0, \cdots, n\}} \left\{ \frac{1}{\alpha!}, \frac{1}{(n-\alpha)!} \right\}$ is equal to $\frac{1}{(\frac{n}{2})!^2}$, if n is even, and to $\frac{1}{(\frac{n+1}{2})! (\frac{n-1}{2})!}$, if n is odd. So, the series (3.3) is absolutely convergent and γ is well-defined. Moreover, from (3.2) and Fubini's theorem,

$$\begin{aligned}
\sum_{n \geq 0} H_{n+1}(s, t; u, v) &= \int_u^s a_1(r, v) \Big(\sum_{n \geq 0} H_n(s, t; r, v) \Big) dr \\
&\quad + \int_v^t a_2(u, w) \Big(\sum_{n \geq 0} H_n(s, t; u, w) \Big) dw,
\end{aligned}$$

which clearly implies (3.1).

Now, to check the derivability of γ, it is enough to show that $H_n(s, t; u, v)$ is derivable for all $n \geq 0$ and that the series of the derivatives is absolutely convergent. First, we will see the derivability respect to u. Obviously H_0 is derivable. From (3.2)

$$(3.5)$$
$$\begin{aligned}
\frac{\partial H_{n+1}(s, t; u, v)}{\partial u} &= -a_1(u, v) H_n(s, t; u, v) + \int_v^t \frac{\partial a_2(u, w)}{\partial u} H_n(s, t; u, w) \, dw \\
&\quad + \int_v^t a_2(u, w) \frac{\partial H_n(s, t; u, w)}{\partial u} \, dw .
\end{aligned}$$

So, $H_n(s, t; u, v)$ is derivable with respect to u for all $n \geq 0$. Furthermore, for all $n \geq 0$, $(u, v) \leq (s, t)$,

$$\left| \frac{\partial H_n(s, t; u, v)}{\partial u} \right| \leq C^{n+1} \sum_{\alpha=0}^{n-1} \left(\sum_{\beta=\alpha}^{n-1} \binom{\beta}{\alpha} \right) \frac{(s - u)^\alpha}{\alpha!} \left[\frac{(t - v)^{n-\alpha}}{(n - \alpha)!} + \frac{(t - v)^{n-1-\alpha}}{(n - 1 - \alpha)!} \right].$$

For $n = 0$, it is obvious. Assume that it is true up to the step n. From (3.4) and (3.5)

$$\left|\frac{\partial H_{n+1}(s,t;u,v)}{\partial u}\right| \le C|H_n(s,t;u,v)| + C \int_v^t |H_n(s,t;u,w)|\, dw$$

$$+ C \int_v^t \left|\frac{\partial H_n(s,t;u,w)}{\partial u}\right| dw$$

$$\le C^{n+1} \sum_{\alpha=0}^{n} \left(\sum_{\beta=\alpha}^{n} \binom{\beta}{\alpha}\right) \frac{(s-u)^\alpha}{\alpha!}\left[\frac{(t-v)^{n+1-\alpha}}{(n+1-\alpha)!} + \frac{(t-v)^{n-\alpha}}{(n-\alpha)!}\right].$$

Thus,

$$\left|\frac{\partial H_{n+1}(s,t;u,v)}{\partial u}\right| \le (2C)^{n+1} \max_{\alpha\in\{0,\cdots,n\}}\left\{\frac{1}{\alpha!}, \frac{1}{(n-\alpha)!}\right\}.$$

Then, the series is absolutely convergent and $\left|\frac{\partial \gamma_{s,t}(u,v)}{\partial u}\right| \le C$, for all $(u,v) < (s,t)$. By the same method, we can study the derivability of γ respect v and the mixed second order derivability with respect to u and v. ∎

Remark. Notice that by continuity we can extend the first order derivatives and the mixed second order partial derivative with respect to u and v of $\gamma_{s,t}(\cdot)$ to $\{(u,v): 0 < u \le s, 0 < v \le t\}$.

The next propositions provide a set of useful properties of γ.

Proposition 3.2 *Fix $(u,v) \in T$. the function $(s,t) \longmapsto \gamma_{s,t}(u,v)$ has bounded derivatives of first order and of mixed second order with respect to s,t on $\{(s,t) \in T : u < s < 1, v < t < 1\}$.*

Proof. First, we will check that $H_n(s,t;u,v)$ is derivable with respect to s for all $n \ge 0$. For $n = 0$, it is obvious. From (3.2)
(3.6)
$$\frac{\partial H_{n+1}(s,t;u,v)}{\partial s} = a_1(s,v)H_n(s,t;s,v) + \int_u^s a_1(r,v) \frac{\partial H_n(s,t;r,v)}{\partial s}\, dr$$
$$+ \int_v^t a_2(u,w) \frac{\partial H_n(s,t;u,w)}{\partial s}\, dw\,.$$

So, by induction, $H_n(s,t;u,v)$ is derivable with respect to s for all $n \ge 0$. Moreover,

$$\left|\frac{\partial H_n(s,t;u,v)}{\partial s}\right| \le C^n \sum_{\alpha=1}^{n} \binom{n}{\alpha} \frac{(s-u)^{\alpha-1}}{(\alpha-1)!} \frac{(t-v)^{n-\alpha}}{(n-\alpha)!}.$$

Indeed, for $n = 1$, it is obvious. From (3.6) and (3.4) it follows

$$\left|\frac{\partial H_{n+1}(s,t;u,v)}{\partial s}\right| \le C^{n+1} \sum_{\alpha=1}^{n+1} \binom{n+1}{\alpha} \frac{(s-u)^{\alpha-1}}{(\alpha-1)!} \frac{(t-v)^{n+1-\alpha}}{(n+1-\alpha)!}\,.$$

Therefore $|\frac{\partial H_n(s,t;u,v)}{\partial s}| \leq (2\,C)^n \ \max_{\alpha\in\{1,\dots,n\}}\left\{\frac{1}{(\alpha-1)!},\ \frac{1}{(n-\alpha)!}\right\}$, ensuring the absolute convergence of the series of the derivatives and $|\frac{\partial\gamma_{s,t}(u,v)}{\partial s}| \leq C$. Analogously $|\frac{\partial\gamma_{s,t}(u,v)}{\partial t}| \leq C$.

Finally, we want to study $\frac{\partial^2\gamma_{s,t}(u,v)}{\partial s\,\partial t}$. the identity (3.6) yields

(3.7)
$$\frac{\partial^2 H_{n+1}(s,t;u,v)}{\partial s\,\partial t} = a_1(s,v)\,\frac{\partial H_n(s,t;s,v)}{\partial t} + a_2(u,t)\,\frac{\partial H_n(s,t;u,t)}{\partial s}$$
$$+ \int_u^s a_1(r,v)\,\frac{\partial^2 H_n(s,t;r,v)}{\partial s\,\partial t}\,dr + \int_v^t a_2(u,w)\,\frac{\partial^2 H_n(s,t;u,w)}{\partial s\,\partial t}\,dw\,.$$

As before, we can check inductively,

$$\frac{\partial^2 H_n(s,t;u,v)}{\partial s\,\partial t} \leq C^n \sum_{\alpha=1}^{n-1}\binom{n}{\alpha}\frac{(s-u)^{\alpha-1}}{(\alpha-1)!}\frac{(t-v)^{n-1-\alpha}}{(n-1-\alpha)!}$$
$$\leq (2\,C)^n \max_{\alpha\in\{1,\dots,n-1\}}\left\{\frac{1}{(\alpha-1)!},\ \frac{1}{(n-1-\alpha)!}\right\}.$$

So, $\frac{\partial^2\gamma_{s,t}(u,v)}{\partial s\,\partial t}$, exists and satisfies $|\frac{\partial^2\gamma_{s,t}(u,v)}{\partial s\,\partial t}| \leq C$. ∎

Remark. Notice that by continuity we can extend the derivatives to $\{(s,t)\in t;\ u\leq s,\ v\leq t\}$.

We finally quote some properties of $\gamma_{s,t}(u,v)$ which are obvious consequences of the results we have been proving so far.

Proposition 3.3 *There exists a universal constant $C>0$ such that*

(3.8) $\displaystyle\sup_{(s,t)\in T}\ \sup_{(u,v)\leq(s,t)}\ |\gamma_{s,t}(u,v)| \leq C\,,$

(3.9) $|\gamma_{s,t}(u,v) - \gamma_{s,t}(\overline{u},\overline{v})| \leq C\,(|u-\overline{u}| + |v-\overline{v}|),\quad (u,v),(\overline{u},\overline{v})\leq(s,t),$

(3.10) $|\gamma_{s,t}(u,v) - \gamma_{\overline{s},\overline{t}}(u,v)| \leq C\,(|s-\overline{s}| + |t-\overline{t}|),\quad (s,t),(\overline{s},\overline{t})\geq(u,v),$

(3.11) $|\gamma_{s,t}(u,v) - \gamma_{\overline{s},t}(u,v) - \gamma_{s,\overline{t}}(u,v) + \gamma_{\overline{s},\overline{t}}(u,v)| \leq C\,(|s-\overline{s}|\cdot|t-\overline{t}|),$
for all $(s,t)\geq(\overline{s},\overline{t})\geq(u,v).$

References

[1] S. Aida, S. Kusuoka and D. Stroock: *On the support of Wiener functionals.* Asymptotic problems in probability theory: Wiener functionals and asymptotics, K.D. Elworthy and N. Ikeda Eds. Pitman Research Notes in Mathematics Series 284. Longman Scientific & Technical, Essex, 1993.

[2] L. Arnold and W. Kliemann: *Qualitative theory of Stochastic Systems.* Academic Press, 1983.

[3] V. Bally, A. Millet and M. Sanz-Solé: *Approximation and support theorem in Hölder norm for parabolic stochastic partial differential equations.* Annals of Probability, to appear.

[4] G. Ben Arous, M. Gradinaru and M. Ledoux: *Hölder norms and the support theorem for diffusions*. Annals de l'Inst. H. Poincaré **30**, 3, 415–436 (1994).

[5] M. Farré and D. Nualart: *Nonlinear stochastic integral equations in the plane*. Stochastic Processes and their Applications **46**, 219–239 (1993).

[6] X. Guyon and B. Prum: *Le théoreme de Girsanov pour une classe de processus à paramètre multidimensionnel*. C. R. Acad. Sc. Paris **285**, 565–567 (1977).

[7] I. Gyöngy: *The stability of stochastic partial differential equations and applications*. Stochastics and Stochastics Reports **27**, 129–150 (1989).

[8] I. Gyöngy and T. Pröhle: *On the approximation of stochastic differential equations and on Stroock-Varadhan's support theorem*. Computers Math. Applic, 65-70 (1990).

[9] I. Gyöngy, D. Nualart and M. Sanz-Solé: *Approximation and support theorems in modulus spaces*. Probability Theory and Related Fields, to appear.

[10] V. Mackevicius: *On the support of the solution of stochastic differential equations*. Lietuvos Matematikos Rinkinys XXXVI (1), 91-98 (1986).

[11] M. Mellouk: *Support des diffusions dans les espaces de Besov-Orlicz*. Preprint, 1994.

[12] A. Millet and M. Sanz-Solé: *A Simple Proof of the Support theorem for Diffusion Processes*. Séminaire de Probabilités XXVIII, J. Azéma, P.A. Meyer, M. Yor eds. Lect. Notes in Math. 1583, 36–48, Springer Verlag, 1994.

[13] A. Millet and M. Sanz-Solé: *The support of the solution to a hyperbolic SPDE* . Probability Theory and Related Fields **98**, 361–387 (1994).

[14] D. W. Stroock and S.R.S. Varadhan: *On the support of diffusion processes with applications to the strong maximum principle*. Proc. Sixth Berkeley Symp. Math. Statist. Prob. III, 333-359, Univ. California Press, Berkeley, 1972.

[15] D. W. Stroock and S.R.S. Varadhan: *On degenerated elliptic-parabolic operators of second order and their associated diffusions*. Comm. on Pure and Appl. Math. Vol XXV, 651-713 (1972).

[16] J.B. Walsh: *An introduction to stochastic partial differential equations*. École d'été de Probabilités de Saint-Flour, XIV. Lectures Notes Math, **1180**, 266–437. Springer Verlag (1986).

Statistical Dynamics with Thermal Noise

R.F. Streater
King's College, Strand,
London WC2R 2LS

Abstract

We consider how to add noise to a non-linear system in a way that obeys the laws of thermodynamics. We treat a class of dynamical systems which can be expressed as a (possibly non-linear) motion through the set of probability measures on a sample space. Thermal noise is added by coupling this random system to a heat-particle distributed according to a Gibbs state. The theory is illustrated by the Brussellator, where it is shown that the noise converts a limit cycle into a global attractor. In the linear case it is shown that every Markov chain with transition matrix close to the identity is obtained by coupling to thermal noise with a bistochastic transition matrix.

1   Motivation

Suppose that we have a dynamical system whose (non-linear) equations depend on a parameter λ, and that there is a bifurcation to periodicity or chaos as we increase λ up to a critical value λ_0. It is then interesting to add noise to the system, and to ask whether this induces the system to bifurcate earlier, at some $\lambda < \lambda_0$, or whether it delays the onset until $\lambda > \lambda_0$. It might on the other hand change the nature of the criticality, of destroy it altogether.

Here we report on a study (made with L. Rondoni [1, 2]) of the chemical system known as the Brussellator [3, 4] modified to follow the activity-led law of mass action. For certain values of the rate constants the system exhibits a limit-cycle, which is converted into a global attractor when the system is coupled to thermal noise in a particular way.

To judge from the literature [5, 6] how the noise is to by added is very much a matter of taste. One can add a random force to Newton's law of motion, to obtain a Langevin equation. One can also try "multiplicative noise", in which the random term multiplies one of the dynamical variables. The nature of the noise is also at our disposal; white noise is the most studied; however, because velocities are finite for real particles, the Ornstein-Uhlenbeck process is a more reasonable choice for the Langevin equation.

White noise is "too hot" to be a good model for noise in many physical and chemical systems; white noise exchanges energy equally fast at all frequencies. The absorption of noise of positive energy, at a rate independent of the energy, can be regarded as caused by the bombardment of the system by the random outpourings of an infinitely hot source, since the Gibbs distribution at $\beta = 0$ is a uniform distribution over positive energies. The negative-energy component of white noise must be interpreted as the emission of energy by the system, equally fast at all frequencies, and so corresponds to the case where the system under study is at infinite temperature. This is not a very general class of systems. A related problem arises in quantum theory; quantum analogues of white noise were introduced by Senitzky [7, 8], which can be made rigorous [9, 10]. But the unphysical aspects were pointed out by Kubo, in a telling intervention in the debate [11]. A model for thermal noise at finite temperature was constructed by Ford, Kac and Mazur [12], and the quantum version by Lewis and Thomas [13].

In this report I take, as the goal, the adding of noise to a system so that

- the extended system obeys the first law of thermodynamics, and

- the extended system obeys the second law of thermodynamics.

In order to discuss the first law of thermodynamics, an energy must be specified. In a random theory, the energy of the system will be a random variable, i.e. a measurable function $\mathcal{E}_c(\omega_c)$ of the configuration ω_c. The set of configurations form the sample space Ω_c of the system. In our model the noise also carries energy, a measurable function $\mathcal{E}_\gamma(\omega_\gamma)$ of the configuration ω_γ of the noise; the set of configurations ω_γ form the sample space Ω_γ. The extended system, including noise, is described by the configuration $\omega = (\omega_c, \omega_\gamma)$, which form the sample space $\Omega = \Omega_c \times \Omega_\gamma$. We assume that the total energy is the random variable $\mathcal{E}(\omega) = \mathcal{E}_c(\omega_c) + \mathcal{E}_\gamma(\omega_\gamma)$. The first law then says that the dynamics conserves the total energy.

In order to discuss the second law, we need an entropy. In a probabilistic treatment, the obvious candidate is the Gibbs-Shannon entropy S. Let $\Sigma(\Omega)$ denote the set of probabilistic measures on Ω, typically denoted p. Then we take

$$S(p) = - \sum_{\omega \in \Omega} p(\omega) \log p(\omega).$$

For this to be finite, Ω must be discrete. One formulation of the second law is that S should not decrease with time.

In setting up the structure of the theory, called *statistical dynamics*, we adopted the **field** point of view. This is in contrast to the **particle** point of view, which gives rise to the Gibbs paradox: entropy fails to be an extensive quantity, contrary to experiment. The issue is how we describe a configuration of the system in a region of space. In the particle point of view, we name the individual particles, $1, 2 \ldots$. Then we list the position x_1 of the first particle, x_2

of the second particle, etc. So a configuration is given by an ordered sequence of points $\{x_1, x_2 \ldots\}$ of points in the space (call it Λ) in which the system sits. In the field point of view, we reverse the order: we look at the point $x_1 \in \Lambda$ and say how many particles $N(x_1)$ are located there; then we look at x_2 and give $N(x_2)$, and so on. So, in the field point of view, a configuration is a map N from Λ to $\{0, 1, 2 \ldots\}$. In 1 c.c. of material, there are 10^{23} molecules, so there are $(10^{23})!$ more particle configurations than field configurations. This gives rise to an enormous extra entropy in a particle theory (called the entropy of mixing by Gibbs) not observed in physics or chemistry. The field point of view gives rise to quantum statistics, whereas the particle point of view leads to Boltzmann statistics (which is appropriate for the study of ball bearings, but not of atoms or molecules). In order for entropy to be extensive, it must be additive over disjoint subsets $\Lambda_1 \subseteq \Lambda$ and $\Lambda_2 \subseteq \Lambda$. For this, the configuration space for the union of the regions must be the product of those for Λ_1 and Λ_2 separately (and the probability distributions for the two regions must be independent). The first requirement leads to the *local structure* postulated in the next section:

$$\Omega(\Lambda) = \prod_{x \in \Lambda} \Omega_x.$$

The configuration space in the field point of view is automatically of this form; this structure is impossible in the particle point of view.

In this article, following [14, 15, 16] I develop the theory of classical statistical dynamics, starting with isolated dynamics in Section (2). This is devised so that the two laws of thermodynamics hold in their usual form: entropy increases and energy is conserved. The theory leads to non-linear dynamical equations. In Section (3), isothermal and isopotential dynamics are described. In isothermal dynamics, heat flows ensure that the temperature is kept constant at all points of Λ, and as a consequence, energy is not a constant of the motion. Instead of entropy's increasing, the Helmholtz free energy F decreases. In isopotential dynamics, matter flows are postulated, which keep a chemical potential constant. Then atomic number is not conserved; usually the temperature is kept fixed as well, and then it is the Gibbs free energy, G, rather than F, which decreases. In these three cases the thermodynamic function, S, F or G, acts as a Lyapunov function for the dynamics, and the system usually converges to equilibrium at large times. When more that one chemical potential is kept fixed, or different parts are kept at different temperatures, we say that we have a *driven* system. A driven system can exhibit varied behaviour, such as limit cycles, convergence to equilibrium or convergence to a stationary but not static limit at large times. This is illustrated in Section (4), where a modified version of the chemical system known as the Brussellator is described; it converges to equilibrium at non-zero temperature, but to a limit cycle at absolute zero. One may interpret the heat as noise, coupled in a particular way so as to be consistent with thermodynamics. We may then say that the limit cycle is destroyed by noise.

2 Statistical Dynamics

We [14, 15] take space to be a finite set Λ and to each $x \in \Lambda$ we associate a sample space Ω_x. Then we take

$$\Omega(\Lambda) = \prod_{x \in \Lambda} \Omega_x.$$

In the same way we define the configuration space for any subset Λ_0 of Λ by restricting the product to $x \in \Lambda_0$. If there are chemical species $A, B \ldots$ then each Ω_x is the product over chemical species:

$$\Omega_x = \prod_A \Omega_{x,A}.$$

Finally, each $\Omega_{x,A}$ is $\{0, 1 \ldots\}$, the "classical Fock space". The integer $j \in \Omega_{x,A}$ gives us the number of molecules of species A at x. It is possible to take another form for $\Omega_{x,A}$, such as $\{0, 1\}$; this particular choice expresses that at most one particle of type A can occupy each point in space. In either case, Ω is a discrete space.

A configuration $\omega \in \Omega$ is a $|\Lambda|$-tuple

$$\omega = (\omega_{x_1}, \omega_{x_2}, \ldots, \omega_{|\Lambda|}).$$

A state of the system is given by a probabilistic measure p on Ω. The set of such measures form a convex set, denoted $\Sigma(\Omega)$. Because of the product structure of Ω, we can define, for any subset Λ_0 of Λ, the marginal map M_{Λ_0} : $\Sigma(\Omega) \to \Sigma(\Omega_{\Lambda_0})$ by

$$M_{\Lambda_0} p(\omega_0) = \sum_{\omega'} p(\omega_0, \omega').$$

Here, $\omega = (\omega_0, \omega')$, with $\omega_0 \in \Omega_{\Lambda_0}$ and $\omega' \in \Omega_{\Lambda - \Lambda_0}$. If Λ_0 consists of a single point x, we write M_x for M_{Λ_0}.

We say that a state p is independent over Λ if

$$p(\omega) = \bigotimes_{x \in \Lambda} M_x(p).$$

In this case the entropy is extensive:

$$S(p) = -\sum_{x \in \Lambda} p_x(\omega_x) \log p(\omega_x) = \sum_{x \in \Lambda} S_x(p_x).$$

We shall denote by $\mathcal{A}$ the abelian C^*-algebra of complex bounded random variables on Ω, furnished with the sup-norm.

Note that the function $f(\omega) = 1$ lies in $\mathcal{A}$ and acts as a unit, which we denote by 1. By introducing $\mathcal{A}$ we make the set-up parallel to the theory of quantum processes, the only difference being that there we choose a non-abelian C^*-algebra. The set Σ can be identified with the subset of normalised, positive elements of $\mathcal{A}^\dagger$, the predual of $\mathcal{A}$; in quantum theory, this is called the set of normal states on $\mathcal{A}$.

Time-evolution is specified by a stochastic map T acting on $\mathcal{A}$; a stochastic map is a normal linear map satisfying [17]

- $T1 = 1$

- If $f(\omega) \geq 0$ for all ω, then $(Tf)(\omega) \geq 0$ for all ω.

We can throw the action of T onto the predual space $\mathcal{A}^\dagger$ by duality; we denote the dual action by $T^\dagger$. Since T is stochastic, $T^\dagger$ maps Σ to itself ([17], Lemma 2.2). Using the notation $\langle \bullet, \bullet \rangle$ for the bilinear form defining the duality, $T^\dagger$ is defined by its property

$$\langle T^\dagger p, f \rangle = \langle p, Tf \rangle \text{ for } p \in \mathcal{A}^\dagger, \ f \in \mathcal{A}.$$

The map $T^\dagger$ gives us one time-step in the discrete-time dynamics:

$$p(t+1) = T^\dagger p(t), \qquad t = 0, 1 \ldots.$$

Now, to obey the second law of thermodynamics, we need $S(T^\dagger p) \geq S(p)$ for all p, and the necessary and sufficient condition for this is that T be *bistochastic*: regarding $p = p(\omega)$ as an element of $\mathcal{A}$, we require that

$$\sum_\omega (Tp)(\omega) = \sum_\omega p(\omega).$$

We restrict our attention to bistochastic maps, thus arriving at the second law. A theorem of Birkhoff [18] asserts that any bistochastic matrix is a convex combination of permutation matrices; indeed, the $n \times n$ permutation matrices are the extreme points of the convex set of $n \times n$ bistochastic matrices. A permutation of the points of Ω is the discrete analogue of a non-dissipative dynamics, since a simple version of Mackey's lifting theorem says that any automorphism of the algebra $\ell^\infty(\Omega)$ is induced by a measure-preserving map on Ω; for a discrete space with counting measure, such a map is a permutation. Thus a Markov chain generated by a bistochastic map can be regarded as a random mixture of conservative dynamics: it is the discrete version of a theory with a random Hamiltonian. It is easy to see that the Birkhoff decomposition into permutations is unique, and that $T^\dagger$ is decomposed into the mixture using the same weights as T, but with the inverses of the permutations that enter T. This result is a version of the law known as *microscopic reversibility*.

To discuss the first law, we must specify an energy function $\mathcal{E}$ on Ω. Given $\mathcal{E}$, let

$$\Omega_E = \{\omega \in \Omega : \mathcal{E}(\omega) = E\}.$$

This is called the energy-shell of energy E. Thus Ω is the disjoint union of energy-shells:

$$\Omega = \bigcup_E \Omega_E.$$

We shall need the assumption $|\Omega_E| < \infty$ for all E. This is not unduly restrictive in finite volume. Let χ_E be the indicator function of Ω_E. We can regard χ_E as an element of $\mathcal{A}$. Then the first law, which says that T must map each energy-shell into itself, can be expressed by the requirement

$$T\chi_E = \chi_E \text{ for all } E.$$

It follows from this, and Birkhoff's theorem, that $T^\dagger \chi_E = \chi_E$, where now χ_E is regarded as the uniform measure on Ω_E. We also have the easy result: if p is a state of finite mean energy, then so is $T^\dagger p$, and

$$\langle p, \mathcal{E} \rangle = \sum_\omega p(\omega)\mathcal{E}(\omega) = \sum_\omega (T^\dagger p)(\omega)\mathcal{E}(\omega) = \langle T^\dagger p, \mathcal{E} \rangle.$$

In this sense we have achieved the first law.

With $\mathcal{E}$ specified, we define the Gibbs canonical state at beta β as usual

$$p_\beta(\omega) = Z_\beta^{-1} e^{-\beta\mathcal{E}(\omega)}.$$

This is also called the thermal state. We need to assume that the partition function

$$Z_\beta = \sum_\omega e^{-\beta\mathcal{E}(\omega)}$$

converges, which puts a condition on the growth of the number of states as energy increases. In particular, the number of points on each energy-shell must be finite. If p_β exists and has finite mean energy, then it has finite entropy, and is the state with the maximum entropy, conditional on the given mean energy. If T conserves energy, then the canonical states are fixed points of $T^\dagger$.

One of the aims of statistical dynamics is to show that the thermal state p_β is the limit of $T^{\dagger n}p$ as $n \to \infty$, for any initial state p of finite energy. In good cases, the parameter β is determined by the requirement that $\langle p_\beta, \mathcal{E} \rangle = \langle p, \mathcal{E} \rangle$, the initial mean energy. The trouble is that an energy-conserving T does not mix up the different energy shells: as time progresses, no probability leaks from one energy-shell to another. If T is ergodic when restricted to an energy-shell, then $T^n p$ converges to the uniform distribution on each energy shell, but the canonical relation between the probabilities of points in shells of different energy will not hold in general. The uniform distribution on a single energy-shell Ω_E is called the microcanonical state, and it expresses the law known as the equipartition of energy.

To explain how we can reach the canonical state, rather than the microcanonical state, in a theory that conserves energy, two philosophies are possible. In the first, use is made of the local structure of the sample space. Suppose that we choose T which exchanges particles between nearest neighbours x, y, while conserving total energy. This provides a link between the energy-shells in Ω_x and Ω_y, and leads to a diffusion term in the dynamical law. By forming chains, we can hope to mix up all the energy in any finite region Λ_0. In this dynamics, the overall energy-shell $\Omega_{\Lambda,E}$, where E is the initial energy, is not mixed with those with a different value of the total energy; the system converges to the overall microcanonical state. The theorem on the "equivalence of ensembles" says [16] that the microcanonical state on $\Omega_{\Lambda,E}$, when restricted to $\mathcal{A}_{\Lambda_0}$, is very close to the canonical state with energy $E|\Lambda_0|/|\Lambda|$, as Λ and E go to infinity with their ratio fixed. That is, the microcanonical state looks, locally, like a canonical state.

The trouble with this philosophy is that it does not include a very interesting situation: this is where the system has a local structure, and very rapidly moves to a state in local thermal equilibrium, so that in a small region it has a well-defined local temperature. We therefore adopt an alternative point of view.

This uses the fact that any theory has hidden degrees of freedom, not explicitly mentioned; for example, when we do the infrared renormalisation of electrodynamics the soft modes of the electromagnetic field are replaced by a single parameter, the energy threshold below which soft photons cannot be detected. As another example, the statistical mechanics of atoms is often studied while ignoring the excitation states with very high energy. The degrees of freedom that are going to be ignored are often called "the fast variables". The idea is that the speed of the ignored reactions is so fast that on the time-scale of the remaining processes, they have already reached their asymptotic limit for large times. Please note, it is not possible to ignore the energy of these modes. It is a tenet of statistical dynamics that it is information that is lost during dissipation, not energy. Otherwise, the first law could not be exact. Of course, dissipated energy reappears as heat. So to take this energy into account, we introduce a sample space, of "heat particles", at each point $x \in \Lambda$, denoted by $\Omega_{\gamma,x}$. We choose a random variable $\mathcal{E}_{\gamma,x}$ to represent the energy of all the hidden degrees of freedom at x. Then we put

$$\mathcal{E}_\gamma = \sum_x \mathcal{E}_{\gamma,x}.$$

and we denote by s_β the canonical state of the heat-particles:

$$s_\beta(\omega_\gamma) = Z^{-1}\exp\{-\beta\mathcal{E}_\gamma(\omega_\gamma)\} \tag{1}$$

In chemical kinetic theory it is assumed that the kinetic energy of the molecules is thermalised, so $\mathcal{E}_\gamma$ then represents all the kinetic energy of all the molecules. It is, in chemistry, far too detailed a description to apportion the heat-energy among the various individual particles; we might apportion it among the different types of particle, for example, if we wanted to describe the dynamics of two gases at different temperatures which are put together. We might also want to distinguish between the heat in phononic modes of different frequency. To allow for these possibilities, it is enough to take each $\Omega_{\gamma,x}$ to be a product of one or more copies of classical Fock space, $\{0, 1, \ldots\}$, where the integer j specifies the number of quanta of heat-particles. In analogy with the quantum oscillator, we take the energy of j quanta to be proportional to j. When we take only one type of heat-particle, of energy κ, we shall denote its probability distribution by $s_x(j)$. Then we put

$$\Omega_x = \Omega_{c,x} \times \Omega_{\gamma,x} \quad \text{so that} \quad \Omega = \Omega_c \times \Omega_\gamma.$$

The suffix c refers to "chemicals" and the suffix γ to the gamma ray or photon, although any other form of heat, such as phonons, are to be included. Thus, a configuration is given by a number of density-fields, $j_{A,x}, j_{B,x} \ldots j_{\gamma,x}$; we treat the heat-particle just like any other species.

It is reasonable to postulate that the total energy is the sum of the chemical energy $\mathcal{E}_c$ and the heat-energy $\mathcal{E}_\gamma$. We do not expect any energy of interaction between them; (there is not usually any interaction between kinetic and potential energy). So we postulate that

$$\mathcal{E} = \mathcal{E}_c + \mathcal{E}_\gamma.$$

In the simplest models, we shall assume that there is no interaction between the different points of Λ, so:

$$\mathcal{E}_c(\omega_c) = \sum_x \mathcal{E}_{c,x}(\omega_{c,x}).$$

Interactions between neighbours, as in the Ising model, do not satisfy this, and are harder to solve. Having decided on the form of the energy, we choose the first step of the dynamics to be $p \mapsto T^\dagger p$, where T is bistochastic and conserves the total energy, potential plus heat. In line with the idea that the stochastic nature of T is to represent the omission of the fast variables, we can interpret T as the scattering process, which is the mean of energy-conserving scattering by random interactions. This is only the first step in constructing the desired dynamics, since it does not mix the energy-shells.

We can achieve the mixing of the energy-shells by adding more randomness to the dynamics, using an idea which is mathematically similar to the Stosszahlansatz of Boltzmann. Since the sample space has the product structure $\Omega = \Omega_c \times \Omega_\gamma$, we can define the marginal maps M_c and M_γ as usual:

$$(M_c p)(\omega_c) = \sum_{\omega_\gamma} p(\omega_c, \omega_\gamma), \text{ with } p \in \Sigma(\Omega),$$

$$(M_\gamma p)(\omega_\gamma) = \sum_{\omega_c} p(\omega_c, \omega_\gamma).$$

We then define the corresponding *Boltzmann map* B to be the non-linear map $B : \Sigma(\Omega) \to \Sigma(\Omega)$

$$Bp = M_c p \otimes M_\gamma p.$$

One time-step in the *isolated dynamics* is then the combined map $\tau p = BT^\dagger p$. It is known that the map B does not decrease the entropy, and so the same is true of τ. More, because the total energy is the sum of potential and heat energy, the map B does not change the mean energy. So if T is energy-conserving, so is τ. The map τ then gives a dynamics obeying the laws of thermodynamics; it is a non-linear map on the set Σ of probabilities. In many cases this leads to the canonical state in the limit of large times [19].

A model constructed in this way carries at any time the detailed distribution of the energy among the states of the heat-particles, and this is far more information than is usual in non-equilibrium theories in Physics and Chemistry. In fact, very little error will be made by a much coarser picture, which is described by fewer parameters. It has been remarked by Fowler [20] that the transfer of heat within a local region x happens 10^{10} times faster than most chemical reactions. So we might expect the state of the heat-particle at x to be almost instantly thermalised. This is known as *the hypothesis of local thermal equilibrium, or LTE*. This can be formulated mathematically as follows. Given a state $s \in \Sigma(\Omega_\gamma)$, it defines what is known as the mean density of kinetic energy, namely, the real-valued function on Λ, $x \mapsto \langle s, \mathcal{E}_{\gamma,x} \rangle$. Let us denote by Q the non-linear map that replaces a state $s \in \Sigma(\Omega_\gamma)$ of the heat-particle by the product of local thermal states at each x:

$$Qs = \otimes_x s_{\beta(x)}.$$

Here, $\beta(x)$ is the beta such that $s_{\beta(x)}$ has the same density of kinetic energy at x as s:

$$\langle Qs, \mathcal{E}_{\gamma,x} \rangle = \langle s_{\beta(x)}, \mathcal{E}_{\gamma,x} \rangle = \langle s, \mathcal{E}_{\gamma,x} \rangle.$$

The map Q is a primitive form of the quasi-free map used in quantum statistical dynamics [21, 22]. Since the entropy of a product state such as Qs is not less than that of any other state with the same marginals, and the entropy of a thermal state is not less than that of any other state with the same mean energy, it follows that the map $Q : \Sigma(\Omega_\gamma) \to \Sigma(\Omega_\gamma)$ does not decrease the entropy. It actually increases the entropy unless $s \in \Sigma(\Omega_\gamma)$ is such that $Qs = s$. It also conserves the mean kinetic energy.

We implement the hypothesis of LTE by combining the map $T^\dagger$ with the map Q. Thus one time-step in the dynamics of the "slow" variables is given by

$$p_c \otimes s \mapsto M_c \left(T^\dagger(p_c \otimes s) \right) \otimes Q M_\gamma \left(T^\dagger(p_c \otimes s) \right). \tag{2}$$

At each time, the mean kinetic energy determines $\beta(x,t)$, the inverse temperature at position x and time t; the variation of β in space-time is determined by T. The amount of each chemical at x is a random variable and its distribution changes with time according to Eqn.(2). According to the way they are constructed, these dynamics conserve the total mean energy, including heat, and increase entropy, which therefore acts as a Lyapunov function for the dynamics. We call the map (2) the isolated dynamics with LTE. Like the isolated dynamics without LTE, it is non-linear, and it converges to equilibrium even faster.

It is possible to construct a sort of "non-linear" stochastic process by making use of the iterates of the time-step given by Eqn. (2); we intend to exploit the mildness of the non-linearity. The equation is quadratic in p_c, the configuration of the system at the previous time. The occurrence of p_c in the second factor, $QM_\gamma \left(T^\dagger (p_c \otimes s) \right)$, tells us how p_c influences the change in s in the time-step, and with the LTE map, this change in s is summarised by the change in $\beta(x,t)$. Having chosen $p^{(0)}$, the state at time $t = 0$, we iterate the map to find $\beta = \beta(x,t)$, and then note that the map given by the first factor

$$p_c \mapsto M_c \left(T^\dagger (p_c \otimes s_\beta) \right)$$

is linear in p_c. By dual action, it therefore defines a stochastic map $T_c(t)$ on $\mathcal{A}_c$ for each t. We can then define an orbit through $\mathcal{A}_c$ by

$$A \mapsto A(t) = T_c(t) T_c(t-1) \cdots T_c(1) A.$$

This can be used, for example, to define the many-time correlation functions of a random variable A as $\langle p^{(0)}, A(t_1) \ldots A(t_n) \rangle$. In the process thus defined, there are no random variables associated with the heat-particle; they are said to be "slave" variables, replaced by the parameters $\beta(x,t)$. It makes little sense to ask for the many-time correlations of random variables involving functions of ω_γ; these have been randomised by the map Q after each time-step.

It is clear that if T factorises, in that $T = T_c \otimes T_\gamma$, then the construction reduces to a Markov chain on each factor $\mathcal{A}_c$ and $\mathcal{A}_\gamma$. If T nearly factorises, we say that the coupling to the noise is weak, and then the non-linear process is nearly that given by the stochastic map T_c.

There is one further randomisation that simplifies the description of the dynamics even more, but which does not spoil the first and second law. In nuclear, atomic and molecular physics there are laws of conservation besides that of energy; charge is conserved, and so is baryon number, and, in molecular physics, atomic number too. Such conserved numbers are called *generalised charges*. We shall represent a typical charge by the random variable $\mathcal{N}$. In describing the statistical dynamics of a system with a conserved charge, the bistochastic map $T^\dagger$ must be chosen to conserve this charge. We do this in the same way as we did to ensure that energy is conserved in the previous paragraph. Thus, we divide Ω_c up into "charge shells", defined in a similar way to energy shells:

$$\Omega_{c,N} = \{\omega_c \in \Omega_c : \mathcal{N}(\omega_c) = N\}.$$

We require that T maps each charge-shell to itself. Then as for energy, we show that the mean value of $\mathcal{N}$ is conserved in time. Since the energy is also to be conserved, T must map the intersection of a charge-shell and an energy-shell to itself. The isolated dynamics obtained from an energy- and charge-conserving map T cannot lead us out of the energy-charge-shell

$$\Omega_{c,E,N} = \Omega_{c,E} \cap \Omega_{c,N},$$

and the isolated dynamics with LTE cannot lead us out of the charge-shell.

In a theory with a conserved charge, the equilibrium state is often described by the *grand canonical state*. This is the probability distribution defined on Ω by two parameters, the beta β and the chemical potential μ by

$$p(\omega_c, \omega_\gamma) = Z_{\beta,\mu}^{-1} \exp\left\{-\beta\left(\mathcal{E}(\omega) - \mu\mathcal{N}(\omega_c)\right)\right\}.$$

The normalisation constant $Z_{\beta,\mu}$ is called the grand partition function, and will exist if each energy-charge-shell has finitely many points and the multiplicity of states for large E and N does not grow too fast. Thus, if the atomic number is unbounded, we require that $\mu < 0$. The grand canonical state is a mixture of states with different values of $\mathcal{N}$, the relative weight given to value n being $e^{\beta\mu n}$. It can be proved to be the state having the maximum entropy among all states with a given mean energy and mean value of $\mathcal{N}$. It is clear that if the initial state has an exact number of particles carrying the quantum number N, then we cannot reach the grand canonical state as t goes to ∞ if T conserves N, unless we do something more. It is believed that the grand canonical state is the correct equilibrium state, and will be reached for large times, whenever each local region exchanges the quantum number $\mathcal{N}$ extremely rapidly with its neighbouring regions, but without any net flow. This can be modelled by introducing more randomness, in a way analogous to the LTE map Q which we now call Q_γ: we apply the map which replaces the marginal state $p_{c,x}$ after any time-step, with the (unique) state of the form

$$p_{\mu(x)}(\omega_{c,x}) = \text{const.} \exp\{\beta(x)\mu(x)\mathcal{N}(\omega_{c,x})\}. \tag{3}$$

having the same mean value of $\mathcal{N}$ as $p_{c,x}$. Here, the local $\beta(x)$ is determined by the density of kinetic energy, as before. Denote by Q_c the map taking $p_{c,x}$ to the state (3). It is clear that Q_c conserves the mean $\mathcal{N}$, and does not decrease the entropy. If, as is usual, the internal energy of the chemical is proportional to the number of particles, then the map Q_c also conserves mean energy. Define the map Q by

$$Q = Q_c \otimes Q_\gamma.$$

The dynamics is then given by the combined map

$$p \mapsto T^\dagger p \mapsto \otimes_x M_x T^\dagger p \mapsto \otimes_x Q M_x T^\dagger p.$$

It T is ergodic on the energy-charge shell, then this map, when iterated, leads to the grand canonical state at large times. At each time, the state is a product of local states of the form (1) and (3), and is described by the values of the fields $\beta(x)$ and $\mu(x)$. This is a much simpler description than is needed for the complete specification of the state $T^\dagger p$.

If there are several quantum numbers $\mathcal{N}_1, \mathcal{N}_2 \ldots$, not all conserved by T, then it is often a good model of the dynamics to assume that they are all conserved by the fast variables, which cause the particles to be exchanged rapidly from region to region without any net flow. Diffusion and reaction in the theory is to be described by the slow variables, whose dynamics is governed by T. We are thus led to generalise the map Q to the case of several charges, $\mathcal{N}_i$; it takes the local state $p_{c,x}$ into the product of states $p_{c,\mu_i(x)}$. This map introduces enough extra randomness to mix up the charge-shells, and leaves us with a tractable theory. It is this simple version that we use to describe the Brussellator in Section 4.

This model has the philosophical difficulty that energy and particle number are only conserved on average. There is a non-zero probability, which is extremely small if Λ is large, that a measurement of the number of particles at time $t + 1$ will not be the same as at time t; this difficulty does not show up when we consider the dynamics of the mean values, which leads to a theory obeying both laws of thermodynamics. We can escape this philosophical dilemma by noting that any system confined within a container has a slightly ambiguous total number of particles, in that atoms adhering to the inside of the walls will enter the system at random, and then will be adsorbed by the walls. The fluctuations in the model can be regarded as describing this.

In the next Section we consider how such a model can be modified so that it describes an isothermal, isopotential or driven system.

3   The Isothermal and Isopotential Dynamics

When chemical or physical reactions take place, heat is produced or used up, tending to change the temperature. Ostwald invented a thermostat, that withdrew heat from, or added heat to, the reactor, keeping the reaction at (approximately) the same temperature. We say that the reactions are then occurring under isothermal conditions. The contraption keeping the temperature fixed is called a heat-bath. Under isothermal conditions, both laws of thermodynamics must be modified, as heat and entropy flow into and out of the system, and must be accounted for.

We can build a very simple mathematical model of a heat-bath by using the idea of a heat-particle. Suppose that at time t the system is in the state $p = p_c \otimes s_\beta$, where s_β is the canonical state of the heat-particles at beta β. Then after the interaction, the state is $T^\dagger p = T^\dagger(p_c \otimes s_\beta)$. This is no longer at thermal equilibrium, and an energy, dq say, has been transferred from Ω_c to Ω_γ. This can happen even when T is energy-conserving, since then it conserves the total energy $\mathcal{E} = \mathcal{E}_c + \mathcal{E}_\gamma$, not each part separately. One time-step in the isolated dynamics with LTE is the obtained by applying the marginal maps and thermalising the heat-particle with the map Q_γ:

$$p(t + 1) = M_c \left(T^\dagger(p_c \otimes s_\beta) \right) \otimes Q_\gamma M_\gamma \left(T^\dagger(p_c \otimes s_\beta) \right). \tag{4}$$

This map is clearly quadratic in p_c. The map Q_γ restores the heat-particle to a canonical state, but at a different beta, to take into account the addition of the energy dq to Ω_γ. Under isothermal conditions, the state is then simply restored, by hand, to the original beta:

$$p \mapsto \tau_\beta p = M_c \left(T^\dagger(p_c \otimes s_\beta) \right) \otimes s_\beta. \tag{5}$$

This defines a linear map τ_β on $\Sigma(\Omega_c)$, and is therefore described by a Markov chain of the usual kind. We show in the appendix that every Markov matrix, near the identity, is obtained in this way from a suitable isolated dynamics. If T is symmetric (in the scalar product given by the counting measure), then τ_β obeys detailed balance in the form advocated by Glauber [23], but in general, where T is bistochastic but not symmetric, a more general form of detailed balance holds [24].

The isothermal dynamics takes place at constant β, in contrast with the isolated dynamics, which takes place at constant $\langle \mathcal{E} \rangle$. We get one from the other by a Legendre transform of the whole orbit. We can explicitly compute dq, the heat that must be subtracted from the system to maintain constant temperature. We then find that there is a change in entropy of at least $-dq/T$ in addition to the entropy created by T [25, 26]; I call this the "free-energy theorem"; it is in agreement with elementary thermodynamics, in which entropy can be defined as the total change $\sum -dq/T$ in creating the current state from absolute zero by adding heat slowly in small doses. In our more detailed theory, we know that

$$dq = -d\langle \mathcal{E}_c \rangle \qquad \text{(by energy conservation)}$$

and this leads to the result that in isothermal dynamics, the thermodynamic function

$$\Psi(t) = S(p(t)) - \beta\langle p(t), \mathcal{E}_c \rangle \tag{6}$$

(related to the Helmholtz free energy F by $F = -\beta\Psi$) increases along the orbit [25]; it is bounded above and is therefore a Lyapunov function. This leads to *the fundamental theorem of chemical kinetics*: in any isothermal system, in which T mixes the charge-energy-shells and the conserved charged have sharp values, the state converges for large time to a canonical state at the beta of its heat-bath. This state is uniform over each charge-energy-shell.

In the same way, we can construct *isopotential* dynamics for each generalised charge $\mathcal{N}$. We start with the sample space of the chemicals, Ω_c, and of the heat particles Ω_γ, and form the product $\Omega = \Omega_c \times \Omega_\gamma$. We assume that Ω_c is the product of several classical Fock spaces, whose elements give the numbers of particles of various types, $A, B \ldots$. Suppose that one of the particles, say A, is allowed to enter or leave the system after each time-step, so that the chemical potential of that species of particle is maintained at a fixed value, say μ_A. Let T be a bistochastic map on Ω conserving energy and the generalised charge $\mathcal{N}$. In our models, $\mathcal{N}$ is a linear sum of the particle numbers $\mathcal{N}_A, \mathcal{N}_B \ldots$, and we assume that $\mathcal{N}_A$ enters $\mathcal{N}$ with coefficient equal to 1. This can always be arranged by scaling if $\mathcal{N}$ contains $\mathcal{N}_A$ with non-zero coefficient. We write $\Omega_c = \Omega_d \times \Omega_A$, where d denotes the sample space of the chemicals omitting A. Choose a chemical potential μ_A and denote by p_A the canonical state

$$p_A(\omega_A) = \text{const.} \exp\{\beta \mu_A \mathcal{N}_A(\omega_A)\}. \tag{7}$$

The isopotential dynamics (at beta β) is then the (non-linear) dynamics on $\Sigma(\Omega)$ given by $\tau_{A,\beta}$:

$$\tau_{A,\beta} p = M_d \left(T^\dagger (p_d \otimes p_A \otimes s_\beta) \right) \otimes p_A \otimes s_\beta. \tag{8}$$

This defines a linear mapping on $\Sigma(\Omega_d)$ by dropping the last two factors. Because the number $\mathcal{N}_A$, whose chemical potential is being kept fixed in the dynamics, is part of the charge $\mathcal{N}$ which is conserved by T, we can show by the same method as for the function F that the Gibbs free energy $G = F + \mu\langle\mathcal{N}\rangle$ is a Lyapunov function for the isopotential dynamics: G is bounded and decreasing along the orbit. Such a system cannot show permanent oscillations, at least if β and μ are finite.

In chemistry, systems in which only energy is exchanged with the outside and the temperature is held fixed, are called "closed"; those in which energy, and one or more types of particle, are exchanged, at a fixed temperature and chemical potential, are called "open". I prefer the less ambiguous words "isothermal" and "isopotential".

Because of the existence of a Lyapunov function for isopotential systems, the fundamental theorem of chemical kinetics holds, and such systems converge at large times to grand canonical states. Such models therefore cannot describe systems of chemicals which show permanent oscillations; the Beloussov-Zabotinsky reaction comes from a clever mixture of organic dyes and other chemicals, and exhibits oscillating concentrations with a period of about one second, and continuing for up to half an hour. Prigogine's group in Brussels have modelled this by a system of four reactions, which have permanent oscillations [3, 4]. This dynamical system is called the Brussellator. From what we have said, it cannot be obtained from an isothermal isopotential chemical system; it is a new sort, a *driven* system. A variant of this model is presented in the next section.

We can get a driven system by keeping two chemical species, which can transmute into each other, at fixed chemical potentials that are not at equilibrium. For example, consider $\Lambda = \{1, 2, 3\}$, and let $\Omega_1 = \Omega_2 = \Omega_3$, all three being classical Fock space of a single particle. Suppose that T causes a hopping from site 1 to site 2 and from site 2 to site 3 and vice versa with equal probability. Left to itself, the system will equilibrate, with the same population at all points of Λ. But if we keep 1 at a higher chemical potential than 3, then we will feed the chemical in at 1 and take it out at 3, and the system will tend to *stasis* for large times, but not to the canonical state. This is a model for diffusion from a region maintained at high density to a region maintained at low density. If the "chemical" is the heat particle, then this is a model for heat conduction. In this model, we could also regard the three variables at the points of Λ as say three isomers of a chemical, which can make transitions into each other and sit in a single cell: the mathematics is the same. The original Brussellator is of this type; it has only one cell and there are six chemical types and four reactions, leaving two conserved quantities. But four of the chemicals are driven, in that their chemical potentials are maintained at fixed values. Since the number of driven variables is greater than the number of conserved variables, we have a more elaborate example of the above situation: the driven variables would, if left alone, reach a particular equilibrium, but are prevented from doing this. This is why the argument leading to a Lyapunov function [25] cannot be completed; it then turns out that the model exhibits permanent oscillations.

We choose to modify this model a bit, because it violates the law of chemistry known as microscopic reversibility: all back reactions are assumed to have zero rate, contrary to the Arrhenius relation. In [1, 2] we have modified the Brussellator in three ways. First, the dynamics is activity-led, rather than density-led. Secondly, we have introduced a heat-particle, making seven chemicals and three conserved charges, including heat. And finally, only two of the original chemicals are driven, which together with the isothermal condition, makes three, the same as the number of conservation laws. This allows us to construct a Lyapunov function, Ψ^{-1}, which is decreasing and bounded below at positive temperatures. The system duly converges to equilibrium. At zero temperature, which allows $p \in \Sigma$ to lie on the boundary of the state space, the putative Lyapunov function Ψ becomes infinite, and it turns out that we get permanent oscillations very similar to those of the original model. It is clear that for driven systems a lot of questions remain open.

4   The Brussellator with Thermal Noise

In the original model [3], Λ consists of a single point, representing a single cell which is well stirred. There are six chemical species A, B, D, E, X, and Y, which react according to the unidirectional processes

$$A \rightarrow X \tag{9}$$
$$B + X \rightarrow Y + D \tag{10}$$
$$2X + Y \rightarrow 3X \tag{11}$$
$$X \rightarrow E \tag{12}$$

The equation of motion was taken to be that given by the law of mass action, which says that the rate of a reaction is proportional to the product of the concentrations of the participating chemicals. This is the "density-led" law. No back-reactions occur in the model, which violates the law of chemistry known as microscopic reversibility; this is well confirmed experimentally, and requires that the forward and backward rates of a reaction should be proportional, related by the Arrhenius factor. The law relates to a system that is not interfered with; in a driven system, the law of Arrhenius might not hold. For example in reactions (10) and (12) the ions D and E might be insoluble, and might be precipitated. Then if they were removed by human hand, the back reaction would not occur. However, it is not easy to prevent the back-reactions in (9) and (11) from occurring. In the Brussellator, A and B are supplied, and are kept at constant concentration, and D and E are removed, and kept at zero concentration. We are left with a dynamical system with two variables, $X(t)$ and $Y(t)$, representing the concentrations of X and Y, and obeying two coupled first-order differential equations with a cubic non-linearity.

To get a model satisfying the laws of chemistry as well the laws of thermodynamics, we modify the Brussellator slightly [1, 2], so that it falls within the class of models discussed in the previous section. We start with the same reactions, but assume that all four reactions are exothermic, so that heat is emitted on the right-hand sides. For simplicity, we assume that the heat of reaction, denoted γ, is the same in all four reactions. In deference to Arrhenius' law, we allow back-reactions to occur; so we consider the reversible reactions

$$A \;\rightleftharpoons\; X + \gamma \tag{13}$$
$$B + X \;\rightleftharpoons\; Y + D + \gamma \tag{14}$$
$$2X + Y \;\rightleftharpoons\; 3X + \gamma \tag{15}$$
$$X \;\rightleftharpoons\; E + \gamma \tag{16}$$

The rates of these reactions are assumed to be given by rate-constants $k_1, k_2, k_3,$ k_4 respectively. We consider the driven isothermal dynamics in which γ is kept in a canonical state at constant beta, and A and B are kept at constant chemical potentials. We found that we did not need to sweep away D and E to get an oscillating system, so we let them evolve without human interference. Thus our model has 7 species including the heat-particle, and four reactions. It therefore has three conserved quantities. And in the dynamics, three quantities are kept constant. This leads to the existence of a Lyapunov function and a free-energy theorem. The model uses the sample space

$$\Omega = \Omega_A \times \Omega_B \times \Omega_D \times \Omega_E \times \Omega_X \times \Omega_Y \times \Omega_\gamma. \tag{17}$$

Here, each factor is classical Fock space $\Omega_A = \{0, 1, 2 \ldots\}$. The state at each time is that given by the product of canonical states

$$p = p_A \otimes p_B \otimes p_D \otimes p_E \otimes p_X \otimes p_Y \otimes s_\beta, \tag{18}$$

where each chemical factor is determined by the corresponding chemical potential according to (7) and s_β is given by (1). One time-step in the isolated dynamics is given by a choice of a bistochastic T that allows each of the four reactions to take place, and also the reverse reaction. We denote a typical element of Ω by its occupation numbers $j_\bullet$:

$$\omega = \left(j_A, j_B, j_D, j_E, j_X, j_Y, j_\gamma \right). \tag{19}$$

For example, if the time-step is Δt, then $T^\dagger$ will possess an entry $k_1 \Delta t$ in the matrix with row

$$\left(j_A - 1, j_B, j_D, j_E, j_X + 1, j_Y, j_\gamma + 1 \right)$$

and column ω given by (19). This expresses that the rate of the forward process $A \to X + \gamma$ is k_1. Similarly, the other non-diagonal entries to the matrix are determined. The diagonal elements are fixed by the requirement that the columns add up to 1. This puts restrictions on the allowed values of the parameters $k_1 \Delta t, \ldots$: they cannot lie outside the *bistochastic region*. By choosing the forward and backward rates to be equal, we get a symmetric stochastic matrix, which is obviously bistochastic. The value of the rate constants $k_1, \ldots$ could depend on the occupation numbers of the sample point. If we choose the simplest case, where they are independent of this, we get the case studied in [27], which gives us the activity-led equations of motion as follows: for each species, put $a = N/(1 + N)$, the activity; then by following the action of $T^\dagger$ by the LTE map Q, and taking an infinitesimal time step, we get

$$\frac{dN_A}{dt} = -k_1 \left(a_A - a_X a_\gamma \right) \tag{20}$$

$$\frac{dN_B}{dt} = -k_2 \left(a_B a_X - a_Y a_D a_\gamma \right) \tag{21}$$

$$\frac{dN_D}{dt} = k_2 \left(a_B a_X - a_Y a_D a_\gamma \right) \tag{22}$$

$$\frac{dN_X}{dt} = k_1 \left(a_A - a_X a_\gamma \right) - k_2 \left(a_B a_X - a_Y a_D a_\gamma \right)$$
$$+ k_3 \left(a_X^2 a_Y - a_X^3 a_\gamma \right) - k_4 \left(a_X - a_E a_\gamma \right) \tag{23}$$

$$\frac{dN_E}{dt} = k_4 \left(a_X - a_E a_\gamma \right) \tag{24}$$

$$\frac{dN_Y}{dt} = k_2 \left(a_B a_X - a_Y a_D a_\gamma \right) - k_3 \left(a_X^2 a_Y - a_X^3 a_\gamma \right) \tag{25}$$

$$\frac{dN_\gamma}{dt} = k_1 \left(a_A - a_X a_\gamma \right) + k_2 \left(a_B a_X - a_Y a_D a_\gamma \right)$$
$$+ k_3 \left(a_X^2 a_Y - a_X^3 a_\gamma \right) + k_4 \left(a_X - a_E a_\gamma \right). \tag{26}$$

At zero temperature, $a_\gamma = 0$ and we prevent the back reactions; with this interpretation, we see that the violation of the Arrhenius relation in the original Brussellator can be regarded as being caused by the choice of absolute zero temperature, rather than by a wrong choice of dynamics. We could also interpret γ as another chemical, which is produced in each of the four reactions. If it is precipitated, and then swept away, its chemical potential will be kept at zero, and the back reactions would not occur. We prefer to interpret γ as thermal noise, and we put $a_\gamma = e^{-\beta\epsilon}$, where ϵ is the energy of one quantum. Then in the isothermal dynamics we ignore Eqn.(26), since it is vitiated by the flows of heat. In our version of the Brussellator, we follow the isopotential dynamics where a_A and a_B are also kept constant, so we ignore Eqns.(20) and (21); we put the constant values of β, a_A and a_B in the remaining four equations. With the choices $a_\gamma = 0.003$, $a_A = 1/8$ and $a_B = 1/4$, the figure shows the time-development of the four remaining variables, drawn on the Apple Mac using Stella. The values of the rate constants were $k_1 = 5/16$, $k_2 = 3/8$, $k_3 = 1/4$ and $k_4 = 5/64$. When $a_\gamma = 0$, linear stability analysis of the fixed point

$$a_A = 1,\ a_B = 1,\ a_X = 1/2,\ a_Y = 3/4$$

shows that the Lyapunov index is purely imaginary, indicating that there are limit cycles. This is borne out by the numerical studies. When the noise is small, as in the diagram, the system makes many rotations about the fixed point, which is very close to $(1/2, 3/4)$, and spirals in very slowly. The real part of the Lyapunov index is small and negative in the presence of small noise, showing that the centre of the oscillations has become a stable fixed point. Indeed, in [1] we construct a relative entropy which is a strict Lyapunov function, showing that the fixed point is a global attractor, provided we avoid the limiting cases where one or more chemicals are zero. An interesting limit is zero temperature; there, the relative entropy becomes infinite, since it contains terms proportional to beta. It is this failure on the boundary of state space which allows oscillating solutions to occur.

In summary, we have described the dynamics of a system of chemicals as a non-linear motion through the set of states, a state being the product of local canonical states, specified by the field of chemical potentials. The noise is just one more chemical, the heat-particle. The fluctuations at any point in energy or density are thermal at any time, this being forced on us by the action of the *LTE* map Q. The isolated dynamics satisfies both laws of thermodynamics; the law of conservation of heat is satisfied on average, and the law of increasing entropy is exact. Isothermal and isopotential dynamics are derived from the isolated dynamics by restoring the beta and the chemical potential to assigned values after each time-step. In this way the dynamics of a driven system is obtained. In the case of the Brussellator, modified by adding noise, and driving two of the six chemicals, there is a Lyapunov function as long as the temperature is not zero. The model therefore show convergence to equilibrium except when the noise is omitted; in this limiting case, permanent oscillations occur (as in the original model) which are destroyed by the noise.

Dissipative Brussellator γ = 0.003

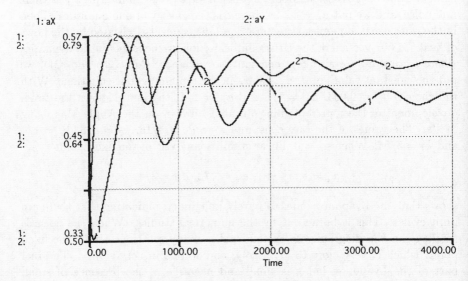

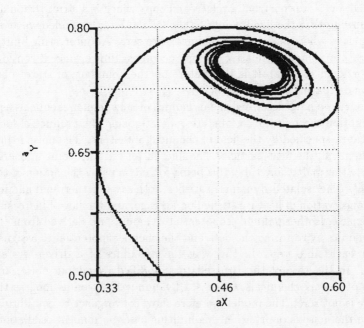

5 Appendix

We now prove that any finite-dimensional Markov chain with a positive fixed point, such that the transition matrix is close enough to the identity, can be considered as the isothermal version of a bistochastic process conserving energy. Earlier results along these lines [14] showed that any such Markov chain can be regarded as a driven system, with a symmetric transition matrix on a much larger space than that used here, and that if it obeys the Glauber form of detailed balance, then the size of the sample space of the heat-particle could be reduced to that of the original chain. The next theorem is a generalisation of the latter.

Theorem

Let $\Omega_c = \{1, 2, \ldots, n\}$. Let $p^{(0)} \in \Sigma(\Omega_c)$ have positive components: $p_j^{(0)} >$ 0, $j = 1, 2, \ldots, n$. Then there exists a neighbourhood $\mathbf{N}$ of the unit $n \times n$ matrix and a sample space Ω_γ, with $|\Omega_\gamma| = n$ and an energy function $\mathcal{E} = \mathcal{E}_c + \mathcal{E}_\gamma$ on $\Omega = \Omega_c \times \Omega_\gamma$ such that for any stochastic matrix $T \in \mathbf{N}$ with $p^{(0)}$ as fixed point, there exists a bistochastic map M on $\mathcal{A}(\Omega)$, conserving $\mathcal{E}$ and such that

$$p_c T(\omega_c) = \sum_{\omega_\gamma} (p_c \otimes s) \, M(\omega_c, \omega_\gamma), \text{ for all } p_c \in \Sigma(\Omega_c) \qquad (27)$$

where s is the canonical state with energy $\mathcal{E}_\gamma$ and $\beta = 1$.

Proof.

Consider the set $\{0\} \cup \Omega_c$, the given sample space with one point adjoined. Define the *fan* of Ω_c, denoted $\text{Fan} \, \Omega_c$, to be the graph whose vertices are the points of $\{0\} \cup \Omega_c$, and whose edges are the pairs $(\omega_c, 0)$. We can regard the edges as chemical bonds between the points of the system and a common heat bath, 0. We define the space Ω_γ of the heat particle to be the set of edges of $\text{Fan} \, \Omega_c$. Then $|\Omega_\gamma| = n$ as required. Define the energy function $\mathcal{E}_c$ on Ω_c by

$$\mathcal{E}_c(\omega_c) = -\beta^{-1} \log p^{(0)}(\omega_c).$$

This is well defined, since the components of $p^{(0)}$ are positive. Then define $\mathcal{E}_\gamma(\omega_c, 0) = -\mathcal{E}_c(\omega_c)$, and the energy function on $\Omega = \Omega_c \times \Omega_\gamma$ to be $\mathcal{E}\left(\omega_c, (\omega_c', 0)\right) = \mathcal{E}_c(\omega_c) - \mathcal{E}_c(\omega_c')$. We now construct a bistochastic matrix M on $\mathcal{A}(\Omega)$ which causes hopping from $i \in \Omega_c$ to $j \in \Omega_c$, by picking up a heat particle from the bond $b_i = (i, 0)$, hopping to j, and releasing a heat particle to the bond $b_j = (j, 0)$. That is, we have the process

$$i + b_i \rightleftharpoons j + b_j.$$

Other bonds and vertices do not interact. Since the transitions occur only on the energy-shell Ω_0 of zero energy, our reaction conserves energy, as claimed. We shall now define an $n \times n$ bistochastic matrix M_1 on $\mathcal{A}(\Omega_0)$. The action of

M will coincide with that of M_1 on the subspace, and will be the unit operator (of size $(n^2 - n) \times (n^2 - n)$) on the remaining points. Since M_1 is bistochastic and conserves energy, so will M be, as it has the form

$$M = \begin{pmatrix} M_1 & 0 \\ 0 & 1 \end{pmatrix}. \tag{28}$$

Let

$$\lambda = \sum_{i,j=1}^{n} e^{\beta(\mathcal{E}(j) - \mathcal{E}(i))} \tag{29}$$

and put

$$\operatorname{diag} p^{(0)} = \begin{pmatrix} p^{(0)}(1) & 0 & \cdots & 0 \\ 0 & p^{(0)}(2) & \cdots & 0 \\ \cdots & \cdots & & \\ 0 & 0 & \cdots & p^{(0)}(n) \end{pmatrix} \tag{30}$$

We define M_1 in terms of T by

$$M_1 = 1 + \lambda \operatorname{diag} p^{(0)} (T - 1). \tag{31}$$

Let

$$p^{(1)} = (1, 1 \ldots, 1)^t. \tag{32}$$

Then $\operatorname{diag} p^{(0)} p^{(1)} = p^{(0)}$. A matrix A is stochastic if its elements are non-negative, and $A p^{(1)} = p^{(1)}$. We now check the second property for M_1 and M_1^t.

$$\begin{aligned} M_1 p^{(1)} &= p^{(1)} + \lambda \operatorname{diag} p^{(0)} T p^{(1)} - \lambda \operatorname{diag} p^{(0)} p^{(1)} \\ &= p^{(1)} + \lambda \operatorname{diag} p^{(0)} p^{(1)} - \lambda \operatorname{diag} p^{(0)} p^{(1)} = p^{(1)} \end{aligned}$$

since T is stochastic. Hence the rows of M_1 add up to 1. Also

$$M_1^\dagger p^{(1)} = p^{(1)} + \lambda (T^\dagger - 1) \operatorname{diag} p^{(0)} p^{(1)} = p^{(1)} + \lambda (T^\dagger - 1) p^{(0)} = p^{(1)} \tag{33}$$

since $p^{(0)}$ is a fixed point of $T^\dagger$. Hence the columns of M_1 add up to 1. We now check the positivity. The off-diagonal elements of M_1 are non-negative, so it remains to check the diagonal elements. It $T = 1$ we may take $p^{(0)} = p^{(1)}$, and $M_1 = 1$. So the diagonal elements are positive for this limiting value. If T_{ik}, $i \neq j$ are small, then $p^{(0)}$ is close to $p^{(1)}$ and $\mathcal{E}(i) - \mathcal{E}(j)$ close to zero. So λ is close to n^2. It follows that the sum of the off-diagonal elements of M_1 along a row is less than 1 if $T_{i,k}$ is less than about n^{-3}. This defines the required neighbourhood $\mathbf{N}$ of the unit matrix, in which M_1 is bistochastic.

It remains to show that we recover the Markov chain generated by T as the isothermal dynamics given by the bistochastic matrix M. Recall that M consists of the matrix M_1 in the block labelled by rows and columns (j, b_j), M

being the unit matrix on the rest of the diagonal, as in Eqn.(28). This may be expressed in terms of matrix elements as:

$$M_{i,b_j;k,b_\ell} = \delta_{i,j}\delta_{k,\ell}\left(\delta_{i,k} + p^{(0)}(i)(T_{i,k} - \delta_{i,k})\right) + (1 - \delta_{i,j}\delta_{k,\ell})\delta_{i,k}\delta_{j,\ell}. \tag{34}$$

One time-step of the isothermal dynamics defined by M is

$$p_c \mapsto p'_c = \sum_{\omega_\gamma} M^\dagger(p_c \otimes s)(\omega_c, \omega_\gamma). \tag{35}$$

Here, ω_c runs over $k \in \{1, 2, \ldots, n\}$, and ω_γ runs over $\ell = b_\ell = (\omega_\ell, 0)$. The heat-particle is in the canonical state

$$s(\ell) = \frac{e^{\beta\mathcal{E}(\ell)}}{\sum_j e^{\beta\mathcal{E}(j)}}. \tag{36}$$

Then $\sum_j s(j) = 1$ and

$$p^{(0)}(i)s(i) = \frac{e^{-\beta\mathcal{E}(i)}}{\sum_k e^{-\beta\mathcal{E}(k)}}\frac{e^{\beta\mathcal{E}(i)}}{\sum_j e^{\beta\mathcal{E}(j)}} = \frac{1}{\lambda}. \tag{37}$$

Both these are used in the following:

$$
\begin{aligned}
p'_c(k) &= \sum_{i,j,\ell} p_c(i)s(j)M_{i,j;k,l} \\
&= \sum_{i,j,\ell} p_c(i)s(j)\delta_{i,j}\delta_{k,l}M_{1,i,k} + \sum_{i,j,\ell} p_c(i)s(j)[\delta_{i,k}\delta_{j,\ell} - \delta_{i,j}\delta_{k,\ell}\delta_{i,k}\delta_{j,\ell}] \\
&= \sum_i p_c(i)s(i)M_{i,k} + p_c(k) - p_c(k)s(k) \\
&= \sum_i p_c(i)s(i)[\delta_{i,k} + \lambda p^{(0)}(i)(T_{i,k} - \delta_{i,k}] + p_c(k) - p_c(k)s(k) \\
&= p_c(k)s(k) + \lambda \sum_i p_c(i)p^{(0)}(i)s(i)T_{i,k} \\
&\quad -\lambda p^{(0)}(k)p_c(k)s(k) + p_c(k) - p_c(k)s(k) \\
&= \sum_i p_c(i)T_{i,k}.
\end{aligned}
$$

Thus, the isothermal dynamics coincides with the given Markov chain. This result seems to be new, and is published here for the first time. The "accidental" cancellation of unwanted terms in the calculation shows that the result is not altogether routine. It means that the "relative entropy" of the Markov chain [28] can be identified as the Hemholtz free energy obtained from an energy-conserving isolated dynamics by a Legendre transform. The free-energy theorem then shows that it decreases along the orbit, a known result, and also leads to new sharp estimates for the rate of decrease, since such estimates are known for the increase in entropy under a bistochastic map.

Birkhoff's theorem shows that a bistochastic map can be decomposed into permutations. Applying this to the matrix M, we get a corresponding decomposition of any stochastic matrix T near the identity having $p^{(0)}$ as a fixed point, into the parts coming from the permutations of $\Omega = \Omega_c \times \Omega_\gamma$ conserving energy. This is a novel interpretation of the decomposition of the convex set of stochastic matrices, having a common fixed point, into extreme points.

In the case of continuous time, we have a semigroup of maps T_t for $t \geq 0$, and it is positivity preserving if and only if the generator,

$$X = \lim_{t \to 0} t^{-1}(T_t - 1)$$

has non-positive off-diagonal elements and the rows add up to zero. The measure $p^{(0)}$ is a fixed point of the semigroup if and only if $Xp^{(0)} = 0$. Let us define M_{1t} by the equation corresponding to (31); we see that the limit of $t^{-1}(M_{1t} - 1)$ also has non-positive off-diagonal elements, and the rows and columns add up to zero. Hence the limit generates a positive semigroup of bistochastic maps. The original Markov process is obtained by summing over the space of the heat-particle. Thus our result can be extended to continuous time.

Acknowledgements

This work was partially supported by the EC "Go East" Fellowship Programme, under grant No.ERBCIPACT931672, and by SERC. I thank Dr. A Etheridge for the kind hospitality at Edinburgh. I am indebted to Prof. R.J. Naftalin for drawing the diagrams.

References

[1] L. Rondoni and R.F. Streater; The Statistical Dynamics of the Brussellator, *Open Systems and Information Dynamics*, to appear 1994.

[2] L. Rondoni and R.F.Streater: The Brussellator with Thermal Noise, *Proceedings of the 1993 Torun Conference on Mathematical Physics*, Edited by M. Michalski, 1994.

[3] R. Lefever and I. Prigogine, Symmetry-breaking Instabilities in Dissipative Systems, *Journal of Chemical Physics*, **48**, 1695 (1968).

[4] L. Reichl; *A Modern Course in Statistical Physics*, University of Texas Press, Austin (1984).

[5] M. J. Freidlin and A.D. Wentzell, *Random Perturbations of a Dynamical System*, Springer Verlag (1984).

[6] Y. Kifer, *Random Perturbations of Dynamical Systems*, Birkhaüser, Boston (1988).

[7] I.R. Senitzky, *Physical Review*, **119**, 670 (1960).

[8] M. Lax, *Physical Review*, **145**, 111-129 (1965).

[9] C. Barnett, R.F. Streater and I.F. Wilde, The Ito- Clifford Integral, *Journal of Functional Analysis*, **48**, 172-212 (1982). Quasi-free Quantum Stochastic Integrals for the CAR and CCR, ibid, **52** 19-47 (1983).

[10] R.L. Hudson and K.R. Parthasarathy, Quantum Ito's Formula and Stochastic Evolution,*Communications in Mathematical Physics*, **93**, 301-323 (1984).

[11] R. Kubo, *Journal of the Physical Society of Japan*, **12**, 570, (1957).

[12] G.W. Ford, M. Kac and P. Mazur, *Journal of Mathematical Physics*, **6**, 504 (1965).

[13] J.T. Lewis and L.C. Thomas, *Annales d'Institute Henri Poincaré*, **22A** 241-248 (1975).

[14] R.F. Streater, Statistical Dynamics, *Reports on Mathematical Physics*, **33**, 203-219 (1993).

[15] R.F. Streater, Statistical Dynamics with Sources and Sinks, in *Quantum Chemistry and Technology in the Mesoscopic Level*, pp. 15-20, edited by H.Hasagawa. The Physical Society of Japan, (1994).

[16] R.F. Streater, Entropy and KMS-norms, in: Proceedings of the Madeira Conference 1993, *Stochastics and White Noise Analysis*, edited by L. Streit, Kluwer.

[17] E.B. Davies, *Quantum Theory of Open Systems*, Academic Press, London, 1976.

[18] G. Birkhoff, *Univ. Nac. Tucuman Rev. Ser.* **A5**,147 (1946).

[19] R.F.Streater, Convergence of the iterated Boltzmann map, *Publications of the Research Institute in Mathematical Sciences,* Kyoto, **20** 913-927, (1984).

[20] R.H. Fowler, *Statistical Mechanics*, 2nd Ed., Cambridge, (1936), p.703.

[21] R.F. Streater, Entropy and the Central Limit Theorem in Quantum Mechanics, *Journal of Physics A, Mathematical and General*, **20**, 4321-4330 (1987).

[22] R.F. Streater, A Boltzmann map for Quantum Oscillators, *Journal of Statistical Physics*, **48**, 753-767 (1987)

[23] R.J. Glauber, Time-dependent Statistics of the Ising Model, *Journal of Mathematical Physics*, **4**, 294-307 (1963).

[24] R.F. Streater, Detailed Balance and Free Energy, in *Proceedings of the Conference on Quantum Information*, Kyoto, (1993). Edited by M. Ohya.

[25] R.F. Streater, The F-Theorem for Stochastic Models, *Annals of Physics*, **218**, 255-278 (1992).

[26] R.F. Streater, The Free-energy Theorem, in *Quantum and Non-commutative Analysis*, pp. 137-147, Edited by H. Araki, K.R. Ito and Kishimoto. Kluwer (1993).

[27] S. Koseki and R.F. Streater, The Statistical Dynamics of Activity-led Reactions, *Open Systems and Information Dynamics*, **2**, 77-94 (1993).

[28] O.Penrose, *The Foundations of Statistical Mechanics*, Pergamon, (1971).

Stochastic Hamilton Jacobi Equations and Related Topics

A. Truman and H.Z. Zhao

Department of Mathematics, University of Wales Swansea, Singleton Park, Swansea SA2 8PP, U.K.

Abstract

In this paper we describe the stochastic Hamilton Jacobi theory and its applications to stochastic heat equations, Schrödinger equations and stochastic Burgers' equations.

Keywords. Stochastic Hamilton Jacobi equation, Stochastic heat equation, Stochastic Burgers' equation, Stochastic harmonic oscillator.

§1. Introduction

It is well-known that the leading term in Varadhan's and Wentzell-Freidlin's large deviation theories (Varadhan (1967), Wentzell and Freidlin (1970)) and Maslov's quasi-classical asymptotics of quantum mechanics (Maslov (1972)) involves the solution of a variational problem which gives a Lipschitz continuous solution of the Hamilton Jacobi equation if it exists (Fleming (1969, 1986)). It was proved by Truman (1977) and Elworthy and Truman (1981, 1982) that before the caustic time the Hamilton Jacobi function which is $C^{1,2}$ gives the exact solutions of the diffusion equations geared to small time asymptotics. The main tools in the theory are classical mechanics and the Maruyama-Girsanov-Cameron-Martin formula. The philosophy of the theory is to choose a suitable drift for a Brownian motion on the configuration space manifold and to employ the MGCM theorem to simplify the Feynman-Kac representation of the solutions for the heat equations. An extended version of this theory to degenerate diffusion equations was obtained in Watling (1992). The Brownian Riemannian bridge process was obtained by this means in Elworthy and Truman (1982). The extension to more general Riemannian manifolds was obtained in Elworthy (1988), Ndumu (1986,1991). The same methods have been applied to travelling waves for nonlinear reaction diffusion equations in Elworthy, Truman and Zhao (1994).

Recently we developed the stochastic Hamilton Jacobi theory in Truman and Zhao (1994A, 1994B). By choosing a suitable measurable drift we obtain the solution for the stochastic heat equation and the corresponding heat kernel on hyperbolic space and a harmonic oscillator potential in Euclidean space, thereby obtaining a stochastic Mehler formula. Applying the Hopf-Cole transformation we get some new rigorous relations for the stochastic Burgers' equation and stochastic Burgers' equation without viscosity. In this paper we shall survey the stochastic Hamilton Jacobi theory and related topics and pose several problems.

§2. The Stochastic Hamilton Jacobi Equations and Burgers' Equations

A. Let M be a n-dimensional Riemannian manifold. For $c \in C^{1,2}([0,+\infty) \times M, R)$ and $k \in C^{1,2}([0,+\infty) \times M, R^m)$ and a m-dimensional Brownian motion w_s on probability space $(\Omega, \mathcal{F}_t, P)$, define $\Phi_s : M \times \Omega \to M$ by the following stochastic mechanical system for each $x \in M$,

$$\begin{cases} d\dot{\Phi}_s(x) = -\nabla c(s, \Phi_s(x))ds - \nabla k(s, \Phi_s(x))dw_s, \\ \dot{\Phi}_0(x) = \nabla S_0(x), \Phi_0(x) = x. \end{cases} \quad (2.1)$$

For $t \geq 0$, $y \in M$ define $S^* : [0,+\infty) \times M \to R$ by the following non-anticipating Itô stochastic integral

$$\begin{aligned} S_t^*(y,\omega) = \frac{1}{2}\int_0^t |\dot{\Phi}_s(y)|^2 ds + S_0(y) \\ - \int_0^t c(s, \Phi_s(y))ds - \int_0^t \langle k(s, \Phi_s(y)), dw_s \rangle. \end{aligned} \quad (2.2)$$

Here by $\langle \cdot, \cdot \rangle$ denote the inner product in R^m. We assume a no-caustic condition: there exists $T > 0$ such that for $0 \leq s \leq T$, $\Phi_s(\omega) : M \to M$ is a diffeomorphism for a.e. $\omega \in \Omega$. With this assumption we define $S(\omega) : [0,T] \times M \to R^1$ for a.e. $\omega \in \Omega$ by

$$S_t(x) = S_t^*(\Phi_t^{-1}(x)). \quad (2.3)$$

Theorem 2.1.

(i) Let Φ be defined by (2.1) and satisfy the no-caustic condition for $0 \leq t \leq T$ and $S_t(x)$ be defined by (2.3) . Then for any $0 \leq t \leq T$, $x \in M$ and a.e. $\omega \in \Omega$

$$\dot{\Phi}_t(x) = \nabla S_t(\Phi_t(x)). \quad (2.4)$$

(ii). For any $0 \leq t \leq T$, $x \in M$ and a.e. $\omega \in \Omega$, S satisfies the following stochastic Hamilton Jacobi equation

$$dS_t(x) + [c(t,x) + \frac{1}{2}||\nabla S_t(x)||^2]dt + \langle k(t,x), dw_t \rangle = 0. \quad (2.5)$$

(iii). Define $\phi : [0,T] \times M \to R$ by $\phi(t,x) = |det T_x \Phi_t^{-1}|$ (using the Riemannian metric of M). Then ϕ satisfies the continuity equation:

$$\frac{\partial \phi}{\partial t}(t,x) + div(\phi(t,x)\nabla S(t,x)) = 0. \quad (2.6)$$

Note that $S(t,x)$ is continuous with respect to t and C^2 with respect to x for $0 \le t \le T$. We call $S(t,x)$ a strong solution of the stochastic Hamilton Jacobi equation (2.5).

Let $\tau(a)$ be first time such that there exist $x_1 \ne x_2$, $x_1, x_2 \in M$ such that $\Phi_\tau(x_1) = \Phi_\tau(x_2) = a$ and $\dot{\Phi}_\tau(x_1) \ne \dot{\Phi}_\tau(x_2)$ for a e. ω, i.e., τ is a caustic time at point a.

Theorem 2.2. *The function $\nabla S(t,x)$ is discontinuous at (τ, a).*

Remark 2.1. *If $k \equiv 0$, one can prove that the assumption that there exist $x_1 \ne x_2$, $x_1, x_2 \in M$ such that $\Phi_\tau(x_1) = \Phi_\tau(x_2)$ implies $\dot{\Phi}_\tau(x_1) \ne \dot{\Phi}_\tau(x_2)$ from the uniqueness of the backward deterministic ordinary differential equation. Therefore $\tau(a)$ is the first time such that there exist $x_1 \ne x_2$, $x_1, x_2 \in M$ such that $\Phi_\tau(x_1) = \Phi_\tau(x_2) = a$.*

B. The philosophy of the stochastic Hamilton Jacobi theory or the stochastic semi-classical analysis is that we can construct a Brownian motion on a Riemannian manifold with a random nonlinear drift A_s satisfying some measurability conditions. Then choosing A_s suitably we use a MGCM transformation to simplify the Feynman-Kac representation for the solution of the following stochastic heat equation

$$\begin{cases} du_t^\mu(x) = [\frac{1}{2}\mu^2 \Delta u_t^\mu(x) + \frac{1}{\mu^2}c(t,x)u_t^\mu(x)]dt + \frac{1}{\mu^2}u_t^\mu(x)\langle k(t,x), \partial w_t\rangle, \\ u_0^\mu(x) = T_0(x)\exp\{-\frac{S_0(x)}{\mu^2}\}, \end{cases}$$

$$(2.7)$$

where $x \in M$, a n-dimensional Riemannian manifold and Δ its Laplace-Beltrami operator, ∂ is the Stratonovich derivative. Given $T_0 : M \to R^1$ a bounded measurable function and $S_0 : M \to R^1$ a C^2 function, let $u_t^\mu(x)$ denote the solution of (2.7). If $c \in C^{1,2}([0,+\infty) \times M, R)$ and $k \in C^{1,2}([0,+\infty) \times M, R^m)$, let $S_t(x)$ be defined by (2.3) for $0 \le t \le T$.

Here we use the Eells-Elworthy formulation of diffusion processes on manifolds (Elworthy (1982)). Let $\Pi : O(M) \to M$ be the orthonormal frame bundle of M. The Levi-Civita connection of M determines a map $\chi : O(M) \times R^n \to TO(M)$ into the tangent space to $O(M)$, which trivializes the horizontal tangent bundle of M. Let $\tilde{A}_t : O(M) \to TO(M)$ be its horizontal lift: so $T\Pi(\tilde{A}_t(\mathcal{U})) = A_t(\Pi(\mathcal{U})), 0 \le t \le \tau, \mathcal{U} \in O(M)$, where $T\Pi : TO(M) \to TM$ is the derivative map of Π.

For $x_0 \in M$ take $\mathcal{U}_0 \in \Pi^{-1}(x_0)$ and consider the Stratonovich stochastic equation for $\mathcal{U} : [0, \xi_A) \times \hat{\Omega} \to O(M)$ with $\mathcal{U}(0, \hat{\omega}) = \mathcal{U}_0$, $0 \le s \le t$

$$d\mathcal{U}_s^{t,\mu} = \mu\chi(\mathcal{U}_s^{t,\mu}) \cdot dB_s + \tilde{A}_s^t(\mathcal{U}_s^{t,\mu})ds, \qquad (2.8)$$

where B_s is an n-dimensional Brownian motion defined on the probability space $(\hat{\Omega}, \hat{\mathcal{F}}, \hat{P})$. Define $\hat{\mathcal{F}}_s = \sigma\{B_{s_1} : 0 \le s_1 \le s\}$. The solution $\mathcal{U}_s^{t,\mu}$ with explosion time $0 < \xi_A \le +\infty$ for all $\hat{\omega} \in \hat{\Omega}$ is $\hat{\mathcal{F}}_s$ measurable. Finally we define $X_s^{t,\mu}$ on M by $X_s^{t,\mu} = \Pi(\mathcal{U}_s^{t,\mu})$. Let $\{B_s^{t,\mu} : 0 \le s \le t\}$ be the solution of (2.8) with $A \equiv 0$. Note that $B_s^{t,\mu}, X_s^{t,\mu} \in \hat{\mathcal{F}}_s$.

Given the initial condition $u_0^\mu(x)$ for equation (2.7), then by the Feynman-Kac formula, we have

$$
u_t^\mu(x) = \hat{E} T_0(B_t^{t,\mu}) \exp \left\{ -\frac{1}{\mu^2} S_0(B_t^{t,\mu}) + \frac{1}{\mu^2} \int_0^t c(t-s, B_s^{t,\mu}) ds \right.
$$
$$
\left. + \frac{1}{\mu^2} \int_0^t \langle k(t-s, B_s^{t,\mu}), dw_{t-s} \rangle \right\}.
$$

Given a drift A_s^t, we get the solution $X_s^{t,\mu}$ to equation (2.8). If the stochastic integral $\int_0^t \langle k(t-s, X_s^{t,\mu}(x)), dw_{t-s} \rangle$ is well defined in Itô's sense, then we can use the Maruyama-Girsanov-Cameron-Martin formula to give

$$
u_t^\mu(x) = \hat{E} T_0(X_t^{t,\mu}(x))
$$
$$
\exp \left\{ -\frac{1}{\mu^2} S_0(X_t^{t,\mu}(x)) + \frac{1}{\mu^2} \int_0^t c(t-s, X_s^{t,\mu}(x)) ds \right.
$$
$$
\left. + \frac{1}{\mu^2} \int_0^t \langle k(t-s, X_s^{t,\mu}(x)), dw_{t-s} \rangle \right\} \mathcal{M}_t^\mu, \tag{2.9}
$$

where

$$
\mathcal{M}_t^\mu = \exp \left\{ -\frac{1}{\mu} \int_0^t \langle A_s^t(X_s^{t,\mu}(x)), \mathcal{U}_s^{t,\mu} dB_s \rangle \right.
$$
$$
\left. - \frac{1}{2\mu^2} \int_0^t \|A_s^t(X_s^{t,\mu}(x))\|^2 ds \right\}. \tag{2.10}
$$

On the other hand if we choose $A_s^t = \nabla Y_s^t$ for the some suitable $Y_s^t : 0 \le s \le t$, then Itô's formula yields

$$
Y_t(X_t^{t,\mu}(x)) - Y_0(x) = \int_0^t \left\{ \langle \nabla Y_s(X_s^{t,\mu}(x)), \mu \mathcal{U}_s^{t,\mu} dB_s \rangle + dY_s(X_s^{t,\mu}(x)) \right.
$$
$$
\left. + [\frac{1}{2} \mu^2 \Delta Y_s(X_s^{t,\mu}(x)) + \langle \nabla Y_s(X_s^{t,\mu}(x)), A_s(X_s^{t,\mu}(x)) \rangle] ds \right\}.
$$
$$\tag{2.11}$$

Substituting $\int_0^t \langle \nabla Y_s(X_s^{t,\mu}(x)), \mathcal{U}_s^{t,\mu} dB_s \rangle$ in (2.10) by solving for it in (2.11), we get

$$
\mathcal{M}_t^\mu = \exp \left\{ \frac{1}{\mu^2} \left(Y_0(x) - Y_t(X_t^{t,\mu}(x)) \right) + \frac{1}{2} \int_0^t \Delta Y_s(X_s^{t,\mu}(x)) ds \right.
$$
$$
\left. + \frac{1}{\mu^2} \int_0^t \left(dY_s(X_s^{t,\mu}(x)) + \frac{1}{2} \|A_s^t(X_s^{t,\mu}(x))\|^2 ds \right) \right\}.
$$

C. Denote $w_s^* = w_{t-s}$ and let $\mathcal{F}_s^*$ be the enlargement of the filtration $\{\mathcal{F}_s^0\}$, where

$$
\mathcal{F}_s^0 = \sigma\{w_r^* : r \le s\}. \tag{2.12}
$$

Then $w_{t-s} = w_s^*$ is $\mathcal{F}_s^*$ measurable and $\mathcal{F}_{s_1}^* \subset \mathcal{F}_{s_2}^*$ if $s_1 \leq s_2$. See Rogers and Williams (1987) for the filtration $\mathcal{F}_s^*$.

If we take $Y_s = -S_{t-s}$ and if ∇S_{t-s} is $\mathcal{F}_s^*$ measurable for fixed t, then we have a $\mathcal{F}_s^*$ measurable solution $X_s^{t,\mu}(x)$ to equation (2.8). Thus we have the following theorem:

Theorem 2.3. *Assume* $c \in C^{2,2}([0,+\infty) \times M, R^1)$, $k \in C^{2,2}([0,+\infty) \times M, R^m)$ *and* $S_0 \in C^1(M \to R)$ *and* $T_0 : M \to R^1$ *bounded and measurable,* w_t *is a m-dimensional Brownian motion on probability space* $(\Omega, \mathcal{F}, P)$. *Let the stochastic flow defined by (2.1) satisfy the no-caustic condition for* $0 \leq t \leq T$ *and let* $S_t(x)$ *be the Hamilton Jacobi function defined by (2.2) and (2.3) and* $X_s^{t,\mu}(x)$ *be the Markov process defined by (2.8) with* $Y_s = -S_{t-s}$. *If* $X_s^{t,\mu}(x)$ *is* $\mathcal{F}_s^*$ *measurable for fixed* t, *then P-a.s.*

$$u_t^\mu(x) = \exp\{-\frac{S_t(x)}{\mu^2}\}\hat{E}T_0(X_t^{t,\mu}(x))\exp\{-\frac{1}{2}\int_0^t \Delta S_{t-s}(X_s^{t,\mu}(x))ds\}.$$

(2.13)

For $0 \leq t \leq T$, let $\psi(t,x) = T_0(\Phi_t^{-1}(x))\sqrt{\phi_t(x)}$. Then we obtain:

Theorem 2.4. *Assume that c and k and* S_0 *are* C^3 *with all the conditions of Theorem 2.3 and also that* $\nabla \log \psi_t$ *is bounded and* ψ_{t-s} *is* $\mathcal{F}_s^*$ *measurable for any* $0 \leq s \leq t$ *and fixed* t. *Then for any* $\mu \neq 0$ *and* $0 \leq t \leq T$, *P-a.s.*

$$u_t^\mu(x) = \exp\{-\frac{1}{\mu^2}S_t(x)\}\psi_t(x)$$

$$\hat{E}\exp\{\frac{1}{2}\mu^2 \int_0^t \psi_{t-s}^{-1}(X_s^{t,\mu}(x))\Delta\psi_{t-s}(X_s^{t,\mu}(x))ds\},$$

(2.14)

where $X_s^{t,\mu}(x)$ *is defined by (2.8) with* $Y_s = -S_{t-s} + \mu^2 \log \psi_{t-s}$.

Remark 2.2. *For* $k \equiv 0$, *we get Formula B of Elworthy and Truman (1982).*

As $\mu \to 0$, $X_s^{t,\mu}(x)$ converges to $z_s^t(x)$ in $\hat{P}$ probability for each $\omega \in \Omega$. As long as $\psi^{-1}\Delta\psi$ is bounded we can apply Lebesgue's dominated convergence theorem and the Taylor expansion theorem to give the following post WKB expansion:

$$u_t^\mu(x) = \exp\{-\frac{1}{\mu^2}S_t(x)\}\psi_t(x)$$

$$[1 + \frac{1}{2}\mu^2 \int_0^t \psi_{t-s}^{-1}(z_s^t(x))\Delta\psi_{t-s}(z_s^t(x))ds + o(\mu^2)].$$

(2.15)

Remark 2.3. *Theorems 2.3 and 2.4 are still true as long as the stochastic integral* $\int_0^t \langle k(t-s, X_s^{t,\mu}), dw_{t-s} \rangle$ *is well defined in Itô sense.*

D. For simplicity in the last of this section we let $M = R^1$. Consider the stochastic heat equation (2.7) with $T_0 \equiv 1$. The well-known logarithmic transformation

$$v^\mu(t, x) = -\mu^2 \nabla \log u^\mu(t, x) \qquad (2.16)$$

given by Hopf in his pioneering work (Hopf (1950)) on Burgers' equation transforms the linear heat equation into a stochastic nonlinear Burgers' equation

$$dv^\mu(t, x) + v^\mu(t, x) \frac{\partial}{\partial x} v^\mu(t, x) dt$$

$$= \frac{1}{2} \mu^2 \frac{\partial^2 v^\mu(t, x)}{\partial x^2} dt + e_1(t, x) dt + e_2(t, x) dw_t, \qquad (2.17)$$

with $v^\mu(0, x) = \nabla S_0(x)$ and $e_1(t, x) = -\nabla c(t, x)$, $e_2(t, x) = -\nabla k(t, x)$. This is the popular KPZ model for the dynamics of the interface introduced by Kardar, Parisi and Zhang (1986). Holden, Lindstrom, Oksendal, Uboe and Zhang (1994), Oksendal (1994) studied the stochastic Burgers' equation with space-time white noise. Newman (1994) introduced an alternative model which is similar to (2.17).

Let $v(t, x) = \nabla S(t, x)$. Then from (2.5) we have

$$dv(t, x) + v(t, x) \frac{\partial}{\partial x} v(t, x) dt = e_1(t, x) dt + e_2(t, x) dw_t. \qquad (2.18)$$

This is the stochastic Burgers' equation without viscosity. Taking the logarithms of (2.13) and differentiating them we get:

Theorem 2.5. *Assume all the conditions of Theorem 2.3 and the stochastic flow defined by (2.1) satisfies the no-caustic condition for $0 \leq t \leq T$ and v^μ is the solution of (2.17) and v is the solution of (2.18) with the same initial conditions. Then for any $\mu \neq 0$ and $0 \leq t \leq T$,*

$$v^\mu(t, x) = v(t, x) - \mu^2 \nabla \log \hat{E} \exp \left\{ -\frac{1}{2} \int_0^t \frac{\partial}{\partial x} v(t - s, X_s^{t,\mu}(x)) ds \right\}, \qquad (2.19)$$

where $X_s^{t,\mu}(x)$ is the same as in Theorem 2.3.

Theorem 2.6. *Assume all the conditions of Theorem 2.4 and let v^μ be the solution of (2.17) and v be the solution of (2.18) with the same initial conditions. Then for any $\mu \neq 0$, $0 \leq t \leq T$,*

$$v^\mu(t, x) = v(t, x) - \mu^2 \nabla \log \psi(t, x)$$

$$- \mu^2 \nabla \log \hat{E} \exp \left\{ \frac{1}{2} \mu^2 \int_0^t \psi^{-1}(t - s, X_s^{t,\mu}(x)) \Delta \psi(t - s, X_s^{t,\mu}(x)) ds \right\} \qquad (2.20)$$

where $X_s^{t,\mu}(x)$ is the same as in Theorem 2.4.

Remark 2.4. *Theorem 2.4 and 2.5 are true for any $\mu \neq 0$. Therefore (2.19) and (2.20) are exact solutions rather than asymptotic solutions of the stochastic Burgers' equation. This means that the stochastic Burgers' equation can be solved if the no-caustic condition is satisfied. It seems that this solution can not be obtained by Maslov canonical operator method (Mishchenko, Shatalov and Sternin (1990)).*

Remark 2.5. *It is trivial to get from Theorem 2.5 that for $0 \leq t \leq T$, $v^{\mu}(t,x) \to v(t,x)$ as $\mu \to 0$ as long as $\nabla \log \psi$ and $\psi^{-1} \Delta \psi$ are bounded.*

Remark 2.6. *Beyond the caustics (2.19) and (2.20) can hardly be true as in general v^{μ} is smooth and v is discontinuous at the caustics by Theorem 2.2. This leads us to pose an open problem:*

Problem. *What change of measure should be used to deduce the detailed asymptotics of the solution beyond the caustics?*

The answer to this problem is not clear even in Elworthy and Truman's deterministic Hamilton Jacobi theory (deterministic here means that the Hamilton Jacobi equation is deterministic).

We now present some of the simplest equations to have stochastic shock wave solutions. First consider

$$dv(t,x) + v(t,x)\frac{\partial}{\partial x}v(t,x)dt = -\varepsilon dw_t, \tag{2.21}$$

with $v(0,x) = 1$ for $x < 0$ and $v(0,x) = 0$ for $x > 0$. Let g be the solution of the Burgers' equation $\frac{\partial}{\partial t}g(t,x) + g(t,x)\frac{\partial}{\partial x}g(t,x) = 0$ with the same initial condition. Then $v(t,x) = g(t,x + \int_0^t \varepsilon w_\sigma d\sigma) - \varepsilon w_t$. From the results on the deterministic equation (Smoller (1983)) we know $v(t,x) = 1 - \varepsilon w_t$ for $x < \frac{1}{2}t - \int_0^t \varepsilon w_\sigma d\sigma$ and $v(t,x) = -\varepsilon w_t$ for $x > \frac{1}{2}t - \int_0^t \varepsilon w_\sigma d\sigma$. This is the simplest example of stochastic shock waves.

We shall demonstrate another shock wave from the harmonic oscillator in §3.

We have not proved the existence of the piecewise continuous solution for the stochastic Burgers' equation (2.18). Nevertheless, we believe:

Conjecture. *Given some bounds on e_1 and e_2 the stochastic Burgers' equation (2.18) has a unique viscosity solution.*

It may not be true if e_1 and e_2 are not bounded. See §3 for a counterexample. For some special cases we can prove: (see also Davies and Truman (1982))

Theorem 2.7. *If $V(t,x) = \int_0^{x - \int_0^t \varepsilon w_\sigma d\sigma} e_1(t,y)dy$ is bounded below and strictly convex in space variable x, then the stochastic Burgers' equation*

$$dv(t,x) + v(t,x)\frac{\partial}{\partial x}v(t,x) = e_1(t,x)dt - \varepsilon dw_t, \tag{2.22}$$

with a bounded initial condition has a unique viscosity solution.

Here we have not striven for the maximum generality. For alternative formulations see Flemming (1969, 1986).

Theorem 2.8. *Consider the solution $u^\mu(t,x)$ of the following stochastic heat equation on R^1*

$$\begin{cases} du_t^\mu(x) = [\frac{1}{2}\mu^2 \Delta u_t^\mu(x) + \frac{1}{\mu^2}c(t,x)u_t^\mu(x)]dt + \frac{1}{\mu^2}\varepsilon x u_t^\mu(x)\partial w_t, \\ u_0^\mu(x) = T_0(x)\exp\{-\frac{S_0(x)}{\mu^2}\}, \end{cases} \tag{2.23}$$

Assume $-c(t, x - \int_0^t \varepsilon w_\sigma d\sigma)$ is bounded below and strictly convex in space variable x. Then the following stochastic Hamilton Jacobi equation

$$dS(t,x) + [\frac{1}{2}|\nabla S(t,x)|^2 + c(t,x)]dt + \varepsilon x dw_t = 0, \tag{2.24}$$

with initial condition $S(0,x) = S_0(x)$ has a unique Lipschitz continuous solution. If T_0 is bounded, then

$$\lim_{\mu \to 0} \mu^2 \log u^\mu(t,x) = -S(t,x), \tag{2.25}$$

uniformly in any compact subset of $[0, +\infty) \times R^1$, P-a.s..

We also believe (2.25) is true in general:

Conjecture. *As long as the stochastic Hamilton Jacobi equation (2.5) has a Lipschitz solution $S(t,x)$, the solution of stochastic heat equation (2.7) satisfies*

$$\lim_{\mu \to 0} \mu^2 \log u^\mu(t,x) = -S(t,x),$$

uniformly in any compact subset of $[0, +\infty) \times M$, P-a.s..

One can easily generalize the above results to R^n. It would be interesting to consider the same problem on a manifold.

§3. The Heat Kernels and the Stochastic Mehler Formula

A. Let M be a n-dimensional Riemannian manifold, with global normal coordinates and a Riemannian metric $r(\cdot, \cdot)$. Consider the following stochastic heat equation

$$du_t = [\frac{1}{2}\Delta u_t + c(t,x)u_t]dt + u_t(\langle \gamma(t), \partial w_t\rangle + \langle x, \varepsilon \partial w_t\rangle), \tag{3.1}$$

where $x \in M$, Δ is a Laplace Beltrami operator, w_t is a m-dimensional Brownian motion, $\varepsilon : R^m \to R^n$ is a linear map, $\gamma : [0, \infty) \to R^m$. Let $p_t^\lambda(x,y)$ be the solution of (3.1) with initial condition

$$p_0^\lambda(x,y) = \frac{1}{(2\pi\lambda)^{\frac{n}{2}}}\exp\left\{-\frac{r^2(x,y)}{2\lambda}\right\}. \tag{3.2}$$

Then the heat kernel of equation (3.1) is given by $p_t(x,y) = \lim_{\lambda \to 0} p_t^\lambda(x,y)$.

Suppose $y \in M$ is a pole for M so that $\exp_y : T_y M \to M$ is a C^∞ diffeomorphism. Denote $x = \exp_y(v)$. Therefore we have a global system of normal coordinates from y in which the geodesics from y are represented as straight lines from the origin. Let $\theta_y(v)$ be the Jacobian determinant of $\exp_y$ at v (Ruses' invariant): $\theta_y(v) = |det_M T_v \exp_y|$, where det_M indicates that the inner product of $T_y M$ and $T \exp_y(v) M$ are used to define the determinant. We shall often write θ for θ_y.

From now on let the stochastic process $X_s^{t,\lambda}$ be defined in our normal coordinates by (2.8) where we take $\mu = 1$ and

$$A_s^{t,\lambda}(x) = \varepsilon w_{t-s} - \frac{v + \int_0^{t-s} \varepsilon w_\sigma d\sigma}{\lambda + t - s} - \frac{1}{2} \nabla \log \theta(v + \int_0^{t-s} \varepsilon w_\sigma d\sigma). \quad (3.3)$$

Let $\xi_s^{t,\lambda} = v_s^{t,\lambda} + \int_0^{t-s} \varepsilon w_\sigma d\sigma$. Then from (2.8) $\xi_s^{t,\lambda}$ satisfies an equation which does not include the stochastic process w. except for the initial condition $\xi_0^{t,\lambda} = v_0 + \int_0^t w_\sigma d\sigma$. Thus we can write $\xi_s^{t,\lambda} = F_s^{t,\lambda}(v_0 + \int_0^t w_\sigma d\sigma)$ for some functional $F_s^{t,\lambda}$ which is independent of w. Recall $v_s^{t,\lambda} = \xi_s^{t,\lambda} - \int_0^{t-s} \varepsilon w_\sigma d\sigma = F_s^{t,\lambda}(v_0 + \int_0^t w_\sigma d\sigma) - \int_0^t \varepsilon w_\sigma d\sigma + \int_{t-s}^t \varepsilon w_\sigma d\sigma$. Thus the stochastic integral $\int_0^t \langle X_s^{t,\lambda}, \varepsilon dw_{t-s} \rangle$ is well defined in Itô's sense.

Lemma 3.1. *As $\lambda \to 0$, $\{X_s^{t,\lambda} : 0 \le s \le t\}$ converges in law $(\hat{P})$ to the process $\{X_s^t(x) : 0 \le s \le t\}$ which is sample continuous, agrees with $X_s^{t,0}$ for $0 \le s < t$ and $X_t^t(x) = y$.*

The process X_s^t is called the stochastic Brownian Riemannian bridge from x to y in time t in M.

Theorem 3.1. *Suppose y is a pole on the n-dimensional Riemannian manifold M. If $\theta^{\frac{1}{2}} \Delta \theta^{-\frac{1}{2}}$ and c are bounded, then the fundamental solution of the equation (3.1) on the Riemannian manifold M is given by*

$$p_t(x,y)$$
$$= \frac{1}{(2\pi t)^{\frac{n}{2}}} \theta^{-\frac{1}{2}} \left(\exp(\exp^{-1}(x) + \int_0^t \varepsilon w_s ds) \right) \exp \left\{ \int_0^t \langle \gamma(s), dw_s \rangle \right.$$
$$+ \frac{1}{2} \int_0^t |\varepsilon w_s|^2 ds + \langle x, \varepsilon w_t \rangle - \frac{r^2 \left(\exp(\exp^{-1}(x) + \int_0^t \varepsilon w_s ds), y \right)}{2t} \right\} \quad (3.4)$$
$$\cdot \hat{E} \exp \left\{ \frac{1}{2} \int_0^t (\theta^{\frac{1}{2}} \Delta \theta^{-\frac{1}{2}}) \left(\exp(v_s^t + \int_0^{t-s} \varepsilon w_\sigma d\sigma) \right) ds \right.$$
$$+ \int_0^t c(t-s, X_s^t) ds \bigg\},$$

where $X_s^t = \exp(v_s^t)$ is the stochastic Brownian Riemannian bridge from x to y at time t on M.

Corollary 3.1. *Assume all the conditions of Theorem 3.1. Then for x, y in any compact subset of M as $t \to 0$, the heat kernel $p_t(x, y)$ of equation (3.1) satisfies*

$$p_t(x, y) = \frac{1}{(2\pi t)^{\frac{n}{2}}} \exp\{-\frac{r^2(x, y)}{2t}\} \left(1 + o((t \log \frac{1}{t})^{\frac{1}{2}})\right), \quad P - a.s.. \quad (3.5)$$

If $\varepsilon \equiv 0$, $\gamma \equiv 0$, equation (3.1) is a deterministic equation on M, and an Elworthy and Truman theorem follows from Theorem 3.1:

Theorem 3.2. *(Elworthy and Truman (1982))*
Suppose y is a pole on the n-dimensional Riemannian manifold M. If $\theta^{\frac{1}{2}}\Delta\theta^{-\frac{1}{2}}$ and c are bounded, then the fundamental solution of the equation (3.1) for $\varepsilon \equiv 0$, $\gamma \equiv 0$ on the Riemannian manifold M is given by

$$p_t(x, y) = \frac{1}{(2\pi t)^{\frac{n}{2}}} \theta^{-\frac{1}{2}}(x) \exp\{-\frac{r^2(x, y)}{2t}\}$$
$$\hat{E} \exp\left\{\frac{1}{2}\int_0^t \theta^{\frac{1}{2}}(X_s^t)\Delta\theta^{-\frac{1}{2}}(X_s^t)ds + \int_0^t c(t - s, X_s^t)ds\right\}, \quad (3.6)$$

where X_s^t is the Brownian Riemannian bridge from x to y at time t on M.

We can give the heat kernel for the stochastic heat equation (3.1) on hyperbolic n-space H^n. The formula for $n = 3$ is particularly nice. On H^3 we have

$$p_t(x, y)$$
$$= \frac{1}{(2\pi t)^{\frac{3}{2}}} \frac{r\left(\exp(\exp^{-1}(x) + \int_0^t \varepsilon w_s ds), y\right)}{R} \frac{1}{\sinh\frac{r\left(\exp(\exp^{-1}(x) + \int_0^t \varepsilon w_s ds), y\right)}{R}}$$
$$\exp\left\{\int_0^t \langle \gamma(s), dw_s \rangle + \frac{1}{2}\int_0^t |\varepsilon w_s|^2 ds + \langle x, \varepsilon w_t \rangle\right.$$
$$\left. - \frac{r^2\left(\exp(\exp^{-1}(x) + \int_0^t \varepsilon w_s ds), y\right)}{2t} - \frac{1}{2R^2}t\right\}.$$
$$(3.7)$$

A natural question to pose is: can one generalize the above results to a more general Riemannian manifold with cut locus? Works of Ndumu (1986, 1991) could be useful here.

B. For simplicity to investigate the stochastic harmonic oscillator we let $M = R^1$. Consider the following stochastic harmonic oscillator in R^1

$$\begin{cases} d\dot{\Phi}_s = -\alpha^2 \Phi_s ds - \varepsilon dw_s, \\ \Phi_0(x) = x, \dot{\Phi}_0(x) = \frac{x}{\lambda}, \end{cases} \quad (3.8)$$

where $\alpha, \lambda > 0, \epsilon \geq 0$ are constants, which is (2.1) in the case $c(s,x) = \frac{1}{2}\alpha^2 x^2$, $k(s,x) = \varepsilon x$ and $S_0(x) = \frac{x^2}{2\lambda}$ and $M = R^1$. (See Markus and Weerasinghe (1988), Albeverio, Hilbert and Zehnder (1990) for some important properties of the stochastic harmonic oscillator.) Solving equation (3.8) we have

$$\Phi_s(x) = x \cos \alpha s + \frac{x}{\lambda \alpha} \sin \alpha s - \frac{\varepsilon}{\alpha} \int_0^s \sin \alpha(s - \sigma) dw_\sigma. \tag{3.9}$$

The map $\Phi_s : R^1 \to R^1$ satisfies the no caustic condition for $s < \frac{1}{\alpha}\text{ctg}^{-1}(-\frac{1}{\alpha\lambda})$ and the inverse function is given by

$$\Phi_t^{-1}(x) = \frac{\lambda(\alpha x + \varepsilon \int_0^t \sin \alpha(t - \sigma)dw_\sigma)}{\alpha\lambda \cos \alpha t + \sin \alpha t}, \text{ for } 0 < t < \frac{1}{\alpha}\text{ctg}^{-1}(-\frac{1}{\alpha\lambda})$$

By the definition of S and careful calculations we have

$S_{t,\lambda}(x)$

$$= [\frac{\sin 2\alpha t}{4\alpha\lambda^2} + \frac{\cos 2\alpha t}{2\lambda} - \frac{1}{4}\alpha \sin 2\alpha t]\frac{\lambda^2 \left(\alpha x + \varepsilon \int_0^t \sin \alpha(t - \sigma)dw_\sigma\right)^2}{(\alpha\lambda \cos \alpha t + \sin \alpha t)^2}$$

$$- \frac{\varepsilon}{\alpha} \int_0^t \cos \alpha(t - \sigma)dw_\sigma \left(\alpha x + \varepsilon \int_0^t \sin \alpha(t - \sigma)dw_\sigma\right) \tag{3.10}$$

$$+ \frac{1}{2}\varepsilon^2 \int_0^t \left[(\int_0^s \cos \alpha(s - \sigma)dw_\sigma)^2 - (\int_0^s \sin \alpha(s - \sigma)dw_\sigma)^2\right] ds$$

$$+ \frac{\varepsilon^2}{\alpha} \int_0^t \int_0^s \sin \alpha(s - \sigma)dw_\sigma dw_s,$$

for $0 < t < T_{\alpha,\lambda} = \frac{1}{\alpha}\text{ctg}^{-1}(-\frac{1}{\alpha\lambda})$. Let us consider Stratonovich stochastic heat equation with harmonic oscillator potential

$$du_t = [\frac{1}{2}\Delta u_t + \frac{1}{2}\alpha^2 x^2 u_t]dt + \varepsilon x u_t \partial w_t. \tag{3.11}$$

Here $x \in R^1$. Let $p_t^\lambda(x)$ be the solution of (3.11) with initial condition $p_0^\lambda(x) = \frac{1}{\sqrt{2\pi\lambda}}\exp\{-\frac{x^2}{2\lambda}\}$ and $p_t(x) = \lim_{\lambda \to 0} p_t^\lambda(x)$. We shall apply Theorem 2.4. Note first $\psi_{t,\lambda}(x) = \sqrt{\frac{\alpha}{2\pi(\alpha\lambda\cos\alpha t + \sin\alpha t)}}$ so that $\nabla\psi_{t-s,\lambda} \equiv 0$ for $0 \leq s \leq t < T_{\alpha,\lambda}$. Having checked the measurability we have

$$p_t^\lambda(x) = \sqrt{\frac{\alpha}{2\pi(\alpha\lambda\cos\alpha t + \sin\alpha t)}}\exp\{-S_{t,\lambda}(x)\}.$$

It turns out by taking the limits $\lambda \to 0$ that

Theorem 3.3. *For* $0 < t < \frac{\pi}{\alpha}$,

$$
\begin{aligned}
p_t(x) =& \sqrt{\frac{\alpha}{2\pi \sin \alpha t}} \exp\left\{ -\frac{\cos \alpha t}{2\alpha \sin \alpha t} \left(\alpha x + \varepsilon \int_0^t \sin \alpha(t - \sigma) dw_\sigma \right)^2 \right. \\
& + \frac{\varepsilon}{\alpha} \int_0^t \cos \alpha(t - \sigma) dw_\sigma \left(\alpha x + \varepsilon \int_0^t \sin \alpha(t - \sigma) dw_\sigma \right) \\
& - \frac{1}{2}\varepsilon^2 \int_0^t \left[(\int_0^s \cos \alpha(s - \sigma) dw_\sigma)^2 - (\int_0^s \sin \alpha(s - \sigma) dw_\sigma)^2 \right] ds \\
& \left. - \frac{\varepsilon^2}{\alpha} \int_0^t \int_0^s \sin \alpha(s - \sigma) dw_\sigma dw_s \right\}.
\end{aligned}
$$

(3.12)

And as $t \to \frac{\pi}{\alpha}$, $p_t(x) \to +\infty$ *P-a.s. for any* $x \neq -\frac{\varepsilon}{\alpha} \int_0^t \sin \alpha(t - \sigma) dw_\sigma$.

Similarly we consider the Stratonovich heat equation with harmonic oscillator potential

$$
du_t = [\frac{1}{2}\Delta u_t - \frac{1}{2}\alpha^2 x^2 u_t]dt + \varepsilon x u_t \partial w_t.
$$

(3.13)

Here $x \in R^1$. Let $q_t^\lambda(x)$ be the solution of (3.13) with initial condition $q_0^\lambda(x) = \frac{1}{\sqrt{2\pi\lambda}} \exp\{-\frac{x^2}{2\lambda}\}$ and $q_t(x) = \lim\limits_{\lambda \to 0} q_t^\lambda(x)$. (Here more general initial conditions can be considered.)

Theorem 3.4. *For any* $t > 0$,

$$
\begin{aligned}
q_t(x) \\
=& \sqrt{\frac{\alpha}{2\pi \sinh \alpha t}} \exp\left\{ -\frac{\cosh \alpha t}{2\alpha \sinh \alpha t} \left(\alpha x + \varepsilon \int_0^t \sinh \alpha(t - \sigma) dw_\sigma \right)^2 \right. \\
& + \frac{\varepsilon}{\alpha} \int_0^t \cosh \alpha(t - \sigma) dw_\sigma \left(\alpha x + \varepsilon \int_0^t \sinh \alpha(t - \sigma) dw_\sigma \right) \\
& - \frac{1}{2}\varepsilon^2 \int_0^t \left[(\int_0^s \cosh \alpha(s - \sigma) dw_\sigma)^2 + (\int_0^s \sinh \alpha(s - \sigma) dw_\sigma)^2 \right] ds \\
& \left. - \frac{\varepsilon^2}{\alpha} \int_0^t \int_0^s \sinh \alpha(s - \sigma) dw_\sigma dw_s \right\}.
\end{aligned}
$$

(3.14)

Corollary 3.2. *For* $\varepsilon = 0$, *then*

$$
p_t(x) = \sqrt{\frac{\alpha}{2\pi \sin \alpha t}} \exp\{-\frac{\alpha \cos \alpha t}{2 \sin \alpha t}x^2\}
$$

(3.15)

and

$$
q_t(x) = \sqrt{\frac{\alpha}{2\pi \sinh \alpha t}} \exp\{-\frac{\alpha \cosh \alpha t}{2 \sinh \alpha t}x^2\}.
$$

(3.16)

Formula (3.15) was given in Elworthy, Truman and Zhao (1994) and (3.16) is the classical Mehler formula (c.f. Berline, Getzler and Vergne (1992)). Therefore we call (3.14) the stochastic Mehler formula. Having dealt with the harmonic oscillator case since the Itô correction term involves only a harmonic oscillator potential we can write the formula for stochastic heat equations of Itô's type.

We shall demonstrate another shock wave for the harmonic oscillator. Consider the deterministic equation first

$$\frac{\partial}{\partial t}g(t,x) + g(t,x)\frac{\partial}{\partial x}g(t,x) = -x, \quad g(0,x) = x. \tag{3.17}$$

From formula (3.10) where we take $\lambda = 1$, $\alpha = 1$, $\varepsilon = 0$ we have

$$g(t,x) = \nabla S_\lambda(t,x) = \frac{\cos t - \sin t}{\cos t + \sin t}x. \tag{3.18}$$

For $t < \frac{3}{4}\pi$, $g(t,x)$ is smooth and as $t \uparrow \frac{3}{4}\pi$, $g(t,x) \to +\infty$ for $x < 0$ and $g(t,x) \to -\infty$ for $x > 0$. Note $\frac{3}{4}\pi$ is the caustic time of the classical mechanics and Theorem 2.2. If we consider the stochastic Burgers' equation

$$dv(t,x) + v(t,x)\frac{\partial}{\partial x}v(t,x)dt = -xdt - \varepsilon dw_t, \quad v(0,x) = x, \tag{3.19}$$

then it follows from (3.10) where $\alpha = \lambda = 1$,

$$v(t,x) = \frac{\cos t - \sin t}{\cos t + \sin t}\left(x + \varepsilon\int_0^t \sin(t-\sigma)dw_\sigma\right) - \varepsilon\int_0^t \cos(t-\sigma)dw_\sigma. \tag{3.20}$$

Therefore for $t < \frac{3}{4}\pi$, $v(t,x)$ is well defined and continuous and as $t \uparrow \frac{3}{4}\pi$, $v(t,x) \to +\infty$ P-a.s. for $x < -\varepsilon\int_0^{\frac{3}{4}\pi}\sin(\frac{3}{4}\pi - \sigma)dw_\sigma$ and $v(t,x) \to -\infty$ P-a.s. for $x > -\varepsilon\int_0^{\frac{3}{4}\pi}\sin(\frac{3}{4}\pi - \sigma)dw_\sigma$. This is also a counter-example of our first conjecture when e_1 is not bounded.

§4. Stochastic Schrödinger Equations

A. We consider the following stochastic Schrödinger equation in R^n

$$i\hbar d\psi_t = -\frac{\hbar^2}{2}\Delta\psi_t dt + \psi_t\langle \varepsilon x, \partial w_t\rangle. \tag{4.1}$$

From the Hamilton Jacobi theory, the heat kernel of (4.1) is given by

$$\psi_t(x,y) = \frac{1}{(2\pi\hbar it)^{\frac{n}{2}}}\exp\left\{\frac{1}{\hbar i}\left[\langle \varepsilon x, w_t\rangle + \frac{1}{2}\int_0^t |\varepsilon^T w_\sigma|^2 d\sigma \right.\right.$$
$$\left.\left. - \frac{r^2(x + \int_0^t \varepsilon^T w_\sigma d\sigma, y)}{2t}\right]\right\}. \tag{4.2}$$

Here $x, y \in R^n$, r is the Euclidean metric in R^n, w_t is a m-dimensional Brownian motion and ε is a $m \times n$ matrix, and $\langle \cdot, \cdot \rangle$ is the inner product in R^m. Formula (4.2) can be obtained from the stochastic Hamilton Jacobi theory. Nevertheless we can show that (4.2) is the solution of the stochastic Schrödinger equation (4.1) by Itô's formula. In fact

$$
\begin{aligned}
d\psi_t(x,y) = \Bigg\{ &-\frac{n}{2t}dt + \frac{1}{\hbar i}\Bigg[\langle \varepsilon x, dw_t\rangle + \frac{1}{2}\langle \varepsilon x, \varepsilon x\rangle dt + \frac{1}{2}|\varepsilon^T w_t|^2 dt \\
&-\frac{\langle x + \int_0^t \varepsilon^T w_\sigma d\sigma - y, \varepsilon^T w_t\rangle_{R^n}}{t}dt \\
&+\frac{r^2(x + \int_0^t \varepsilon^T w_\sigma d\sigma, y)}{2t^2}dt\Bigg]\Bigg\} \psi_t(x,y) \\
= \frac{1}{\hbar i}\Bigg\{ &\Bigg[\frac{n\hbar^2}{2t\hbar i} + \frac{1}{2}|\varepsilon^T w_t|^2 - \frac{\langle x + \int_0^t \varepsilon^T w_\sigma d\sigma - y, \varepsilon^T w_t\rangle_{R^n}}{t} \\
&+\frac{r^2(x + \int_0^t \varepsilon^T w_\sigma d\sigma, y)}{2t^2}\Bigg]dt + \langle \varepsilon x, \partial w_t\rangle\Bigg\} \psi_t(x,y).
\end{aligned}
$$

Here $\langle \cdot, \cdot \rangle_{R^n}$ denotes the inner product in R^n. Moreover

$$
\nabla \psi_t(x,y) = \frac{1}{\hbar i}(\varepsilon^T w_t - \frac{x + \int_0^t \varepsilon^T w_\sigma d\sigma - y}{t})\psi_t(x,y).
$$

Therefore

$$
\begin{aligned}
&-\frac{\hbar^2}{2}\Delta\psi_t(x,y) \\
&= -\frac{\hbar^2}{2}\Bigg[\frac{1}{-\hbar^2}|\varepsilon^T w_t - \frac{x + \int_0^t \varepsilon^T w_\sigma d\sigma - y}{t}|^2 + \frac{1}{\hbar i}(-\frac{n}{t})\Bigg]\psi_t(x,y) \\
&= \Bigg[\frac{1}{2}|\varepsilon^T w_t - \frac{x + \int_0^t \varepsilon^T w_\sigma d\sigma - y}{t}|^2 + \frac{n\hbar^2}{2t\hbar i}\Bigg]\psi_t(x,y).
\end{aligned}
$$

B. An immediate application to the above of the heat kernel formula (4.2) is that we can easily calculate the wave function of the particle at time t subject to the white noise potential for the initial condition

$$
\psi_0(x) = \exp\left\{-\frac{\langle p_0, x\rangle_{R^n}}{\hbar i} - \frac{|x - a|^2}{2\delta^2}\right\}. \tag{4.3}
$$

In the limit $h \to 0$ this boundary condition is equivalent to fixing the initial momentum of the wave-particle to be p_0.

Theorem 4.1. Let $\psi_t(x)$ be the solution of (4.1) with initial condition (4.3). Then

$$\lim_{\hbar \to 0} \exp \left\{ -\frac{1}{\hbar i} [\langle \varepsilon x, w_t \rangle + \frac{1}{2} \int_0^t |\varepsilon^T w_\sigma|^2 d\sigma \right.$$

$$\left. - \langle p_0, x + \int_0^t \varepsilon^T w_\sigma d\sigma \rangle_{R^n}] \right\} \psi_t(x) \qquad (4.4)$$

$$= \exp \left\{ -\frac{|x + \int_0^t \varepsilon^T w_\sigma d\sigma - a - p_0 t|^2}{2\delta^2} \right\}.$$

Theorem 4.1 tells us a lot of information about the quantum mechanical particle subjected to an external force field associated with the white noise potential. One sees that the centre of the wave packet and its velocity evolve in time like an integral of the Wiener process and the Wiener process, respectively.

As a simple consequence of Theorem 4.1, we have for fixed t

$$\lim_{\hbar \to 0} \int_{R^n} |\psi_t(x)|^2 dx = \int_{R^n} |\psi_0(x - p_0 t + \int_0^t \varepsilon^T w_\sigma d\sigma)|^2 dx$$

$$= \int_{R^n} |\psi_0(x)|^2 dx. \qquad (4.5)$$

For the simplicity in the following we take R^1 as our configuration space. (Every thing carries over to R^n). In the case of the configuration R^1, we have for any $\beta_1 < \beta_2$,

$$\lim_{\hbar \to 0} \int_{\beta_1 + p_o t - \int_0^t \varepsilon^T w_\sigma d\sigma}^{\beta_2 + p_o t - \int_0^t \varepsilon^T w_\sigma d\sigma} |\psi_t(x)|^2 dx$$

$$= \int_{\beta_1 + p_o t - \int_0^t \varepsilon^T w_\sigma d\sigma}^{\beta_2 + p_o t - \int_0^t \varepsilon^T w_\sigma d\sigma} |\psi_0(x - p_0 t + \int_0^t \varepsilon^T w_\sigma d\sigma)|^2 dx \qquad (4.6)$$

$$= \int_{\beta_1}^{\beta_2} |\psi_0(x)|^2 dx.$$

It follows that the probability of the quantum mechanical particle being in $[\beta_1, \beta_2]$ at time 0 equals the probability of the quantum mechanical particle being in $[\beta_1 + p_o t - \int_0^t \varepsilon^T w_\sigma d\sigma, \beta_2 + p_o t - \int_0^t \varepsilon^T w_\sigma d\sigma]$ in time t. Thus, the support of the wave function is concentrated in a neighbourhood of the flow of the following stochastic Hamiltonian system

$$d\dot{\Phi}_t = -\varepsilon^T dw_t, \Phi_0(x) = x, \dot{\Phi}_0 = p_0, \qquad (4.7)$$

which is a special case of the stochastic Hamiltonian system (2.1). In this sense stochastic quantum mechanics (corresponding to the stochastic Schrödinger equation) converges to stochastic classical mechanics (corresponding to the stochastic Hamiltonian system) as $\hbar \to 0$. The deterministic version of this result was first proved in Truman (1977).

Acknowledgements

We wish to thank the conference organizer, Alison Etheridge, for her kind invitations and hospitality during the conference and the referee for his/her useful comments to this paper. We would also like to acknowledge the support from SERC grant GR/J34095.

References

[1] S. Albeverio, A. Hilbert, E. Zehnder (1990), Hamiltonian systems with stochastic forces: nonlinear versus linear and a Girsanov formula, Preprint.

[2] N. Berline, E. Getzler and M. Vergne (1992), Heat kernels and Dirac operators, Berlin, Heidelberg: Springer-Verlag.

[3] I. Davies and A. Truman (1982), Laplace Asymptotic Expansions of Conditional Wiener Integrals and Generalised Mehler Kernel Formulas, J. Math. Phys. Vol. 23, 2059-2070.

[4] K. D. Elworthy and A. Truman (1981), Classical mechanics, the diffusion (heat) equation and the Schrödinger equation on a Riemannian manifold, J. Math. Phys., Vol. 22, No.10, 2144-2166.

[5] K. D. Elworthy and A. Truman (1982), The diffusion equation and classical mechanics: an elementary formula, in: Stochastic Processes in Quantum Physics, ed. S. Albeverio et al, Lecture Notes in Physics, No.173, Springer-Verlag, 136-146.

[6] K. D. Elworthy, A. Truman and H.Z. Zhao (1994), Approximate travelling waves for the generalized KPP equations and classical mechanics, Proceedings of the Royal Society of London, Series A, Vol. 446, 529-554.

[7] K. D. Elworthy (1982), Stochastic differential equations on manifolds, London Mathematical Society Lecture Notes Series 70, Cambridge University Press.

[8] K. D. Elworthy (1988), Geometric aspects of diffusions on manifolds. LNM 1362, pp. 277-425, Springer-Verlag.

[9] W. H. Fleming (1986), Controlled Markov processes and viscosity solution of nonlinear evolution equations, Lezioni Fermiane, Scuola Normale Superiore di Pisa.

[10] W. H. Fleming (1969), The Cauchy problem for a nonlinear first order partial differential equations, J. Diff. Equn. Vol. 5, 515-530.

[11] H. Holden, T. Lindstrom, B. Oksendal, J. Uboe and T. S. Zhang (1994), The Burgers' equation with a noise force and the stochastic heat equations, Comm. PDE., Vol. 19, 119-141.

[12] E. Hopf (1950), The partial differential equation $u_t + u u_x = \mu u_{xx}$, Comm. Pure and Appl. Math. 3(3), 201-230.

[13] M. Kardar, G. Parisi and Y.C. Zhang (1986), Dynamical scaling of growing interfaces, Phys, Rev. Letter, Vol. 56, 889-892.

[14] L. Markus & A. Weerasinghe (1988), Stochastic oscillators, J. Diff. Equn., Vol. 21, 288-314.

[15] V.P. Maslov (1972), Theorie des perturbations et methods asymptotiques, Dunod, Paris.

[16] A. S. Mishchenko, V. E. Shatolov and B. Yu. Sternin (1990), Lagrangian manifolds and the Maslov operator, Berlin, Heidelberg: Springer-Verlag.

[17] M.N. Ndumu (1986), An elementary formula for the Dirichlet heat kernel on Riemannian manifolds, in: From Local Times to Global Geometry, Control and Physics (ed. K. D. Elworthy), Pitman Research Notes in Mathematics Series, 150, Longman and Wiley, 320-328.

[18] M.N. Ndumu (1991), The heat hernel formula in a geodesic chart and some applications to the eigenvalue problem of 3-sphere, Probability Theory and Related Fields, Vol. 88, No.3, 343-361.

[19] T. J. Newman (1994), Exact results for a model of interface growth, submitted to Phys. Rev. E..

[20] B. Oksendal (1994), Stochastic partial differential equations and applications to hydrodynamics, Preprint series 9 of Institute of Mathematics, University of Oslo.

[21] L.C.G. Rogers & D. Williams (1987), Diffusions, Markov processes and martingales, Vol. 2. Itô Calculus, John Wiley & Sons.

[22] J. Smoller (1983), Shock waves and reaction-diffusion equations, New York:
Springer-Verlag.

[23] A. Truman (1977), Classical mechanics, the diffusion (heat) equation, and the
Schrödinger equation, J. Math. Phys., Vol.18, No.12, 2308-2315.

[24] A. Truman and H.Z. Zhao (1994A), The stochastic Hamilton Jacobi equation, stochastic heat equation and Schrödinger equation, submitted to Stochastic Processes and Applications.

[25] A. Truman and H.Z. Zhao (1994B), The stochastic diffusion equations and the stochastic Burgers' equations, (preprint).

[26] R.S.R. Varadhan (1967), Diffusion process in a small time interval, Comm. Pur. and Appl. Math. XX, 659-685.

[27] K. D. Watling (1992), Formulae for solutions to (possibly degenerate) diffusion equation exhibiting semi-classical asymptotics, in: Stochastics and Quantum Mechanics (ed. A. Truman & I.M. Davies), World Scientific, 248-271.

[28] A.D. Wentzell and M.I. Freidlin (1970), On small random perturbations of dynamical systems, Russian Math. Surveys, Vol.25, 1-55.

On backward filtering equations for SDE systems (direct approach)

Veretennikov, A.Yu.
Institute for Information Transmission Sciences
Russian Academy of Sciences
19 Ermolovoy str., 101447, Moscow, Russia *
e-mail: veretenn@ippi.ac.msk.su

Abstract

We propose a direct approach to linear backward filtering equations for SDE systems. This approach is a development of a Krylov's idea of a backward linear partial stochastic equation for an ordinary SDE solution. The author's hope is that this method can be applied to a vide variety of filtering problems.

1 Main result

Filtering theory is, in fact, one of the sources of stochastic partial differential equations (SPDE's).

We consider the filtering problem for the most simple two-dimensional stochastic differential equation system

$$dX_t = f(X_t)dt + dw_t^1, \quad X_0 = x,$$
$$dY_t = h(X_t)dt + dw_t^2, \quad Y_0 = y, \tag{1}$$

where functions f and h are smooth and bounded, namely, we assume that both of them have *three* bounded derivatives, w^1 and w^2 are independent standard Wiener processes on some probability space $(\Omega, F, (F_t, t \geq 0), P)$. The problem is to describe the estimate of unobservable process X_t via observable component Y_t, $0 \leq s \leq t$, which is optimal in a regular mean–square sense, i.e., $m_t = E[X_t|F_t^Y]$, $t \geq 0$. The answer is, in fact, known, even for a

*This work was fulfilled, to great extent, due to the support of the organizers of the Conference on SPDE's in Edinburgh University, and Grant # R82000 from the International Science Foundation

much more general situation, see [6]. In particular, m_t may be represented via backward stochastic differential equations. Backward equations make sense if we are interested in an optimal estimation for some fixed time t; in this case we should find the solution of our backward SPDE and substitute there the path of our observable process. The idea of filtering problem representation via linear equation for non-normalized density was first introduced by Zakai [9] and then Rozovskii [7]. Backward filtering equations were first studied by Kushner [4] and Pardoux [5]. Kushner proposed the equation (3) (see below) without any proof, but he has investigated an approximation method for that equation instead. Pardoux was the first who has given a rigorous proof of this equation. Both proofs, Pardoux's and Rozovskii's ones, are based on the preliminary study of this SPDE: they show existence and uniqueness of its solution in some classes of functions, and then prove that this solution coincides with m_T with the help of Ito's formula.

In this paper we propose a *direct* approach to such a representation using a similar idea for an equation without filtering which was introduced by Krylov (see [2], [3], [6], [8]). This method was used earlier in the problem of first integrals and stochastic Liouville equations for diffusion processes. There's no need to study existence and uniqueness using this approach. It seems that the application of that idea to filtering equations is new; see [3], though. Actually, the origin of the very idea is closely connected with Kolmogorov's backward equations for Markov process densities. We consider the case of unit diffusions in both components which is included in [5] and [6], but we do hope that the same method can be used for degenerate case also. This may be regarded as one more "justification" for this paper.

Due to Girsanov's theorem, the process Y_t, $0 \leq t \leq T$ is a Wiener process on some probability space $(\Omega, F, (F_t, 0 \leq t \leq T), \tilde{P})$ (see below).

Theorem 1 *Process m_t may be represented as follows:*

$$m_T = \frac{v(0, x)}{v^1(0, x)}, \tag{2}$$

where functions v and v^1 satisfy the following linear backward stochastic differential equation:

$$-dv(t, x) = [(1/2)v_{xx}(t, x) + f(x)v_x(t, x)]dt + h(x)v(t, x) \star dY_t, \quad 0 \leq t \leq T, \tag{3}$$

with initial data

$$v(T, x) = x, \quad x \in R^1 \tag{4}$$

on probability space $(\Omega, F, (F_t, 0 \leq t \leq T), \tilde{P})$.

Here in (3), which is understood in an integral form, $\star dY_t$ means "backward" stochastic Ito integral, i.e. "regular" stochastic Ito integral with inverse time, see [3], [6]. Function v^1 has it's "initial data" $v^1(T, x) \equiv 1$ and satisfies the same SPDE (3). Recall that random function $v(t, x)$ is, in fact, $F_{t,T}^{w^1,w^2}$ adapted, so that such an integral makes sense. See [6], Theorem 6.3.1.

Before the proof we recall the following Krylov's result concerning multi-dimensional SDE (see [3], [6], [8]):

Consider the family of processes $(Z_t^{s,z}, t \geq s, s \geq 0, z \in R^d)$ which satisfy the following *multidimensional* SDE's:

$$dZ_t^{s,z} = b(Z_t^{s,z})dt + \sigma(Z_t^{s,z})dw_t, \quad t \geq s, \quad Z_s^{s,z} = z, \tag{5}$$

where b is a d-dimensional vector, σ is a matrix $d \times d_1$, $b, \sigma \in C_b^3$, w_t is a d_1-dimensional Wiener process. We'll use the following notations:

$$Z_t^{s,z} \equiv Z(s,t,z),$$

and for $t = T$ also

$$Z_T^{s,z} = u(s,z).$$

Recall that T is always fixed.

Theorem 2 *(Krylov) Function $u(s,z)$ satisfies SPDE*

$$-du(t,z) = [(1/2)(\sigma\sigma^*)_{ij}(z)u_{z_iz_j}(t,z) + b^i(z)u_{z_i}(t,z)]dt$$
$$+\sigma(z)_{ij}u_{z_i}(t,z) \star dw_t^j \tag{6}$$

with "initial data"

$$u(T,z) \equiv z. \tag{7}$$

here σ^* means σ transposed; equation (6) holds true for each component of vector $u(t,z)$, i.e. it's a vector equation.

This fact was discovered by N.Krylov in the seventies, see his exposition in [3]; the direct approach to this result may be found in [8].

We'll use also one well-known lemma which is, in fact, nothing more but Bayes formula for conditional expectations, see [6].

Lemma 1

$$m_T = \frac{\tilde{E}[X_T\rho^{-1}|F_T^Y]}{\tilde{E}[\rho^{-1}|F_T^Y]},$$

where $\tilde{E}$ is the expectation with respect to measure $\tilde{P}$: $d\tilde{P} = \rho dP$, with

$$\rho \equiv \rho_{0T} = \exp(-\int_0^T h(X_t)dw_t^2 - (1/2)\int_0^T |h(X_t)|^2 dt).$$

Recall that due to Girsanov's theorem the process $(Y_t, 0 \leq t \leq T)$ is a Wiener process on the probability space $(\Omega, F, (F_t, 0 \leq t \leq T), \tilde{P})$, in general, with a nonzero starting value. Denote $\tilde{w}_t = Y_t$. Then our system (1) on the space $(\Omega, F, (F_t, 0 \leq t \leq T), \tilde{P})$ has the form

$$dX_t = f(X_t)dt + dw_t^1, \quad X_0 = x,$$
$$dY_t = d\tilde{w}_t, \quad Y_0 = y. \tag{8}$$

2   Proof of theorem 1

The main idea is to use a partition of the time interval and then to apply directly the evolutionary property for markovian SDE's. To make the exposition clearer we first use directly the evolutionary property while in the following steps we rather apply Theorem 2. The non-trivial point is then to change correctly sigma-fields in conditional expectations.

1. Denote

$$v(s,x) = \tilde{E}[X_T^{s,x} \rho_{s,T}^{-1} | F_{s,T}^Y].$$

Actually, we'd like to obtain the equality

$$v(0,x) - v(T,x)$$
$$= \int_0^T [(1/2)v_{xx}(t,x) + f(x)v_x(t,x)]dt + \int_0^T h(x)v(t,x) \star d\tilde{w}_t, \qquad (9)$$

Let us use the identity

$$v(0,x) - v(T,x) = \sum_{i=1}^N (v(t_{i-1},x) - v(t_i,x))$$

for any partition $0 = t_0 < t_1 < \ldots < t_N = T$. Consider one term from this sum: we have

$$v(t_{i-1},x) - v(t_i,x)$$
$$= \tilde{E}[\rho_{t_{i-1},T}^{-1}X(t_{i-1},T,x)|F_{t_{i-1},T}^Y] - \tilde{E}[\rho_{t_i,T}^{-1}X(t_i,T,x)|F_{t_i,T}^Y]$$
$$= \tilde{E}[\rho_{t_{i-1},T}^{-1}X(t_{i-1},T,x)|F_{t_{i-1},T}^{\tilde{w}}] - \tilde{E}[\rho_{t_i,T}^{-1}X(t_i,T,x)|F_{t_i,T}^{\tilde{w}}].$$

2. Now, due to the evolution property (or strong uniqueness plus Markov property plus continuity properties) of the family $X(s,T,x)$ with respect to all variables [1] we get (a.s.)

$$X(t_{i-1},T,x) = X(t_i,T,X_{t_i}^{t_{i-1},x})$$

$$= X(t_i,T,x) + X_x(t_i,T,x)(X_{t_i}^{t_{i-1},x} - x) + (1/2)X_{xx}(t_i,T,x)(X_{t_i}^{t_{i-1}} - x)^2 + \alpha_i^1$$

$$= X(t_i,T,x) + X_x(t_i,T,x)(f(x)\Delta t_i + \Delta w_{t_i}^1) + (1/2)X_{xx}(t_i,T,x)\Delta t_i + \alpha_i^2,$$

where $\Delta t_i = t_i - t_{i-1}$, $\Delta w_{t_i}^j = w_{t_i}^j - w_{t_{i-1}}^j$, and $|\alpha_i^1| + |\alpha_i^2| = o(\Delta t_i)$ in a regular mean-square sense.

Indeed, by virtue of Taylor's expansion formula in an integral form we get

$$X(t_i,T,x') - X(t_i,T,x) = X_x(t_i,T,x)(x' - x) + (1/2)X_{xx}(t_i,T,x)(x' - x)^2$$

$$+ \int_0^1 \int_0^1 \int_0^1 X_{xxx}(t_i,T,\alpha\beta\gamma x' + (1 - \alpha\beta\gamma)x)\alpha^2\beta d\alpha d\beta d\gamma (x' - x)^3.$$

Thus,

$$\alpha_i^1 = \int_0^1 \int_0^1 \int_0^1 X_{xxx}(t_i, T, \alpha\beta\gamma X_{t_i}^{t_{i-1},x} + (1 - \alpha\beta\gamma)x)\alpha^2\beta d\alpha d\beta d\gamma (X_{t_i}^{t_{i-1},x} - x)^3.$$

One can easily obtain the estimates

$$E|X(t_i, T, x) - x|^m \le C(T, m) < \infty$$

for any $m \ge 0$. Moreover, the derivatives of $X(t, T, x)$ in x satisfy the equations

$$X_x(s, t, x) = 1 + \int_s^t f'(X(s, r, x))X_x(s, r, x)dr,$$

$$X_{xx}(s, t, x) = \int_s^t f'(X(s, r, x))X_{xx}(s, r, x)dr + \int_s^t f''(X(s, r, x))X_x^2(s, r, x)dr,$$

$$X_{xxx}(s, t, x) = \int_s^t f'(X(s, r, x))X_{xxx}(s, r, x)dr$$

$$+3\int_s^t f''(X(s, r, x))X_x(s, r, x)X_{xx}(s, r, x)dr + \int_s^t f'''(X(s, r, x))X_x^3(s, r, x)dr.$$

So, random functions X_x, X_{xx}, X_{xxx} are bounded by some nonrandom constant (remind that T is fixed and $f \in C_b^3$). Moreover, for any $m \ge 0$ the inequalities

$$E|X_s^{t,x} - x|^m \le C_m|t - s|^{m/2} \qquad (|t - s| \le 1)$$

are well known. It's true also under less restrictive linear growth conditions on the coefficients. Now, we get

$$E|\alpha_i^1|^{2m} \le CE|X_{t_i}^{t_{i-1},x} - x|^{6m} \le C(t_i - t_{i-1})^{3m},$$

or, equivalently,

$$(E|\alpha_i^1|^{2m})^{1/2} \le C(t_i - t_{i-1})^{3m/2}$$

(here C may be different). Since f is smooth and bounded (actually, the linear growth condition for f is also sufficient), the difference $\alpha_i^1 - \alpha_i^2$ satisfies a similar inequality,

$$(E|\alpha_i^1 - \alpha_i^2|^{2m})^{1/2} \le C(t_i - t_{i-1})^{5m/2} \le C(t_i - t_{i-1})^{3m/2} \qquad (t_i - t_{i-1} \le 1).$$

Thus, we get the same bound for α_i^2,

$$(E|\alpha_i^2|^{2m})^{1/2} \le C(t_i - t_{i-1})^{3m/2}.$$

3. We have

$$\tilde{E}[\rho_{t_{i-1},T}^{-1} X(t_{i-1}, T, x)|F_{t_{i-1},T}^{\tilde{w}}]$$

$$= \tilde{E}[\rho_{t_{i-1},T}^{-1} X(t_i, T, X_{t_i}^{t_{i-1},x})|F_{t_{i-1},T}^{\tilde{w}}]$$

$$= \tilde{E}[\rho_{t_{i-1},T}^{-1}\{X_T^{t_i,x} + X_x(t_i, T, x)f(x)\Delta t_i$$

$$+X_x((t_i, T, x)\Delta w_{t_i}^1 + (1/2)X_{xx}(t_i, T, x)\Delta t_i\}|F_{t_{i-1},T}^{\tilde{w}}] + \alpha_i^3.$$

Here again $E|\alpha_i^3|^{2m} = O(\Delta t_i)^{3m/2}$ due to Hölder's inequality for conditional expectations and bounds like $(E\rho^{-4m})^{1/2} \leq C_m$.

4. Now, we'd like to replace $\rho_{t_{i-1},T}^{-1}$ by $\rho_{t_i,T}^{-1}$. For this aim we apply theorem 2 to the process $(X_t^{s,x}, \rho_{s,t}^{-1}, t \geq s)$. Let us note that this two-dimensional process satisfies the following SDE system:

$$dX_t^{s,x} = f(X_t^{s,x})dt + dw_t^1, \quad X_s^{s,x} = x,$$
$$d\rho_{s,t}^{-1} = h(X_t^{s,x})\rho_{s,t}^{-1}d\tilde{w}_t, \quad \rho_{s,s}^{-1} = 1, \quad s \leq t \leq T. \tag{10}$$

Indeed, $\rho_{s,t}^{-1}$ has the following representation:

$$\rho_{s,t}^{-1} = \exp(\int_s^t h(X_r^{s,x})d\tilde{w}_r - (1/2)\int_s^t |h(X_r^{s,x})|^2 dr).$$

Let us consider a bit more general set of processes $\{(X_t^{s,x}, \rho_{s,t}^{-1,\xi})\}$ which satisfy SDE's

$$dX_t^{s,x} = f(X_t^{s,x})dt + dw_t, \quad X_s^{s,x} = x,$$
$$d\rho_{s,t}^{-1,\xi} = h(X_t^{s,x})\rho_{s,t}^{-1,\xi}d\tilde{w}_t, \quad \rho_{s,s}^{-1,\xi} = \xi, \quad s \leq t \leq T, \tag{11}$$

with $\xi \in R$. In fact, $X_t^{s,x}$ here is the same as earlier, and $\rho_t^{-1,\xi}$ has the following representation:

$$\rho_{s,t}^{-1,\xi} = \xi \exp(\int_s^t h(X_r^{s,x})d\tilde{w}_r - (1/2)\int_s^t |h(X_r^{s,x})|^2 dr) = \xi\rho_{s,t}^{-1}.$$

Due to Blagovesh'enski & Freidlin's results [1] the process $\rho_{s,t}^{-1,\xi}$ is twice continuously differentiable in x variable. Then due to theorem 2 we get

$$-d_s\rho_{s,t}^{-1,\xi} = [(1/2)h^2(x)(\rho_{s,t}^{-1,\xi})^2(\rho_{s,t}^{-1,\xi})_{\xi\xi} + (1/2)(\rho_{s,t}^{-1,\xi})_{xx}$$

$$+f(x)(\rho_{s,t}^{-1,\xi})_x]dt + (\rho_{s,t}^{-1,\xi})_x \star dw_t^1 + h(x)(\rho_{s,t}^{-1,\xi})_\xi \star d\tilde{w}_t.$$

Note that, in fact, $(\rho_{s,t}^{-1,\xi})_\xi = \rho_{s,t}^{-1}$ and $(\rho_{s,t}^{-1,\xi})_{\xi\xi} = 0$. Hence,

$$-d_s\rho_{s,t}^{-1,\xi} = [(1/2)(\rho_{s,t}^{-1,\xi})_{xx} + f(x)(\rho_{s,t}^{-1,\xi})_x]dt$$

$$+(\rho_{s,t}^{-1,\xi}) \star dw_t^1 + h(x)\rho_{s,t}^{-1} \star d\tilde{w}_t.$$

So, similarly to bounds for α_i^1 and α_i^2, we get

$$\rho_{t_{i-1},T}^{-1,\xi} - \rho_{t_i,T}^{-1,\xi} \equiv -\Delta\rho_{t_i}^{-1,\xi}$$

$$= [(1/2)(\rho_{t_i,T}^{-1,\xi})_{xx} + f(x)(\rho_{t_i,T}^{-1,\xi})_x]\Delta t_i + \rho_{t_i,T}^{-1,\xi}\Delta w_{t_i}^1 + h(x)\rho_{t_i,T}^{-1,\xi}\Delta \tilde{w}_{t_i} + \alpha_i^4,$$

with similar inequality for α_i^4 as for previous α_i (below we use this assertion with $\xi = 1$). Indeed, we can use the remark about the bounds under the linear growth conditions. It's the very case. So we have

$$(E|\alpha_i^4|^{2m})^{1/2} \le C(t_i - t_{i-1})^{3m/2}.$$

5. Now, we obtain

$$\tilde{E}[\rho_{t_i,T}^{-1}X(t_{i-1},T,x)|F_{t_{i-1},T}^{\tilde{w}}]$$
$$= \tilde{E}[\{X_T^{t_i,x} + (f(x)X_x(t_i,T,x) + (1/2)X_{xx}(t_i,T,x))\Delta t_i$$
$$+X_x(t_i,T,x)\Delta w_{t_i}^1\}\{\rho_{t_i,T}^{-1} + ((1/2)(\rho_{t_i,T}^{-1})_{xx} + f(x)(\rho_{t_i,T}^{-1})_x)\Delta t_i$$
$$+(\rho_{t_i,T}^{-1})_x\Delta w_{t_i}^1 + h(x)\rho_{t_i,T}^{-1}\Delta \tilde{w}_{t_i}\}|F_{t_i,T}^{\tilde{w}}] + \alpha_i^5 \qquad (12)$$

with

$$(E|\alpha_i^5|^{2m})^{1/2} \le C(t_i - t_{i-1})^{3m/2}, \quad m > 0$$

due to Hölder's inequality.

In particular, for $m = 1$ we obtain the bound

$$(E|\alpha_i^5|^2)^{1/2} \le C(t_i - t_{i-1})^{3/2}.$$

6. Now, note that $F_{t_{i-1},T}^{\tilde{w}} = F_{t_{i-1},t_i}^{\tilde{w}} \bigvee F_{t_i,T}^{\tilde{w}}$ and, moreover, this σ-field is independent from w^1. Using the regular calculus for conditional expectations (cf. [6]), we get

$$\tilde{E}[X_T^{t_i,x}\rho_{t_i,T}^{-1}|F_{t_{i-1},T}^{\tilde{w}}] = \tilde{E}[X_T^{t_i,x}\rho_{t_i,T}^{-1}|F_{t_i,T}^{\tilde{w}}],$$

and in just the same manner we can replace σ-fields $F_{t_{i-1},T}^{\tilde{w}}$ by $F_{t_i,T}^{\tilde{w}}$ in all expressions of (12). Hence, we obtain

$$\tilde{E}[X_T^{t_{i-1},x}\rho_{t_{i-1},T}^{-1}|F_{t_{i-1},T}^{\tilde{w}}]$$

$$= \tilde{E}[X_T^{t_i,x}\rho_{t_i,T}^{-1}|F_{t_{i-1},T}^{\tilde{w}}]$$
$$+\tilde{E}[(1/2)X_T^{t_i,x}(\rho_{t_i,T}^{-1})_{xx} + X_x(t_i,T,x)(\rho_{t_i,T}^{-1})_x + (1/2)X_{xx}\rho_{t_i,T}^{-1}|F_{t_{i-1},T}^{\tilde{w}}]\Delta t_i$$

$$+\tilde{E}[f(x)(X(t_i,T,x)_x\rho_{t_i,T}^{-1} + X(t_i,T,x)(\rho_{t_i,T}^{-1})_x)|F_{t_{i-1},T}^{\tilde{w}}]$$
$$+\tilde{E}[X(t_i,T,x)h(x)\rho_{t_i,T}^{-1}\Delta \tilde{w}_{t_i}|F_{t_{i-1},T}^{\tilde{w}}] + \alpha_i^5$$

$$= \tilde{E}[X(t_i,T,x)\rho_{t_i,T}^{-1}|F_{t_i}^{\tilde{w}}] + \tilde{E}[(1/2)(X(t_i,T,x)\rho_{t_i,T}^{-1})_{xx}|F_{t_i,T}^{\tilde{w}}]$$
$$+\tilde{E}[f(x)(X(t_i,T,x)\rho_{t_i,T}^{-1})_x|F_{t_i,T}^{\tilde{w}}] + \Delta \tilde{w}_{t_i}\tilde{E}[h(x)X(t_i,T,x)\rho_{t_i,T}^{-1}|F_{t_i,T}^{\tilde{w}}] + \alpha_i^5$$
$$= v(t_i,x) + (1/2)v_{xx}(t_i,x)\Delta t_i + f(x)v_x(t_i,x)\Delta t_i + h(x)v(t_i,x)\Delta \tilde{w}_{t_i} + \alpha_i^5.$$

Last equality here holds true because of the possibility to change the order of integration and derivation with respect to x variable.

7. Thus, we get as $\sup \Delta t_i \to 0$ the equality

$$v(0, x) - v(T, x)$$

$$= \sum \{(1/2)v_{xx}(t_i, x) + f(x)v_x(t_i, x)\}\Delta t_i + \sum h(x)v(t_i)\Delta \tilde{w}_{t_i} + \alpha^6$$

with $\alpha^6 = o(1)$, $\sup \Delta t_i \to 0$. As $\sup \Delta t_i \to 0$, we get the desired integral equality (9).

Acknowledgement. The author is thankful to the anonymous referee for helpful comments and remarks.

References

[1] Blagovesh'enski, Yu.A., Freidlin, M.I. Certain properties of diffusion processes depending on a parameter. Soviet Math. Dokl., 1961, vol. 2, 633-636.

[2] Krylov, N.V., Rozovskii, B.L. On the first integrals and Liouville equations for diffusion processes. Lecture Notes in Control and Information Sciences, 1981, vol. 36, 117-125.

[3] Krylov, N.V. On explicit formulas for solutions of evolutionary SPDE's (a kind of introduction to the theory). Lecture Notes in Control and Information Sciences, 1992, vol. 176, 153-164.

[4] Kushner, H.J. Probability methods for approximations in stochastic control and for elliptic equations. N.Y., Acad. Press, 1977.

[5] Pardoux, E. Stochastic partial differential equations and filtering of diffusion processes. Stochastics, 1979, vol. 3, 127-167.

[6] Rozovskii, B.L. Stochastic Evolution Systems. Kluwer Acad. Publ.: Dordreht et al., 1990.

[7] Rozovskii, B.L., Shirjaev, A.N. Reduced form of non-linear filtering equations. In: Suppl. to prepr. of IFAC Symp. on Stochastic control. Budapest, 1974, 59-61.

[8] Veretennikov, A.Yu. Inverse Diffusion and Direct Derivation of Stochastic Liouville Equations. Soviet Math. Notes, 1983, vol. 33, no. 5-6, p. 397-400.

[9] Zakai, M. On the optimal filtering of diffusion processes. Zeitschrift für Wahr. und verw. Geb. 1969, vol.11, 230-243.

Ergodicity of Markov Semigroups

Boguslaw Zegarlinski

Department of Mathematics

Imperial College

London, UK

Abstract: *We present a method for study of the ergodicity problem of Markov semigroups on infinite dimensional spaces. We answer also affirmatively a question, raised during the conference, whether it is possible to apply this method to the case of a Markov semigroup with generator given by a nonsymmetric Dirichlet form.*

1. An Introduction .

Let $P_t \equiv e^{t\mathcal{L}}$, $t \in [0, \infty)$ be a Markov semigroup on the space $\{\mathcal{C}(\Omega), \|\cdot\|_u\}$ of continuous functions on a Polish space Ω, with the supremum norm. Let μ be a probability measure on (Ω, Σ) which is invariant with respect to the semigroup P_t, i.e. satisfying

$$\mu(P_t f) = \mu(f) \tag{1.1}$$

for every $f \in \mathcal{C}$, where by μF we have denoted the expectation of a function F. We would like to discuss the following question:

Ergodicity Problem

Given a set $\mathcal{A} \subset \mathcal{C}(\Omega)$, what is the best estimate of the form

$$\|P_t f - \mu f\| \leq C\gamma(t)\|\|f\|\| \tag{1.2a}$$

with some norm $\|\cdot\|$ on $\mathcal{C}(\Omega)$, or estimate of the form

$$|P_t f - \mu f|(\omega) \leq C(\omega)\gamma(t)\|\|f\|\| \tag{1.2b}$$

with a positive rate function $\gamma(t) \to_{t \to \infty} 0$ independent of $f \in \mathcal{A}$, some positive constants C or $C(\omega)$ which are also (essentially) independent of f (i.e. possibly dependent on some its localization properties), and with some seminorm $\|\|\cdot\|\|$ on $\mathcal{A}$.

Of course the strongest estimate we get when the norm on the left hand side of (1.2a) equals to the supremum norm. However if the underlying space Ω

is noncompact, typically the best that one can get is the pointwise estimate (1.2b). In the first part of this lecture we will discuss these two cases. We will present a powerful method which allows us to get these strongest ergodicity properties. This method is based on use of the Hypercontractivity property of the Markov semigroup or equivalently of Logarithmic Sobolev inequality for the corresponding generator. It has been introduced in [**HS1**]; see also [**SZ2**] and [**Z4**] for some extensions. A general description of this method, extended to include nonsymmetric Markov semigroups, will be given in Section 2. In Section 3 we show how one can use a knowledge about certain conditional measures to prove useful coercive inequalities including Logarithmic Sobolev inequality. Some interesting examples of applications will be given in Section 4.

In many situations the best that we can expect to get is the L_2 ergodicity, i.e. the estimate (1.2a) with the $L_2(\mu)$ norm on the left hand side. We will show that some ideas used to solve the strong ergodicity problem can be useful also in this more complicated situation. This will be described in the last section together with a simple example of application.

2. A Strategy for Strong Ergodicity.

Let us suppose that by an appropriate choice of coordinates in the space Ω we can identify it with a (subset of a) product space. (This can be done, as we have chosen Ω to be a wonderful space.) In many applications this is in fact what we start with. However, as we will see later, the usefulness of a particular choice of coordinates depends on the problem. Thus Ω will be regarded as a space of some sequences $\omega \equiv (\omega_i)_{i \in \Gamma}$, ω_i belonging to a space $\mathbf{M}$ which is either a smooth, connected, (boundaryless), and finite dimensional Riemannian manifold or just a finite set, and with Γ being a countable index set, as e.g. $I\!N$ or $\mathbb{Z}^d$, $d \in I\!N$. We will concentrate on the first case; for a description of the second see [**Z6**] and references therein. For $\Lambda \subset \Gamma$, let Σ_Λ be the smallest σ-algebra such that all the coordinate mappings $\omega \longmapsto \omega_{\mathbf{j}}, \mathbf{j} \in \Lambda$, are measurable. In particular if $\Lambda = \Gamma$, the corresponding subscript will be omitted. By E_Λ we will denote a (regular) conditional expectation with respect to Σ_{Λ^c} associated to the measure μ. Frequently one can construct explicitly a family $\mathcal{E} \equiv \{E_\Lambda\}_{\Lambda \in \mathcal{F}}$, called a local specification, consisting of the probability kernels on (Ω, Σ), indexed by elements of the family $\mathcal{F}$ of all finite subsets of Γ and compatible in the sense that: if $\Lambda_1 \subset \Lambda_2$, then

$$E_{\Lambda_2} E_{\Lambda_1}(F) = E_{\Lambda_2}(F) \tag{2.1}$$

Let us mention that for a finite set $\Lambda \in \mathcal{F}$, the corresponding kernel E_Λ can be regarded essentially as a probability measure on a finite dimensional space $(\mathbf{M}, \mathcal{B}_{\mathbf{M}})^\Lambda$. The local specification can be used to define a probability measure μ on (Ω, Σ), called a Gibbs measure for $\mathcal{E}$, as a solution of the following equation

$$\forall \Lambda \in \mathcal{F} \qquad \mu E_\Lambda(F) = \mu F \tag{DLR}$$

Given a local specification $\mathcal{E}$ on (Ω, Σ) we would like to construct and study a Markov semigroup $P_t \equiv e^{t\mathcal{L}}$ which preserves every Gibbs measure for $\mathcal{E}$. One possible way is the following. We define first local generators $\mathcal{L}_{\mathbf{j}}$ acting only on the coordinate $\omega_{\mathbf{j}}$, respectively, by requiring that

$$E_{\mathbf{j}}\left(g(-\mathcal{L}_{\mathbf{j}}f)\right) = E_{\mathbf{j}}\left(\nabla_{\mathbf{j}}g \cdot \nabla_{\mathbf{j}}f\right) + E_{\mathbf{j}}\left(g\alpha_{\mathbf{j}} \cdot \nabla_{\mathbf{j}}f\right) \tag{2.2}$$

where $\nabla_{\mathbf{j}}$ denotes the gradient operator with respect to the coordinate $\mathbf{j}$, induced by the one given on $\mathbf{M}$, $\alpha_{\mathbf{j}}$ denotes a (sufficiently smooth) vector field and the dot has been used to denote the scalar product inherited from the tangent space $T\mathbf{M}$, (i.e. in some coordinate system we have $\alpha_{\mathbf{j}} \cdot \nabla_{\mathbf{j}}f \equiv \sum_{k,l=1,..,dim\mathbf{M}} \mathbf{h}_{kl}\alpha_{\mathbf{j}}^{(k)}\frac{\partial}{\partial\omega_{\mathbf{j}}^{(l)}}f$, with $\mathbf{h}_{kl}$ denoting the components of the metric tensor, and similarly for $\nabla_{\mathbf{j}}g \cdot \nabla_{\mathbf{j}}f$). In case when the probability measure $E_{\mathbf{j}}$ restricted to $\Sigma_{\mathbf{j}}$ is absolutely continuous with respect to the Riemannian measure and has sufficiently smooth density, one can see that on twice differentiable functions we have

$$\mathcal{L}_{\mathbf{j}}f \equiv \Delta_{\mathbf{j}}f - \beta_{\mathbf{j}} \cdot \nabla_{\mathbf{j}}f + \alpha_{\mathbf{j}} \cdot \nabla_{\mathbf{j}}f \equiv \mathcal{L}^0{}_{\mathbf{j}}f + \alpha_{\mathbf{j}} \cdot \nabla_{\mathbf{j}}f \tag{2.3}$$

where $\Delta_{\mathbf{j}}$ is the Laplace-Beltrami operator with respect to $\omega_{\mathbf{j}}$ and $\beta_{\mathbf{j}}$ is some vector field uniquely determined by the conditional expectation $E_{\mathbf{j}}$. Let us note that in this situation the corresponding adjoint operator $\mathcal{L}_{\mathbf{j}}^*$ in $L_2(E_{\mathbf{j}})$ is given by

$$\mathcal{L}_{\mathbf{j}}^* f = \mathcal{L}_{\mathbf{j}}^0 f - \alpha_{\mathbf{j}} \cdot \nabla_{\mathbf{j}}f + \left(-\beta_{\mathbf{j}} \cdot \alpha_{\mathbf{j}} + div_{\mathbf{j}}\,\alpha_{\mathbf{j}}\right)f \tag{2.4}$$

In order to guarantee later the invariance property for every Gibbs measure we require that the following condition is true

$$-\beta_{\mathbf{j}} \cdot \alpha_{\mathbf{j}} + div_{\mathbf{j}}\,\alpha_{\mathbf{j}} = 0 \tag{2.5}$$

for every $\mathbf{j} \in \Gamma$. Let us note that if this condition is satisfied, the quadratic form (2.2) used to define the operator $\mathcal{L}_{\mathbf{j}}$ can be written as follows

$$E_{\mathbf{j}}\left(g(-\mathcal{L}_{\mathbf{j}}f)\right) = E_{\mathbf{j}}\left(\nabla_{\mathbf{j}}g \cdot \nabla_{\mathbf{j}}f\right) + E_{\mathbf{j}}\left((g - E_{\mathbf{j}}g)\alpha_{\mathbf{j}} \cdot \nabla_{\mathbf{j}}f\right) \tag{2.6}$$

Now for $\Lambda \in \mathcal{F}$ and $\omega \in \Omega$ we define generators

$$\mathcal{L}_{\Lambda,\omega}f \equiv \delta_{\omega|\Lambda^c}\sum_{\mathbf{j}\in\Lambda}\mathcal{L}_{\mathbf{j}}f \equiv \delta_{\omega|\Lambda^c}\left(\mathcal{L}^0{}_{\Lambda}f + \sum_{\mathbf{j}\in\Lambda}\alpha_{\mathbf{j}} \cdot \nabla_{\mathbf{j}}f\right) \tag{2.7}$$

where $\delta_{\omega|\Lambda^c}$ denotes the restriction of the Dirac delta at the point ω to the σ-algebra Σ_{Λ^c}. It is not difficult to see that the corresponding semigroups $P_t^{\Lambda,\omega} \equiv e^{t\mathcal{L}_{\Lambda,\omega}}$ are well defined on $\mathcal{C}(\mathbf{M}^{\Lambda})$. Moreover, the following simple computation shows that the semigroup is positivity preserving:

$$P_t^{\Lambda,\omega}f^2 - (P_t^{\Lambda,\omega}f)^2 = \int_0^t ds\left(\frac{d}{ds}P_s^{\Lambda,\omega}\left(P_{t-s}^{\Lambda,\omega}f\right)^2\right)$$

$$= \int_0^t ds P_s^{\Lambda,\omega}\left(\mathcal{L}_{\Lambda,\omega}(P_{t-s}^{\Lambda,\omega}f)^2 - 2(P_{t-s}^{\Lambda,\omega}f)\mathcal{L}_{\Lambda,\omega}(P_{t-s}^{\Lambda,\omega}f)\right)$$

$$= \int_0^t ds P_s^{\Lambda,\omega}\left(\mathcal{L}^0{}_{\Lambda,\omega}(P_{t-s}^{\Lambda,\omega}f)^2 - 2(P_{t-s}^{\Lambda,\omega}f)\mathcal{L}^0{}_{\Lambda,\omega}(P_{t-s}^{\Lambda,\omega}f)\right) \qquad (2.8)$$

$$= \int_0^t ds P_s^{\Lambda,\omega}|\nabla_\Lambda P_{t-s}^{\Lambda,\omega}f|^2 \geq 0,$$

where we have set

$$|\nabla_\Lambda F|^2 \equiv \sum_{j\in\Lambda}|\nabla_j F|^2, \qquad (2.9)$$

Because of our assumption (2.5), for the operator $\mathcal{L}_{\Lambda,\omega}^*$, the adjoint of $\mathcal{L}_{\Lambda,\omega}$ in $L_2(E_\Lambda^\omega)$, we have also

$$\mathcal{L}_{\Lambda,\omega}^*1 = 0 \qquad (2.10)$$

Thus the semigroup $P_t^{\Lambda,\omega}$ is a Markov semigroup preserving the conditional measure E_Λ^ω. Under some mild assumptions on α_j's and β_j's (see Section 4.) one can simply extend the constructions of [**SZ2**] (in the compact case) and [**Z4**] (in noncompact case) and show that the following limit exists

$$P_t f = \lim_{\Lambda\to\Gamma} P_t^{\Lambda,\omega}f \qquad (2.11)$$

and defines a Markov semigroup on $\mathcal{C}(\Omega)$, for which every Gibbs measure for $\mathcal{E}$ is invariant. The limit (2.11) is independent of external conditions $\omega \in \Omega$. Suppose the set Γ is equipped with a metric $d(\cdot,\cdot)$. and suppose additionally that the system is of finite range $R \in \mathbb{N}$ in the sense that

$$\nabla_k\alpha_j = 0 \quad \text{and} \quad \nabla_k\beta_j = 0 \qquad (2.12)$$

for $d(\mathbf{k},\mathbf{j}) > R$, then it is possible to have the following strong control on the convergence of the sequence of finite volume semigroups.

Exponential Approximation Property

For any $A \in (0,\infty)$ and any $f \in \mathcal{C}(\Omega)$ which is Σ_X-measurable for some $X \in \mathcal{F}$, we have

$$|P_t f(\eta) - P_t^{\Lambda,\omega}f(\eta)| \leq B(\eta)e^{-At}|||f||| \qquad (2.13)$$

with some constant $B \equiv B(\eta) \in (0,\infty)$ dependent only on $\eta \in \Omega$ and $X \in \mathcal{F}$, and with some seminorm $||| \cdot |||$, provided that

$$d(f,\Lambda^c) \equiv d(X,\Lambda^c) \geq Ct \qquad (2.14)$$

with a constant $C \in (0, \infty)$ dependent only on the choice of A.

This property will be the first ingredient in our strategy.

Remarks:
Typically when the space Ω is compact one can choose the constant B independent of η. Strictly speaking the assumption of finite range and so of the exponential approximation property is not necessary for our strategy to work. It would work even if we would have a weaker approximation property with the exponential factor replaced by the one converging to zero more slowly. In this case however, we would get a slower decay to equilibrium.

The construction of the Markov semigroups on infinite dimensional product space and the exponential approximation property have been well known for a long time for spin systems with compact configuration space, see [**Li1**] and references therein. For the noncompact case an appropriate general construction together with the corresponding approximation property has been given recently in [**Z4**], (see also [**Di**] for some special cases by a different method).

The second ingredient is the following property.

<u>Weak Ultracontractivity:</u>

There is $T \in (0, \infty)$ and a positive function $I(\Lambda, \eta)$ such that

$$|P_T^{\Lambda, \eta} f(\eta)| \leq I(\Lambda, \eta) \left(E_\Lambda^\eta (f^2) \right)^{\frac{1}{2}} \tag{2.15}$$

and for any $\epsilon > 0$ we have

$$\lim_{d(0, \Lambda^c) \to \infty} e^{\epsilon d(0, \Lambda^c)} \log I(\Lambda, \eta) = 0 \tag{2.16}$$

for any sequence of Λ's with the volume growing polynomially with the distance from the origin.

In case of compact configuration space one can get a stronger property with a function I independent of η and growing exponentially in the cardinality of the set Λ, and with $L_1(E_\Lambda^\omega)$ norm on the right hand side. One can derive it there as a consequence of Classical Sobolev inequalities via the use of Nash's inequality or using an appropriate Feynman - Kac formula. One gets trivially the similar property when the space **M** is finite and the corresponding conditional measures are strictly positive. It is the case of noncompact configuration space which requires the dependence on η, see [**Z4**]. Finally let us mention that the estimate (2.15) one can also get for some infinite dimensional situations (encountered in the Euclidean field theory). Therefore one may expect that our strategy can be extended to this case.

The third and the last ingredient in our strategy is the following property of a Markov semigroup:

Hypercontractivity

A Markov semigroup $P_t \equiv e^{t\mathcal{L}}$, $t \geq 0$ *preserving a probability measure* μ *is called Hypercontractive iff*

$$||P_t f||_{L_q(\mu)} \leq ||f||_{L_2(\mu)} \tag{H}$$

for $q \equiv q(t)$ *defined by*

$$q(t) \equiv 1 + e^{\frac{2}{c}t} \tag{2.17}$$

with some constant $c \in (0, \infty)$.

Abbreviating $f_t \equiv P_t f$ and $||f_t||_q \equiv ||f_t||_{L_q(\mu)}$, with $q = q(t)$ as in (2.17), by classical calculation of Gross [**G**], with a positive function f we get

$$\frac{d}{dt} \log ||f_t||_q = \frac{\frac{d}{dt}q}{q||f_t||_q^q} \left\{ \mu f_t^q \log \frac{f_t}{||f_t||_q} + \frac{q}{2(q-1)} c\mu(f_t^{q-1}(\mathcal{L}f_t)) \right\} \tag{2.18}$$

Thus the hypercontractivity property **H** is equivalent to the following family of inequalities

$$\mu f^q \log \frac{f}{||f||_q} \leq \frac{q}{2(q-1)} c\mu(f^{q-1}(-\mathcal{L}f)) \tag{2.19}$$

Since we have

$$\mu(f^{q-1}\alpha_{\mathbf{j}} \cdot \nabla_{\mathbf{j}} f) = \frac{1}{q}\mu(\alpha_{\mathbf{j}} \cdot \nabla_{\mathbf{j}} f^q) = \frac{1}{q}\mu(E_{\mathbf{j}}\alpha_{\mathbf{j}} \cdot \nabla_{\mathbf{j}} f^q) = 0, \tag{2.20}$$

where in the last step we have applied our invariance condition (2.5), we get that

$$\mu(f^{q-1}(-\mathcal{L}f)) = \mu(f^{q-1}(-\mathcal{L}^0 f)) \tag{2.21}$$

Hence we conclude that the semigroup $P_t = e^{t\mathcal{L}}$ is hypercontractive provided that we have for every $q \in [2, \infty)$ the following inequality

$$\mu f^q \log \frac{f}{||f||_q} \leq \frac{q}{2(q-1)} c\mu(f^{q-1}(-\mathcal{L}^0 f)) \tag{2.22}$$

By classical arguments of [**G**] the condition (2.22) with the symmetric generator on the right hand side is equivalent to the condition that the probability measure μ satisfies the following <u>Logarithmic Sobolev</u> inequality

$$\mu f \log \frac{f}{\mu f} \leq 2c\mu |\nabla f^{\frac{1}{2}}|^2 \tag{LS}$$

with a coefficient $c \in (0, \infty)$ independent of a positive function f for which the right hand side is finite; here $|\nabla f^{\frac{1}{2}}|^2$ is defined as in (2.9) with $\Lambda = \Gamma$. One can slightly generalize the situation described above in a way which can allow us to assume slightly less smoothness on α's and β's. For this it is useful to remark that by the reversibility requirement (2.5) for local operators $\mathcal{L}_j$ and its consequence (2.6) together with the fact that μ is a Gibbs measure, we have for smooth local functions

$$\mu(g(-\sum_{j \in \Gamma} \alpha_j \cdot \nabla_j f)) = -\sum_{j \in \Gamma} \mu E_j ((g - E_j g)\alpha_j \cdot \nabla_j f) \qquad (2.23)$$

Let us suppose that the following condition is true
(**A.1**)

$$|E_j ((g - E_j g)\alpha_j \cdot \nabla_j f)| \leq \bar{\alpha} \left(E_j (g - E_j g)^2 \right)^{\frac{1}{2}} \left(E_j |\nabla_j f|^2 \right)^{\frac{1}{2}} \qquad (2.24)$$

with some constant $\bar{\alpha} \in (0, \infty)$ independent of $j \in \Gamma$ and the functions f and g.

Remark: In particular the condition (2.24) is satisfied if

$$\sup_j || \, |\alpha_j| \, ||_u < \infty \qquad (2.25)$$

(although it can be realized also in more general situation).

Using (2.24) and (2.23), by application of Schwartz inequality, we get

$$\left| \mu(g(-\sum_{j \in \Gamma} \alpha_j \cdot \nabla_j f)) \right| = \left| \sum_{j \in \Gamma} \mu E_j ((g - E_j g)\alpha_j \cdot \nabla_j f) \right|$$

$$\leq \bar{\alpha} \sum_{j \in \Gamma} \left(\mu E_j (g - E_j g)^2 \right)^{\frac{1}{2}} \left(\mu |\nabla_j f|^2 \right)^{\frac{1}{2}} \qquad (2.26)$$

Let us suppose also that the following inequality is true:
(**A.2**)

$$m_0 E_j (g - E_j g)^2 \leq E_j |\nabla_j g|^2 \qquad (2.27)$$

with some constant $m_0 \in (0, \infty)$ independent of $j \in \Gamma$.

Remark: In applications this is a very mild assumption. For example it is always satisfied for spin systems on a lattice with compact spin space, as e.g. considered in [**SZ1-2**] where E_j is an absolutely continuous with respect to the Riemannian measure with density which is uniformly bounded from below and above. It is also satisfied for the case of spin systems on a lattice with noncompact spin space, as e.g. considered in [**Z4**] where E_j has a density with respect to some log-concave measure, which is uniformly bounded from above and below.

Using (**A.1**) and (**A.2**) we arrive at the following sectorial condition

$$\left| \mu(g(-\sum_{j\in\Gamma}\alpha_j\cdot\nabla_j f)) \right| \leq \bar{\alpha}\cdot m_0^{-\frac{1}{2}}\left(\mu|\nabla g|^2\right)^{\frac{1}{2}}\left(\mu|\nabla f|^2\right)^{\frac{1}{2}} \tag{2.28}$$

Summarizing we have.

Theorem:
*Suppose the conditions (**A.1**) and (**A.2**) are satisfied. Then the nonsymmetric semigroup $P_t = e^{t\mathcal{L}}$ preserving a Gibbs measure μ is hypercontractive, if the measure μ satisfies Logarithmic Sobolev inequality. Moreover, as a consequence of that, we have*

$$\mu(P_t f - \mu f)^2 \leq e^{-\frac{2}{c}t}\mu(f - \mu f)^2 \tag{2.29}$$

o

The inequality (2.29) follows from the fact that **LS** with $q = 2$, implies the following Spectral Gap inequality for $\mathcal{L}$

$$\frac{1}{c}\mu(f - \mu f)^2 \leq \mu(f(-\mathcal{L}f)) \tag{SG}$$

This completes the discussion of the third ingredient necessary for our proof of strong ergodicity.

To make the presentation of its idea as short as possible, we will assume additionally that the exponential approximation property is uniform in η and also that for any $\Lambda \in \mathcal{F}$, the Radon-Nikodym derivative of the restriction $\mu_{|\Sigma_\Lambda}$ of the measure μ to the σ-algebra Σ_Λ satisfies a similar bound as the factor $I(\Lambda, \eta)$ in the weak ultracontractivity property. This situation is true in the case of spin systems on the lattice with the configuration space constructed with a compact Riemmanian manifold **M** considered in [**SZ2**]; (see also [**HS1**]). The more general case (together with some other variants of that strategy) can be found in [**Z4**].

First of all, assuming the exponential approximation property, for a function localized in a finite region $X \in \mathcal{F}$, we have

$$|P_t f(\eta) - \mu f| \leq |P_t^{\Lambda,\eta}f(\eta) - \mu f| + \mathcal{O}(e^{-At}) \tag{2.30}$$

with exponentially small error, provided $d(X, \Lambda^c) \geq Ct$, for some constant $C > 0$ dependent only on the choice of A. Next, for arbitrary $q \in (0, \infty)$ and $T \in (0, \infty)$ such that the weak ultracontractivity is true, we get for $t > T$

$$|P_t^{\Lambda, \eta} f(\eta) - \mu f| = \left(|P_T^{\Lambda, \eta} P_{t-T}^{\Lambda, \eta} f(\eta) - \mu f|^q \right)^{\frac{1}{q}} \leq \left(P_T^{\Lambda, \eta} |P_{t-T}^{\Lambda, \eta} f - \mu f|^q (\eta) \right)^{\frac{1}{q}}$$

$$\leq I(\Lambda, \eta)^{\frac{1}{q}} \cdot \left(E_\Lambda^\eta |P_{t-T}^{\Lambda, \eta} f(\cdot) - \mu f|^{2q} \right)^{\frac{1}{2q}} \tag{2.31}$$

We will chose later q so that we have

$$2q = 1 + e^{\frac{2}{c}[(1-\theta)t - T]} \tag{2.32}$$

with some $\theta \in (0, 1)$. Then the first factor on the right hand side (2.31) converges to one, (by our weak ultracontractivity assumption), provided that we change Λ with time in the way that $d(X, \Lambda^c) \geq Ct$. Therefore the factor $I(\Lambda, \eta)^{\frac{1}{q}}$ is irrelevant for further considerations. Since by our assumption we also have

$$\lim_{t \to \infty} \left(\frac{dE_\Lambda^\eta |\Sigma_\Lambda}{d\mu |\Sigma_\Lambda} \right)^{\frac{1}{2q}} = 1, \tag{2.33}$$

instead of the L_{2q} norm with the conditional measure E_Λ^η we can consider the norm with the measure μ. Then using again the exponential approximation property we get

$$\left(\mu |P_{t-T}^{\Lambda, \eta} f(\cdot) - \mu f|^{2q} \right)^{\frac{1}{2q}} \leq \left(\mu |P_{t-T} f(\cdot) - \mu f|^{2q} \right)^{\frac{1}{2q}} + \mathcal{O}(e^{-At}) \tag{2.34}$$

Now we make use of the hypercontractivity of our Markov semigroup as follows. By semigroup property we split $P_{t-T} = P_{(1-\theta)t - T} P_{\theta t}$ for some $\theta \in (0, 1)$. Then the application of the hypercontractivity to the first part give us the following bound

$$\left(\mu |P_{t-T} f(\cdot) - \mu f|^{2q} \right)^{\frac{1}{2q}} = \left(\mu |P_{(1-\theta)t - T} P_{\theta t} f(\cdot) - \mu f|^{2q} \right)^{\frac{1}{2q}} \leq$$

$$\leq \left(\mu |P_{\theta t} f(\cdot) - \mu f|^2 \right)^{\frac{1}{2}} \leq e^{-\frac{1}{c} \theta t} \left(\mu |f(\cdot) - \mu f|^2 \right)^{\frac{1}{2}} \tag{2.35}$$

where on the left hand side we have used the hypercontractivity property with (2.32). Thus combining all our considerations we arrive at the following strong ergodicity estimate (in the considered special case of compact configuration space)

Theorem:

$$||P_t f - \mu f||_u \leq C e^{-\frac{1}{c} \theta t} \left(\mu |f(\cdot) - \mu f|^2 \right)^{\frac{1}{2}} + \mathcal{O}(e^{-At}) \tag{2.36}$$

For the case of noncompact configuration space consult [**Z4**].
Let us mention also some other interesting approaches to the ergodicity problem via use of stochastic differential equations as e.g. in [**AKT**], [**AKR**], [**DZ**] and also in [**Di**] (in the last case with an extra use of cluster expansion for the measure representing the semigroup); see also references given there. At the moment these other methods can be applied to much more restrictive situations then ours, but it would be very interesting to develop them further.

In the next section we discuss a general method developed in recent years, [**Z1-7**], [**SZ1-3**], for the proof of Logarithmic Sobolev inequalities, the last ingredient of the above presented strategy.

3. The Multiconditioning Strategy.

In this section we would like to sketch a general idea which can be used to prove some important inequalities such as the Logarithmic Sobolev inequality, Spectral Gap inequality or General Nash's inequality and Logarithmic Nash's inequality. First of all let us observe that the considered inequalities have the following form

$$\mu \mathbf{Q}_\mu(f) \leq \mathbf{\Gamma}_\mu(f) \tag{$*$}$$

with $\mathbf{Q}_\mu$ satisfying

$$\mu \mathbf{Q}_\mu(f) \geq 0,$$
$$\mathbf{Q}_\mu(\text{const}) = 0, \tag{3.1}$$
$$\nu \mathbf{Q}_\mu(f) \leq \nu \mathbf{Q}_\nu(f) + \mathbf{Q}_\mu(\nu f)$$

for any probability measure ν, and $\mathbf{\Gamma}_\mu(f) \geq 0$ being a nonnegative functional vanishing only on constants.

How to prove ($*$): A Multiconditioning procedure.

For the proof of ($*$) we need some additional (local) structure:
– Suppose we are given a family of σ-algebras $\{\Sigma_a, \Sigma'_a \subset \Sigma\}_{a \in \mathbb{I}}$, with $\mathbb{I}$ a countable index set, such that

$$\forall a \in \mathbb{I} \qquad \Sigma_a \cup \Sigma'_a \text{ generate } \Sigma$$
$$\text{and } \bigcup_{a \in \mathbb{I}} \Sigma_a \text{ generate } \Sigma \tag{3.2}$$

Let for $\omega \in \Omega$

$$E_a^\omega f \equiv E_\mu(f|\Sigma'_a)(\omega), \tag{3.3}$$

be the corresponding regular conditional expectations.

Remark:
Let us recal that if $I\!\!I$ is partially ordered by some relation $\preceq$ and we have a compatibility condition in the sense that

$$a_1 \preceq a_2 \Rightarrow E_{a_2}^\omega \left(E_{a_1}^{\cdot} f \right) = E_{a_2}^\omega f, \tag{3.4}$$

the family $\mathcal{E} \equiv \{E_a^{\cdot}\}_{a \in I\!\!I}$, is called a local specification and a probability measure μ satisfying

$$\mu(E_a^{\cdot} f) = \mu f \tag{3.5}$$

is called a Gibbs measure for $\mathcal{E}$. ○

Given a sequence $\{a_n \in I\!\!I\}_{n \in I\!\!N}$ and an integrable function f, we define a sequence f_n, $n \in \mathbb{Z}^+$, setting $f_0 \equiv f$ and for $n \geq 1$

$$f_n \equiv E_{a_n} f_{n-1} \tag{3.6}$$

Due to properties of the function $\mathbf{Q}.$, for any sequence $\{a_n \in I\!\!I\}_{n \in I\!\!N}$, we have

$$\mu \mathbf{Q}_\mu(f) \leq \sum_{n \in I\!\!N} \mu \left(E_{a_n} \mathbf{Q}_{E_{a_n}} (f_{n-1}) \right) \tag{3.7}$$

provided that

$$\lim_{n \to \infty} f_n = \mu f \quad , \qquad\qquad \mu - a.e. \tag{3.8}$$

— Suppose we have following inequalities involving conditional expectations

$$E_a \mathbf{Q}_{E_a}(f) \leq \mathbf{\Gamma}_{E_a}(f) \tag{3.9}$$

with some nonnegative functionals $\mathbf{\Gamma}_{E_a}(f)$.
— Using (3.7) together with the inequalities (3.9) for conditional measures, we get

$$\mu \mathbf{Q}_\mu(f) \leq \mu \left(\sum_{n \in I\!\!N} \mathbf{\Gamma}_{E_{a_n}} (f_{n-1}) \right) \tag{3.10}$$

If the following "sweeping out" relations are satisfied

$$\mu \mathbf{\Gamma}_{E_{a_n}} (f_{n-1}) \leq \sum_{k \leq n} \gamma_{a_n, a_k} \mu \mathbf{\Gamma}_{E_{a_k}}(f) \tag{3.13}$$

with some coefficients γ_{a_n, a_k} such that

$$\sum_{n \geq k} \gamma_{a_n, a_k} \equiv c_k < \infty \tag{3.14}$$

we can turn (3.10) into (∗) with

$$\Gamma_\mu(f) \geq \sum_k c_k \mu \Gamma_{E_{a_k}}(f) \tag{3.15}$$

The above idea has been introduced in [**Z1-3,5**] for the proof of **LS** and **SG**, and has been successfully exploited later in [**SZ1-3**] and more recently in [**MO**], [**LY**], [**La**], [**Z5-7**]. Let us mention that there is no canonical choice of the family of σ-algebras in proving the inequality (∗). Dependent on the situation one can make different choices. In fact in the case of **LS** for spin systems on a lattice, three different choices have been used: (compare one group in [**Z1-4**], [**SZ1-3**], [**La**], [**Z5-6**], the second in [**MO1-2**] and the third in [**LY**]).

Also a concrete choice of the functional Γ_μ can depend on the purpose. We illustrate this in the next two sections.

4. Strong Decay to Equilibrium: Some Examples.

In this section we describe shortly some results concerning strong ergodicity obtained by use of the ideas described above.

Example 1

Let

$$\Omega = \mathbf{M}^{\mathbb{Z}^d}, \quad \mathbf{M} = a \ smooth, \ compact, \ connected \ Riemannian \ manifold \tag{4.1}$$

Let $\mathcal{F}$ denote the family of all finite subsets of $\mathbb{Z}^d$. For $\Lambda \in \mathcal{F}$ and $\omega \in \Omega$, let us define

$$E_\Lambda^\omega(f) = \delta_\omega \frac{\mu_{0|\Lambda}(e^{-U_\Lambda} f)}{\mu_{0|\Lambda}(e^{-U_\Lambda})}, \tag{4.2}$$

where $\mu_{0|\Lambda}$ denotes the restriction to Σ_Λ of the product measure

$$\mu_0 = \rho^{\mathbb{Z}^d}, \quad \rho \text{ the normalized Riemannian measure on } \mathbf{M}, \tag{4.3}$$

and

$$U_\Lambda \equiv \sum_{X \cap \Lambda \neq \emptyset} \Phi_X \tag{4.4}$$

with potential $\Phi \equiv \{\Phi_X \in \mathcal{C}^2(\Omega) \cap \Sigma_X - \text{measurable}\}_{X \in \mathcal{F}}$ of finite range R, i.e.

$$\Phi_X \equiv 0, \quad \text{diam}(X) > R \tag{4.5}$$

We will consider the following <u>Markov generators</u>

$$\mathcal{L}_\Lambda^\omega f = \delta_{\omega|\Sigma_{\Lambda^c}} \left(\sum_{j \in \Lambda} \mathcal{L}_j f \right) \tag{4.6}$$

where

$$\mathcal{L}_j \equiv \Delta_j - \beta_j \cdot \nabla_j \tag{4.7}$$

with

$$\beta_j \equiv \nabla_j U_j \tag{4.8}$$

The following result has been obtained in [**SZ2**] (see also [**SZ3**] for the case of discrete spins).

Theorem:

The following conditions are equivalent:

a) *Dobrushin-Shlosman Mixing*

$\exists \kappa > 0 \quad \forall \Lambda \in \mathcal{F}, \omega \in \Omega$

$$|E_\Lambda^\omega(f,g)| \le e^{-\kappa d(f,g)} |||f||| \cdot |||g||| \tag{4.9}$$

with some seminorms $||| \cdot |||$

b) *Uniform SG*

$$\inf_{\Lambda,\omega} gap_{L_2(E_\Lambda^\omega)}(\mathcal{L}_\Lambda^\omega) > 0 \tag{4.10}$$

c) *Uniform LS*

$$E_\Lambda^\omega f \log \frac{f}{E_\Lambda^\omega f} \le c E_\Lambda^\omega |\nabla_\Lambda f^{\frac{1}{2}}|^2 \tag{4.11}$$

with some $c \in (0,\infty)$ *independent of* $\Lambda \in \mathcal{F}$ *and* $\omega \in \Omega$, *for all positive functions* f *for which the right hand side is finite.*

The first condition is a statement about the conditional measures and it appears to be equivalent to the complete analyticity condition of Dobrushin-Shlosman, which means that the map

$$\Phi \longrightarrow E_{\Lambda,\Phi}^\omega \tag{4.12}$$

is (uniformly in Λ and ω) analytic. The second is spectral information about the generators of the corresponding stochastic dynamics. It is easy to see, (because of the exponential approximation property of the semigroups described in Section 2), that this L_2 information implies the strong decay of correlations described in (a). The Dobrushin-Shlosman mixing implies some "sweeping out" property involving in the present case $\Gamma_i(f) \equiv |\nabla_i f|^2$ and a sequence of conditional expectations E_{X+j_n}, where $X + j_n$ is a sequence of translations of some sufficiently large cube $X \in \mathcal{F}$. This property is used, in the way described in Section 3, to prove Logarithmic Sobolev inequalities (with a coefficient $c \in (0, \infty)$ uniform in volume and external conditions). Using **LS** , or better to say its equivalent condition - hypercontractivity, one can solve the strong ergodicity problem by the method presented in Section 2. Let us mention also that the **LS** inequality, by abstract arguments, implies spectral gap property of size $m \geq c^{-1}$. In general the converse need not to be true. Thus implication (b) $\Longrightarrow$ (c) may be slightly surprising. In the present situation we see that the ergodicity problem in L_2, if uniform in volume and external configuration of spins, is equivalent to the strongest one in supremum norm.

Let us mention that as shown recently [**SZ4**], in fact for the solution of strong ergodicity problem for the considered spin systems it is sufficient that one has **LS** for one infinite volume Gibbs measure and we have the following.

Theorem: [SZ4]
Let

$$P_t \equiv e^{t\mathcal{L}} \tag{4.13}$$

and

$$\mu(E_\Lambda f) = \mu f \tag{4.14}$$

If $\exists c \in (0, \infty)$

$$\mu f \log \frac{f}{\mu f} \leq 2c\mu |\nabla f^{\frac{1}{2}}|^2 \tag{4.15}$$

then

$$\|P_t f - \mu f\|_u \leq e^{-mt} A_m |||f||| \tag{4.16}$$

for any

$$0 < m < \text{gap}_2(\mathcal{L}) \tag{4.17}$$

with some positive constant A_m and some seminorm $|||\cdot|||$.

For related ergodicity results for spin systems with a compact configuration space see also [**HS1,2**], [**SZ3**], [**MO-II**], [**La**], [**LY**]. As shown in [**Z4**] it is possible to extend this result also to unbounded spin systems.

For the solutions of the strong ergodicity problem by some other methods see also [**Li1**], [**AH**], [**MO-I**].

One important feature of solving the ergodicity problem via use of Logarithmic Sobolev inequality is the fact that once we have got it for one stochastic dynamic, we can get it easily for other preserving a given measure provided its generator has the quadratic form equivalent to the form of the generator of the former stochastic dynamic. In particular it is very useful for discrete spin systems for which one can have a lot of different (symmetric) Markov generators which quadratic forms are mutually equivalent. (This feature is not present for example in the elegant method of [**AH**].) From the discussion of Section 2 we see that all the above mentioned results extends to the stochastic dynamics with nonsymmetric generator. For example we could consider a stochastic dynamics given by the following generators

$$\mathcal{L}' \equiv \mathcal{L} + \sum_{\mathbf{j} \in \mathbb{Z}^d} \alpha_{\mathbf{j}} \cdot \nabla_{\mathbf{j}} \tag{4.18}$$

with the coefficients $\alpha_{\mathbf{j}}$ defined by

$$\alpha_{\mathbf{j}} \equiv e^{U_{\mathbf{j}}} \alpha_{\mathbf{j}}^0 \tag{4.19}$$

where $\alpha_{\mathbf{j}}^0$ satisfies

$$div_{\mathbf{j}} \, \alpha_{\mathbf{j}}^0 = 0 \tag{4.20}$$

If $\alpha_{\mathbf{j}}^0$ and $\beta_{\mathbf{j}}$'s satisfy also the following condition

$$\left. \begin{array}{c} \sup_{\mathbf{j}} \sum_{\mathbf{k}} |||\nabla_{\mathbf{k}} \alpha_{\mathbf{j}}^0| \, |||_u < \infty \\ \sup_{j} \sum_{\mathbf{k}} |||\nabla_{\mathbf{k}} \beta_{\mathbf{j}}| \, |||_u < \infty \end{array} \right\} \tag{4.21}$$

then it is easy to construct the semigroup $P_t' \equiv e^{t\mathcal{L}'}$, which is Feller and such that for any function dependent on a finite number of coordinate we have

$$|||P_t'f||| \equiv \sum_{\mathbf{j} \in \mathbb{Z}^d} |||\nabla_{\mathbf{j}} P_t'f| \, |||_u < \infty \tag{4.22}$$

provided that the left hand side is finite at $t = 0$. (In fact for existence of P_t' little less smoothness would also do the job.) Combining the arguments of the previous sections together with the technique of [**SZ2**] (for the proof of the Logarithmic Sobolev inequality), we arrive at the following ergodicity result for the nonsymmetric case.

Theorem:

Suppose (4.21) is true and that for some sufficiently large cube Λ the Dobrushin - Shlosman Mixing condition is satisfied. Then the corresponding Gibbs measure μ satisfies Logarithmic Sobolev inequality and the following strong ergodicity estimate is true for the stochastic dynamics P_t' with nonsymmetric generator

$$||P_t'f - \mu f||_u \le e^{-mt} A_m |||f||| \tag{4.23}$$

for any $0 < m < gap_2\mathcal{L}$ with some constant A_m and a cylinder function f.

Example 2

In this example we describe some new results for <u>unbounded spin</u> systems. Now a natural configuration space to consider is the following

$$\Omega = \mathcal{S}'(Z^d) \equiv \text{ the space of tempered sequences in } I\!\!R^{Z^d} \qquad (4.24)$$

Let us note that Ω is <u>not</u> a product space, as was the case before. Now we would like to consider a local specification defined as follows.

$$E_\Lambda^\eta(f) \equiv \frac{E_{\Lambda,\mu_\mathbf{C}}^\eta(e^{-U_\Lambda}f)}{E_{\Lambda,\mu_\mathbf{C}}^\eta(e^{-U_\Lambda})} \qquad (4.25)$$

where

(i) $$E_{\Lambda,\mu_\mathbf{C}}^\eta(f) \equiv E_{\mu_\mathbf{C}}(f \mid \Sigma_{\Lambda^c})(\eta) \qquad (4.26)$$

is a regular conditional expectation with respect to Σ_{Λ^c} associated to a Gaussian measure $\mu_\mathbf{C}$ with mean zero and a covariance $\mathbf{C}$ s.t.

$$\mathbf{C}_{\mathbf{ij}}^{-1} = 0, \quad d(\mathbf{i},\mathbf{j}) > R \qquad (4.27)$$

and

(ii) $$U_\Lambda(\omega) = \sum_{\mathbf{i}\in\Lambda} U(\omega_{\mathbf{i}}) \qquad (4.28)$$

with a semibounded U.

Similarly as before we define the generator of a stochastic dynamics:

$$\mathcal{L} = \sum_{\mathbf{i}} \mathcal{L}_{\mathbf{i}} \qquad (4.29)$$

so that
$$E_{\mathbf{i}}|\nabla_{\mathbf{i}}f|^2 = E_{\mathbf{i}}(f(-\mathcal{L}_{\mathbf{i}}f)) \qquad (4.30)$$

Since now our configuration space is noncompact, the problem of constructing corresponding infinite volume Markov semigroup $P_t \equiv e^{t\mathcal{L}}$, $t \geq 0$ was much more complicated. It has been solved in [**Z4**], where between other things, the following theorem has been proved.

Theorem: [Z4]
Let
$$U = V + W \qquad (4.31)$$

with
$$\left.\begin{array}{l} \|W\|_u, \|W'\|_u < \infty \\ V \in \mathcal{C}^2 \text{ convex} \end{array}\right\} \qquad (4.32)$$

Suppose the measures E_Λ^ω have uniform in Λ and ω mixing property. Then the unique Gibbs measure μ satisfies for $f \geq 0$

$$\mu f \log \frac{f}{\mu f} \leq 2c\mu |\nabla f^{\frac{1}{2}}|^2 \qquad (4.33)$$

with some $c \in (0, \infty)$ and therefore for any $\eta \in \Omega$ and any cylinder function f we have

$$|P_t f(\eta) - \mu f| \leq C(\eta) e^{-mt} |||f||| \qquad (4.34)$$

with some $C(\eta) \in (0, \infty)$, some seminorm $||| \cdot |||$ and any

$$0 < m < \mathrm{gap}_2(\mathcal{L}) \qquad (4.35)$$

In particular if $d = 1$, one has **LS** *and so the exponential decay to equilibrium for any local potential U described above. Moreover, if $d \geq 2$ there are potentials for which the uniform mixing property is satisfied and so the assertion given above remains true, but the Bakry-Emery condition for* **LS** *fails.*

Let us mention that in the general system described above, it is not necessarily true that one has ultracontractivity estimate in a finite volume, essentially used in previous works for compact single spin space. Nevertheless it was possible to generalize the previous strategy to get a successful solution of the ergodicity problem. As a matter of fact such a strategy can also be extended to treat continuous systems, where the lattice is replaced by Euclidean space. The work [**Z4**] contains also hints how to prove **LS** in this interesting case for essentially nonconvex local potentials. (For convex one dimensional case see also [**DZ**].)

Let us mention also that the strong ergodicity problem for noncompact spin systems has been solved by some other methods in [**AKT**] and [**DZ**], but only for potentials for which one can show **LS** by the Bakry-Emery condition; see also [**Di**] for another way of studying ergodicity based on application of a cluster expansion. In [**DZ**] the authors consider also nonsymmetric case. As we can see from our previous discussion one can handle this case also by our method.

5. Weak Decay to Equilibrium in Some Critical Spin Systems.

Let μ be a probability measure on a configuration space (Ω, Σ) and let $P_t \equiv e^{t\mathcal{L}}$ be a Markov semigroup. In this section we would like to study a weak decay to equilibrium, i.e. we are interested in getting an estimate of the form

$$||P_t(f - \mu f)||_{L_2(\mu)} \leq \gamma(t) |||f||| \qquad (5.1)$$

with some rate function $\gamma(t)$ and a seminorm $||| \cdot |||$. If the semigroup is symmetric, one can view this problem as a problem from the spectral analysis of the selfadjoint operator $\mathcal{L}$. It is especially interesting if its spectrum extends to zero. Clearly we would get an estimate of the form (5.1) by finding a potential $G \equiv G(\mathcal{L}^{-1})$, (which in general is an unbounded operator), such that $f - \mu f$ belongs to its domain $\mathcal{D}(G)$. Then we get

$$
\begin{aligned}
||P_t(f - \mu f)||_{L_2(\mu)} &= ||P_t G^{-1}(G(f - \mu f))||_{L_2(\mu)} \\
&\leq ||P_t G^{-1}|| \cdot ||(G(f - \mu f))||_{L_2(\mu)}
\end{aligned}
\tag{5.2}
$$

i.e. we obtain the estimate (5.1) with

$$
\gamma(t) = ||P_t G^{-1}|| = \sup_{\lambda \in \sigma(\mathcal{L})} |e^{t\lambda} G^{-1}(\lambda^{-1})|
\tag{5.3}
$$

where $\sigma(\mathcal{L})$ denotes the spectrum of the selfadjoint operator $\mathcal{L}$, and with a seminorm

$$
|||f||| = ||G(f - \mu f)||_{L_2(\mu)}
\tag{5.4}
$$

However such a solution, although in principle possible, is seldom available, especially in the infinite dimensional context.

We would like to argue that one can get very good results by a kind of approximate spectral analysis, based on the multiconditioning idea. For this let us consider first a rather simple situation when we can represent our probability measure μ as follows

$$
\mu = \otimes_{k \in \mathbb{N}} E_k
\tag{5.5}
$$

with E_k being some probability measures on (Ω, Σ_k) for some sub-σ-algebras $\Sigma_k \subset \Sigma$, chosen in such a way that we have the following decomposition of the semigroup of interests to us

$$
P_t = \prod_k P_t^{(k)}
\tag{5.6}
$$

with some semigroups $P_t^{(k)} \equiv e^{t\mathcal{L}^{(k)}}$ acting in $L_2(E_k)$, respectively. (In general even if such decomposition exists, it does not need to be unique.) Given this structural decomposition, let us observe that we have

$$
\begin{aligned}
\mu(P_t f - \mu f)^2 &= \mu(P_t f - \mu P_t f)^2 = \mu E_1(P_t f - \mu P_t f)^2 \\
&= \mu E_1(P_t f - E_1 P_t f)^2 + \mu(E_1 P_t f - \mu P_t f)^2 \\
&\leq \mu E_1\left(P_t^{(1)} f - E_1 f\right)^2 + \mu(E_1 P_t f - \mu P_t f)^2 \\
&= \mu E_1\left(P_t^{(1)} f - E_1 f\right)^2 + \mu(P_t E_1 f - \mu E_1 f)^2
\end{aligned}
\tag{5.7}
$$

Now we can apply the same arguments with some other conditional expectation to the last term on the right hand side of (5.7). By applying this procedure inductively we arrive at the following inequality

$$\mu\left(P_t f - \mu f\right)^2 \leq \sum_{n \in I\!N} \mu E_n \left(P_t^{(n)} f_{n-1} - E_n f_{n-1}\right)^2 \tag{5.8}$$

where f_n is defined by setting $f_0 \equiv f$ and for $n \in I\!N$

$$f_n \equiv E_n f_{n-1} \tag{5.9}$$

and we have assumed that

$$L_2(\mu) - \lim_{n \to I\!N} f_n = \mu f \tag{5.10}$$

Suppose now that for every $k \in I\!N$ we have the following Spectral Gap inequality

$$m_k E_k (f - E_k f)^2 \leq E_k(f(-\mathcal{L}^{(k)} f)) \tag{5.11}$$

Then using (5.8) and (5.11), we get

$$\mu\left(P_t f - \mu f\right)^2 \leq \sum_{n \in I\!N} e^{-2m_n t} \mu\left(f_{n-1} - E_n f_{n-1}\right)^2 \tag{5.12}$$

Hence, choosing a sequence of numbers $a_n \in (0, \infty)$, we could for example arrive at the weak ergodicity estimate (5.1) with the following rate function

$$\gamma(t) \equiv \left(\sum_{n \in I\!N} e^{-2m_n t} a_n^2\right)^{\frac{1}{2}} \tag{5.13}$$

and the seminorm

$$|||f||| \equiv \sup_{n \in I\!N} a_n^{-1} \left(\mu\left(f_{n-1} - f_n\right)^2\right)^{\frac{1}{2}} \tag{5.14}$$

One could choose also some other norms. Dependent on this choice one would get in general different rate functions.

We would like to describe also another way of analysing the situation described above, using more directly the multiconditioning method of Section 3. First of all we note that for a conditional expectation E_l, we have

$$\mu(f - \mu f)^2 = \mu E_l(f - E_l f)^2 + \mu(E_l f - \mu f)^2 \tag{5.15}$$

Using (5.15) inductively we get

$$\mu(f - \mu f)^2 = \sum_{n \in I\!N} \mu E_n (f_{n-1} - E_n f_{n-1})^2 \tag{5.16}$$

provided that (5.10) is true. (In the context of proving a spectral gap inequality such the decomposition has been considered first in [**Z5**]; see also [**LY**].) Now we can use (5.11) to estimate every term in the sum on the right hand side of (5.16). In this way we get

$$\mu(f - \mu f)^2 \leq \sum_{n \in I\!N} \mu \left(m_n^{-1} E_n(f_{n-1}(-\mathcal{L}^{(n)} f_{n-1})) \right) \qquad (5.17)$$

Hence by Hölder inequality with some $q \in (1, \infty)$, one obtains

$$\mu(f - \mu f)^2 \leq \left(\sum_{n \in I\!N} \mu E_n(f_{n-1}(-\mathcal{L}^{(n)} f_{n-1})) \right)^{\frac{1}{q}} \mathbf{A}_q(f) \qquad (5.18)$$

with a functional

$$\mathbf{A}_q(f) \equiv \left(\sum_{n \in I\!N} m_n^{-p} \mu \left(E_n(f_{n-1}(-\mathcal{L}^{(n)} f_{n-1})) \right)^p \right)^{\frac{1}{p}} \qquad (5.19)$$

where p satisfies $\frac{1}{p} + \frac{1}{q} = 1$. Setting

$$|\nabla_n F|^2 \equiv \frac{1}{2} \left(\mathcal{L}^{(n)} F^2 - 2F \mathcal{L}^{(n)} F \right) \qquad (5.20)$$

and taking an advantage of the product representation for our semigroup, we get

$$E_n \left(f_{n-1}(-\mathcal{L}^{(n)} f_{n-1}) \right) = E_n |\nabla_n f_{n-1}|^2 \leq E_n ... E_1 |\nabla_n f|^2 \qquad (5.21)$$

(In a more general situation we would need to use an appropriate "sweeping out" inequality as in Section 3.) Combining (5.18)-(5.21) we obtain the following <u>Nash's inequality</u>

$$\mu(f - \mu f)^2 \leq \left(\mu |\nabla f|^2 \right)^{\frac{1}{q}} \mathbf{A}_q(f) \qquad (5.22)$$

where

$$\mu |\nabla f|^2 = \sum_{n \in I\!N} \mu |\nabla_n f|^2 = \mu(f(-\mathcal{L}f)) \qquad (5.23)$$

Let us remark that choosing inequalities different than (5.11) as a basis of out inductive procedure, we could get another Nash's inequalities with a functional different than $\mathbf{A}_q$.

Now we can use our Nash's inequality in a classical way to get an information about the decay to equilibrium as follows. Let us observe that, in the situation described above, the functional $\mathbf{A}_q$ is monotone in the sense that

$$\mathbf{A}_q(f_t) \leq \mathbf{A}_q(f) \qquad (5.24)$$

Thus from (5.22) we get for $f_t \equiv P_t f$

$$\mu(f_t - \mu f)^2 \leq (\mu(f_t(-\mathcal{L}f_t)))^{\frac{1}{q}} \mathbf{A}_q(f) \qquad (5.25)$$

whence we obtain

$$\mathbf{A}_q(f)^{-q} \left(\mu(f_t - \mu f)^2 \right)^q \leq \mu(f_t(-\mathcal{L}f_t)) \qquad (5.26)$$

Now we note that

$$\frac{d}{dt}\mu(f_t - \mu f)^2 = 2\mu(f_t(\mathcal{L}f_t)) \leq -2\mathbf{A}_q(f)^{-q} \left(\mu(f_t - \mu f)^2 \right)^q \qquad (5.27)$$

Solving this differential inequality we conclude that

$$\mu(f_t - \mu f)^2 \leq \left(2(q-1)\mathbf{A}_q(f)^{-q} \cdot t + \left(\mu(f - \mu f)^2 \right)^{-(q-1)} \right)^{\frac{1}{q-1}} \qquad (5.28)$$

Before we give a simple example of application, we would like to make the following remarks. First of all let us note that the last two methods has been based essentially on a use of some inequalities and, contrary to the considerations based on the spectral analysis given at the beginning of this section, no assumption of symmetry of the Markov semigroup has been used. Thus one can apply them also in a nonsymmetric case. We have noticed also that in the last method the special assumption of the product decomposition could be substituted by an assumption of weak dependence between some degrees of freedom. Therefore one may hope also for a more general applications, which will be considered elsewhere.

As a simple example of application let us consider here a critical Gaussian model on a lattice defined by a Gaussian probability measure $\mu_{\mathbf{C}}$ on the space $\Omega \equiv I\!\!R^{\mathbb{Z}^d}$, with mean zero and covariance $\mathbf{C} \equiv \{\mathbf{C_{ij}}\}_{i,j \in \mathbb{Z}^d}$ given by

$$\mathbf{C_{ij}} \equiv \frac{1}{(2\pi)^d} \int_{(-\pi,\pi)^d} d\mathbf{q} e^{i\mathbf{q} \cdot (\mathbf{i}-\mathbf{j})} \hat{\mathbf{C}}(\mathbf{q}) \qquad (5.29)$$

with the Fourier transform $\hat{\mathbf{C}}$ satisfying

$$\lambda_1 |\mathbf{q}|^{-\theta d} \leq \hat{\mathbf{C}}(\mathbf{q}) \leq \lambda_2 |\mathbf{q}|^{-\theta d} \qquad (5.30)$$

with some positive constants λ_1, λ_2 and $\theta \in (0,1)$. We will be interested in the stochastic dynamic $P_t \equiv e^{t\mathcal{L}}$ defined with the generator $\mathcal{L}$ satisfying

$$\mu_{\mathbf{C}}(f(-\mathcal{L}f)) = \mu_{\mathbf{C}}|\nabla f|^2 \qquad (5.31)$$

for all functions for which the right hand side is finite. It is well known that for such stochastic dynamic one can find explicitly $P_t f$ for linear functions f, i.e. functions of the following form

$$\varphi(h) \equiv \sum_{\mathbf{j} \in \mathbb{Z}^d} h_{\mathbf{j}} \varphi_{\mathbf{j}} \tag{5.32}$$

with $h \equiv \{h_{\mathbf{j}} \in I\!\!R\}_{\mathbf{j} \in \mathbb{Z}^d}$ being a tempered sequence and $\varphi_{\mathbf{j}}$ being the natural coordinate functions on Ω given by

$$\Omega \ni \omega \longmapsto \varphi_{\mathbf{j}}(\omega) \equiv \omega_{\mathbf{j}} \tag{5.33}$$

Using this and some additional property called attractivity, the L_2-decay to equilibrium for more general functions has been determined in [**De**]. (In this interesting work one can find also a solution of the weak ergodicity problem by some different method for other critical lattice models for which one has some attractivity and linearity properties. Let us mention also that another attractive models have been considered in [**Li2**] using Nash's inequalities.) In our method we do not assume attractivity.

For our purposes let us recal that the Dirichlet form from the right hand side of (5.31) can be written as follows

$$\mu |\nabla f|^2 = \sum_{k \in I\!\!N} \mu |\nabla_{g_k} f|^2 \tag{5.34}$$

where $\{g_k\}_{k \in I\!\!N}$ is an arbitrary O-N base in $l_2(\mathbb{Z}^d)$ and

$$\nabla_{g_k} f(\omega) \equiv \frac{d}{dt} f(\omega + t g_k)_{|t=0} \tag{5.35}$$

We would like to represent the measure $\mu_{\mathbf{C}}$ as follows

$$\mu_{\mathbf{C}} = \otimes_{n \in I\!\!N} \mu_{\mathbf{C}_n} \tag{5.36}$$

with $\mu_{\mathbf{C}_n}$ being a Gaussian measure with mean zero and a covariance $\mathbf{C}_n$, $n \in I\!\!N$, which Fourier transform is given for $n = 1$ by

$$\hat{\mathbf{C}}_1(\mathbf{q}) \equiv \chi(\varepsilon_1 \le |\mathbf{q}|) \hat{\mathbf{C}}(\mathbf{q}) \tag{5.37a}$$

and for $n > 1$ by

$$\hat{\mathbf{C}}_n(\mathbf{q}) \equiv \chi(\varepsilon_n \le |\mathbf{q}| < \varepsilon_{n-1}) \hat{\mathbf{C}}(\mathbf{q}) \tag{5.37b}$$

where $\{\varepsilon_n\}_{n \in I\!\!N}$ is a positive sequence decreasing to zero. By a judicious choice of the O-N base in $l_2(\mathbb{Z}^d)$ we get the following representation

$$\mathcal{L} = \sum_{n \in I\!\!N} \mathcal{L}_n \tag{5.38}$$

with the $\mathcal{L}_n$ being the Dirichlet operators in $L_2(\mu_{\mathbf{C}_n})$, respectively. Let us mention that in the present situation we have

$$m_n \mu_{\mathbf{C}_n}(f - \mu_{\mathbf{C}_n} f)^2 \leq \mu_{\mathbf{C}_n}(f(-\mathcal{L}_n g)) \tag{5.39}$$

where

$$m_n \geq \lambda \varepsilon_n^{\theta d} \tag{5.40}$$

with a positive constant λ independent of n. To have some fun we would like to apply the formula (5.12) in the present situation with the (simplest possible but still interesting) function $f = \varphi_{\mathbf{j}}$. Since we have

$$\mu(f_{n-1} - f_n)^2 = \mu_{\mathbf{C}_n}(\mu_{\mathbf{C}_{n-1}} f - \mu_{\mathbf{C}_n} f)^2 = \mathbf{C}_n(0) \tag{5.41}$$

Using our assumption about $\mathbf{C}_n$ we have

$$\mathbf{C}_n(0) \leq a(\varepsilon_{n-1}^{(1-\theta)d} - \varepsilon_n^{(1-\theta)d}) \tag{5.42}$$

with some positive constant a independent of n. Thus using (5.12) we obtain

$$\mu(P_t \varphi_{\mathbf{j}} - \mu\varphi_{\mathbf{j}})^2 \leq a \sum_{n \in I\!N} e^{-\lambda \varepsilon_n^{\theta d} \cdot t} (\varepsilon_{n-1}^{(1-\theta)d} - \varepsilon_n^{(1-\theta)d}) \tag{5.43}$$

To get an idea what is the time dependence of the right hand side of (5.43) one should observe that in the limit when supremum of the distance between consecutive ε_n's converges to zero, the formula of interests to us converges to some integral. A simple analysis of this integral gives us the following estimate

$$\mu(P_t \varphi_{\mathbf{j}} - \mu\varphi_{\mathbf{j}})^2 \leq C t^{-\frac{1-\theta}{\theta}} \tag{5.44}$$

with some constant $C \in (0, \infty)$. The application of the second method based on the Nash's inequality we leave for the reader as an exercise.

Finally let us mention that similar analysis can be easily carried out also for nonGaussian measures constructed by perturbing $\mu_{\mathbf{C}}$ with the use of the following additive functional

$$U_\Lambda(\varphi) \equiv \sum_{\mathbf{j} \in \Lambda} \left(\sum_n V_n(\varphi_{\mathbf{j}}^{(n)}) \right) \tag{5.45}$$

with $\varphi^{(n)}$ being the Gaussian field on the lattice with covariance $\mathbf{C}_n$, respectively, and with V_n being convex functions (or at least such that using them we can get a probability measure satisfying spectral gap inequality). One could also consider a limiting case obtained with continuous decomposition of our original Gaussian field.

Acknowledgements: The author would like to thank Alison Etheridge for organizing a very nice and interesting Conference. He would like also to thank Sergio Albeverio, Terry Lyons and Michael Röckner for questions and discussions.

References

[AH] Aizenman M. and Holley R.A. , Rapid Convergence to Equilibrium of Stochastic Ising Models in the Dobrushin-Shlosman Regime , Percolation Theory and Ergodic Theory of Infinite Particle Systems, Ed.Kesten H., Springer-Verlag (1987) 1-11

[AKT] Albeverio S., Kondratiev Yu.G. and Tsycalenko T.V. , Stochastic dynamics for quantum lattice systems and stochastic quantization I: Ergodicity , Preprint 1993

[AKR] Albeverio S., Kondratiev Yu.G. and Röckner M. , Dirichlet Operators and Gibbs Measures , BiBoS Preprint 541 / 1992

[AKR] Albeverio S., Kondratiev Yu.G. and Röckner M. , Dirichlet Operators via Stochastic Analysis , BiBoS Preprint 571 / 1993

[AKR] Albeverio S., Kondratiev Yu.G. and Röckner M. , Infinite Dimensional Diffusions, Markov Fields, Quantum Fields and Stochastic Quantization , BiBoS Preprint 631 / 4 / 1994

[AR] Albeverio S. and Röckner M. , Dirichlet Form Methods for Uniqueness of Martingale Problems and Applications , BiBoS Preprint 615 / 2 / 1994

[BE] Bakry D. and Emery M. , Hypercontractivite de semi - groupes des diffusion , C.R.Acad.Sci. Paris Ser. I29 9 (1984) 775-777 , Diffusions hypercontractives pp 177-206 , Sem. de Probabilites XIX, Azema J. and Yor M.(eds.) **LNM** 1123

[CS] Carlen E.A. and Stroock D.W. , An application of the Bakry - Emery criterion to infinite dimensional diffusions, Sem.de Probabilites XX, Azema J. and Yor M. (eds.) **LNM** 1204 341-348

[DZ] Da Prato G. and Zabczyk J. , Convergence to Equilibrium for Spin Systems , Preprint 1994

[DeS] Deuschel J.-D. and Stroock D.W. , Hypercontractivity and Spectral Gap of Symmetric Diffusions with applications to the Stochastic Ising Model , J. Func. Anal. **92** (1990) 30-48

[De] Deuschel J.-D. , Algebraic L_2 Decay of Attractive Critical Processes on the Lattice , Ann. Prob. **22** (1994) 264-283

[Di] Dimock J. , A Cluster Expansion for Stochastic Lattice Fields , J. Stat. Phys. **58** (1990) 1181 - 1207

[G] Gross L. , Logarithmic Sobolev inequalities, Amer. J. Math. **97** (1976) 1061–1083

[HS1] Holley R. and Stroock D.W., Logarithmic Sobolev inequalities and stochastic Ising models, J. Stat. Phys. **46** (1987) 1159–1194

[HS2] Holley R. and Stroock D.W., Uniform and L_2 Convergence in One Dimensional Stochastic Ising models , Commun. Math. Phys. **123** (1989) 85 - 93

[La] Laroche E., Sur des Inégalités de Corrélation et sur les Inégalités de Sobolev Logarithmiques en Mécanique Statistique, Thése Toulouse 1993

[Li1] Liggett Th. M. , Infinite Particle Systems, Springer - Verlag , Grundlehren Series # 276, New York (1985)

[Li2] Liggett Th. M. , L_2 Rates of Convergence for Attractive Reversible Nearest Particle Systems: The Critical Case , Ann. Probab. **19** (1991) 935 - 959

[LY] ShengLin Lu and Horng - Tzer Yau , Spectral Gap and Logarithmic Sobolev Inequality for Kawasaki and Glauber Dynamics, Preprint 1993

[MO] Martinelli F. and Olivieri E. , Approach to Equilibrium of Glauber Dynamics in the One Phase Region: I. The Attractive case/ II. The General Case , Preprints 1992 /1993

[SZ,1] Stroock D.W. and Zegarlinski B. , The Equivalence of the Logarithmic Sobolev Inequality and the Dobrushin–Shlosman Mixing Condition , Commun. Math. Phys. **144** (1992) 303 - 323

[SZ,2] Stroock D.W. and Zegarlinski B. , The Equivalence of the Logarithmic Sobolev Inequality and the Dobrushin–Shlosman Mixing Condition , Commun. Math. Phys. **144** (1992) 303 - 323

[SZ3] Stroock D.W. and Zegarlinski B. , The Logarithmic Sobolev Inequality for Discrete Spin Systems on a Lattice , Commun. Math. Phys. **149** (1992) 175 - 193

[SZ4] Stroock D.W. and Zegarlinski B. , MIT 1993

[Z1] Zegarlinski B. , On log-Sobolev inequalities for infinite lattice systems, Lett. Math. Phys. **20** (1990) 173–182

[Z,2] Zegarlinski B. , Log-Sobolev inequalities for infinite one-dimensional lattice systems , Comm. Math. Phys. **133** (1990) 147–162

[Z,3] ————, Dobrushin uniqueness theorem and logarithmic Sobolev inequalities , J. Func. Anal. **105** (1992) 77–111

[Z,4]———— , The Strong Decay to Equilibrium for the Stochastic Dynamics of Unbounded Spin Systems on a Lattice , Preprint 1992, revised 1994

[Z,5]———— , Gibbsian Description and Description by Stochastic Dynamics in the Statistical Mechanics of Lattice Spin Systems with Finite Range

Interactions , in Proc. of IIIrd International Conference: Stochastic Processes, Physics and Geometry, Locarno, Switzerland, June 24 - 29, (1991)

[Z,6]————, Ergodicity of Markov Semigroups on Infinite Dimensional Spaces , in Proc. of International Conference: Dirichlet Forms and Stochastic Processes, Beijing, China, October 25 - 31, (1993) , Ed.Ma Z.M. and Röckner M.

[Z,7]————, Strong Decay to Equilibrium in One Dimensional Random Spin Systems , J. Stat. Phys. **77** (1994) 717 - 732